BIOTECHNOLOGY IN GROWTH REGULATION

UNIVERSITY LIBRARY LIVERPOOL

BIOTECHNOLOGY IN GROWTH REGULATION

editors

R.B.Heap

C.G.Prosser

G.E.Lamming

International Symposium
held at
AFRC Institute of Animal Physiology
and Genetics Research
Babraham
Cambridge CB2 4AT
United Kingdom

September 18th-20th, 1988

Butterworths

London Boston Singapore Sydney Toronto Wellington

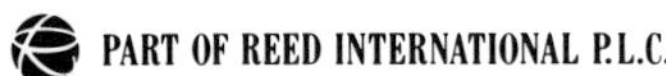

All rights reserved. No part of this publication may be reproduced or transmitted in any form or by any means (including photocopying and recording) without the written permission of the copyright holder except in accordance with the provisions of the Copyright Act 1956 (as amended) or under the terms of a licence issued by the Copyright Licensing Agency Ltd, 33–34 Alfred Place, London, England, WC1E 7DP. The written permission of the copyright holder must also be obtained before any part of this publication is stored in a retrieval system of any nature. Applications for the copyright holder's written permission to reproduce, transmit or store in a retrieval system any part of this publication should be addressed to the Publishers.

Warning: The doing of an unauthorised act in relation to a copyright work may result in both a civil claim for damages and criminal prosecution.

This book is sold subject to the Standard Conditions of Sale of Net Books and may not be re-sold in the UK below the net price given by the Publishers in their current price list.

First published 1989

© **Butterworth & Co. (Publishers) Ltd, 1989**

British Library Cataloguing in Publication Data
Biotechnology in growth regulation
1. Livestock. Production. Growth hormones
I. Heap, R. B. II. Prosser, C. G.
III. Lamming, G. E.
636.08′52
ISBN 0-407-01473-X

Library of Congress Cataloging-in-Publication Data
Biotechnology in growth regulation/edited by R. B. Heap, C. G. Prosser, G. E. Lamming.
p. cm.
Bibliography: p.
Includes index.
ISBN 0-407-01473-X:
1. Somatotropin–Physiological effect. 2. Growth–Regulation.
I. Heap, R. B. (Robert Brian), 1935- . II. Prosser, C. G.
III. Lamming, G. E. (George Eric), 1927-
QP572.S6B56 1989
599′.01927–dc19

Printed and bound in Great Britain by
Courier International Ltd, Tiptree, Essex

PREFACE

Advances in biotechnology and the application of molecular biology techniques have stimulated new investigations into the regulation of growth in animals including man. The purpose of the meeting reported here was to bring together leading scientists studying the mechanisms of action of growth hormone and growth factors and how growth responses may be modified by immunological and transgenic procedures. The primary emphasis of the meeting was on the fundamental aspects of this burgeoning aspect of biological science. Separate sessions were devoted to an examination of the practical application of one of the products of biotechnology, notably bovine somatotropin, and to the ethical and sociological implications of the adoption of such innovations capable of enhancing the efficiency of growth and lactation.

With the new era of highly purified genetically engineered somatotropin, applications in clinical medicine as in the treatment of growth disorders herald investigations which were previously impracticable. During the meeting the death was announced of Professor Sir Frank Young, FRS, a former Professor of Biochemistry in the University of Cambridge who made major contributions to growth hormone research and especially to our understanding of its mechanisms of action in diabetes and metabolism. Further developments described in this book relate to the evolution and structure of the growth hormone receptor structure and to growth factor action, while new studies using transgenic models provide insights into the regulation of copies of genes inserted into different hosts.

The meeting was held in The Conference Centre of the Institute of Animal Physiology and Genetics Research, Babraham, Cambridge, and we are indebted to the Agricultural and Food Research Council for permission to use its facilities for the first conference on this topic. There was an overwhelming response with over 200 delegates from more than 15 countries wishing to attend the meeting. Accommodation and hospitality were provided at Jesus College, Cambridge and this gave participants a taste not only of college life but of a historic city which rightly claims to be the home of biotechnology.

In today's climate scientists have been accused of secrecy and elitism. With regard to the present Conference, two innovations were introduced. An opportunity was provided for four major biotechnology companies to present some of their exciting new work in open session. A growing investment in studies of novel biotechnological compounds resides in the province of such companies. Regrettably, only one responded to the invitation and the presented work is reported in the paper of Peel and colleagues. The second innovation was an invitation to members of the scientific and quality press to attend the open session which was given over to certain issues of immediate public concern.

We are indebted to colleagues at this Institute who helped in many different ways to make the meeting so successful scientifically and socially enjoyable. In particular we thank Simone Prosser, Janet Hood, Janet Tickner, Maureen Hamon, Janet Pendleton, Matthew Hannah, Ivan Fleet, Arthur Davis and Kevin Turner. We also thank Jad and Peter James for preparing the manuscripts for publication with care and efficiency, and our sponsors for their financial support. Professor J.M. Payne was closely involved in the initial stages of this meeting and we greatly regret that he did not live to see its outcome.

LIST OF SPONSORS

We wish to thank the following contributors for financial support.

American Cyanamid
Animal Biotechnology Cambridge Ltd.
Beecham Group Plc.
Cambio Cambridge
Cambridge Research Biochemicals
Celltech Ltd.
Coopers Animal Health
Enzymatix Ltd.
Elanco
Heffers Cambridge
Hoechst Animal Health
Intervet International and Intervet (U.K.)
Johnson and Johnson
KabiVitrum
Merck Sharp & Dohme Research Laboratories
Monsanto Europe SA/NV
Norden Laboratories
Pfizer Central Research
Scientific Generics Inc.
Serono Laboratories
Upjohn International Inc.

CONTENTS

GROWTH HORMONE

SPECIES SPECIFICITY AND STRUCTURE-FUNCTION RELATIONSHIPS OF GROWTH HORMONE

M. Wallis

Biochemistry Laboratory, School of Biological Sciences, University of Sussex, Falmer, Brighton BN1 9QG, U.K.

INTRODUCTION

Growth hormone (GH) is a protein hormone derived from the anterior pituitary gland. It comprises a single polypeptide chain of about 190 amino acids, containing two intra-chain disulphide bridges. In some species there is a tendency for the polypeptide chain to associate to form dimers or larger aggregates, but it is not clear whether this has any biological significance. GHs are found in all vertebrate groups except possibly the primitive, jawless fishes (Agnatha). There are considerable differences between GHs obtained from different groups, however, in both structure and biological properties.

In this paper I shall survey current knowledge about the species specificity of GHs, concentrating mainly on the differences between the primary structures for which information is available. Some aspects of the relationship between structure and function in GH will also be considered.

MICROHETEROGENEITY OF GROWTH HORMONE

Although for any one species it is normal to refer to GH as a single component, the hormone in many species is known to exist in a number of different forms. As a consequence, GH isolated from anterior pituitaries usually shows microheterogeneity. Such heterogeneity may arise from a number of causes, including the occurrence of more than one gene for the hormone, variant forms of mRNA due to different processing pathways for the mRNA precursors, allelic variation in the population of animals from which the hormone is derived, variable glycosylation (although in most species GH is not glycosylated), limited enzymic cleavage, N-terminal heterogeneity and deamidation. Heterogeneity has been most extensively studied in the case of human GH (20), but occurs in GHs from most other species also. Its physiological significance, if any, is not fully understood, but its possible importance should be borne in mind when species differences and structure-function relationships are discussed.

SPECIES SPECIFICITY OF GROWTH HORMONES: VARIATIONS IN AMINO ACID SEQUENCE

The first GH for which the amino acid sequence was reported was the human hormone (25). Subsequent work on GHs from a number of species produced the sequences of ovine, bovine, horse and pig GHs (summarized in Ref. 40) and revision of the human GH sequence. In 1977 application of the techniques of recombinant DNA allowed the sequence of rat GH to be deduced from the corresponding cDNA/mRNA nucleotide sequence (32). Since then application of both recombinant DNA techniques and conventional protein sequencing has increased the number of complete GH sequences available to at least 20 including 2 primates, 9 non-primate mammals, 2 birds, 1 amphibian and 7 teleost fish. Some of these sequences are summarized in Fig. 1; some are based on protein sequencing, some on cDNA/genomic DNA sequencing and some on both.

The availability of sequences for GHs from such a considerable number of species enables detailed comparisons to be made and assessment of possible pathways of molecular evolution in this protein-hormone family. These have been discussed previously (26, 27

```
                10        20        30        40        50        60        70        80        90        100
Anc. p.m.   AFPAMPLSSLFANAVLRAQHLHQLAADTYKEFERAYIPEGQRYS-IQNAQAAFCFSETIPAPTGKDEAQQRSDMELLRFSLLLIQSWLGPVQFLSRVFTNS
pig         ———————————————————————————————————————————————————————————————————————V————————————————————
horse       ————————————————————————————————————————————————————————————————————————————————————L———————
fin whale   ——————————————————————————————————————————————————————————————————N————V——L————————L———————
rat         ——————————————————————————————————————————————————————————————————E————T——————————————I———
ox          ——S—G——————————F———T——————T-V——————N——K—L——I——————L———————
man         —TI—R—D—M—HR———F—Q—E——KE-K—FL—P-TSL——S—T-SNRE-T—K-NL——I———————E———RS—A—
chicken     ————N——————L—Q————T———D—T-NK-S—Y——————————D—K—G————V————T—Y—K———N
bullfrog    —Q-S—N—T—I————MV———RDY—T———D—FK-QTLLISVY-Y—————D—NTH—K—ID—————T-L——MT-IQIVN——G-N

                110       120       130       140       150       160       170       180       190
Anc. p.m.   LVFGTSD-RVYEKLKDLEEGIQALMRELEDGSPRAGQILKQTYDKFDTNMRSDDALLKNYGLLSCFKKDLHKAETYLRVMKCRRFVESSCAF
pig         ——————————————————————————————————————————————L——————————————————————————————
horse       ————————R—————————————————————————————————————L——————————————————————————————
whale       —————————————————————————————————————————————————————————————————————————————
rat         -M——————————————Q————I—————A————————————————————————————A———
ox          ———————————L————T——————————————————R———T————————G-A———
man         —Y-A—SN—DL——————T—GR————T——F——S———SHN———————Y—R—MD-V—F—IVQ——S—G—G-
chicken     ————F———————————R——GP—L—RP————IHL-NE———————————————V——K————G—N-TI
bullfrog    Q——NI———DR-R—D—LHI-I——D—NV·NYGV·TF————V·L—EEGRA————————M—V———K——————N·T·
```

41). The additional sequences that have been described in the past few years provide confirmation of the main features recognized previously, particularly with respect to apparently variable rates of evolution.

When the amino acid sequences of non-primate mammalian GHs are compared it is clear that they are very similar. However, when any of these sequences (or a consensus sequence derived from them) is compared with the sequence of human GH the difference is marked (41). The various orders of placental mammals are thought to have diverged at about the same time, 70 million years ago. The divergence time for rat and human is therefore about the same as that for rat and pig; however, rat GH differs from human GH at ~ 35% of all residues but from pig GH at ~ 5% of all residues. Such observations suggest that the rate of evolution of GHs in the primates has been much greater than that in non-primate mammals. A variable rate of evolution within a protein family is relatively unusual - normally it is found that the rate of evolution for any individual protein is remarkably constant (though rates vary from one protein to another; eg Ref. 16). The rate of evolution can be quantified by determining a most probable sequence for the GH of the ancestor of the placental mammals (from sequence data available for exisiting mammals) (41). This is very similar to the sequences of pig, fin whale and horse GHs (Fig. 1). The number of amino acid differences observed between the existing GH sequences and this ancestral sequence can then be used to determine the rates of evolution for GH in each of the groups concerned (Table 1).

Table 1 Rates of evolution for growth hormones

Evolutionary line			Rate of evolution (APMs)*
'Placental ancestor'**	---------→	pig GH	1.5
'Placental ancestor'	---------→	fin whale GH	3.0
'Placental ancestor'	---------→	horse GH	2.3
'Placental ancestor'	---------→	rat GH	6.0
'Placental ancestor'	---------→	bovine GH	13.8
'Placental ancestor'	---------→	ovine/goat GH	14.2
'Placental ancestor'	---------→	human GH	52.0

* Accepted point mutations/100 residues/10^8 yr

** 'Placental ancestor' represents the hypothetical GH sequence of the common ancestor of the placental mammals.

Fig. 1 The amino acid sequences of some tetrapod GHs (see opposite)

A hypothetical (best fit sequence for the GH of the common ancestor of the placental mammals (Anc. p.m.) was derived from available GH sequences, using methods described previously (41). This is shown, using one-letter code. Sequences of various tetrapod GHs are compared with this. A solid line indicates sequence identity; differences are shown using one-letter code; - indicates a gap. Note that the sequences of pig, horse and whale GHs are very similar to the ancestral sequence but those of man, chicken and bullfrog differ from it substantially. Original data for the sequences shown may be found in Refs. 32 (rat), 34 (chicken), 33 (pig), 38 (whale), 29 (bullfrog) or references cited in Ref. 41 (horse, man, ox).

The apparent rapid evolution of GH in the primates makes sequence studies on GH for primates other than man of particular interest. The sequence of GH from the rhesus monkey has been reported (23). This is very similar to that of human GH, differing at only 4 residues, suggesting that the rate of evolution of GH since the divergence of old-world monkeys and great apes (approximately 20 million years ago) has been slow. The implication is that GH underwent a period of remarkably rapid evolution early in primate evolution. An alternative explanation is that the primate and non-primate GH genes are not strictly homologous, implying that a second gene (or family of genes) is present in the human genome, homologous with non-primate GHs (but not expressed as a major form in the pituitary) and a second gene is present in the non-primate mammalian genome, homologous with human GH (but again not expressed in the pituitary). The situation is complicated by the occurrence of a cluster of GH-like genes in man (eg Ref. 12). More information is needed, especially about GH (and GH genes) in other primates. However, it is clear that the nature of GH in primates and non-primates is remarkably different and this must be taken into account when considering the biological actions of the hormone in different species.

The rate of evolution of GHs also appears to have increased, though to a less marked extent, during the evolution of the ruminants (Table 1). The sequences of sheep, ox and goat GHs have all been determined; they show marked similarities. Sequences of any of these ruminant GHs differ from the sequences of rat or pig GHs at about 11% of all residues, whereas the rat, horse, fin whale and pig hormones differ at only 1-5% of all residues.

Most previous discussion of GH evolution has concentrated on mammalian GHs. Sequences of several non-mammalian GHs are now available. In general they are fairly similar to the sequences of the non-primate mammalian GHs. Sequences of GHs from bird (chicken and duck) (5, 34) are more similar to the non-primate GHs than the latter are to human GH, confirming the very rapid evolution seen in the case of primate GHs. GHs from teleost fish show extensive sequence differences from those of mammals, correlating with the lack of activity of these proteins in mammals. On the other hand, partial sequence data available for an elasmobranch fish (21) suggests that here the differences from mammalian GHs are less marked.

Comparison of GH sequences will be influenced markedly by the recent report of the tertiary structure of recombinant pig GH (1). About 55% of the polypeptide chain is in the form of α-helix, folded to give 4 helices in an antiparallel, twisted helical bundle. Sequence comparisons indicate that the α-helical regions are rather more conserved than other parts of the molecule, suggesting that a similar conformation is found in GHs from other species, and possibly in other proteins of the GH family.

On the basis of sequence comparisons, GHs are thought to be members of a family of homologous proteins along with prolactin and placental lactogens. Comparison of sequences of bovine GH and prolactin (eg Ref. 40) shows identity at about 25% of all residues, enough to establish homology but less similarity than is seen within the GHs (though teleost and mammalian GHs show quite low homology; eg 32% identity for GHs from man and tuna - Ref. 30). Distinct GHs and prolactins are known to occur in most fish groups and it seems likely that GH and prolactin diverged following a gene duplication early in vertebrate evolution.

The relationship of GHs to placental lactogens is more complex. The best-characterized placental lactogen is the human protein. This has an amino acid sequence very similar to that of human GH, differing at only about 15% of all residues. Indeed, human GH resembles human placental lactogen much more closely than it does any of the non-primate GHs. This implies that human placental lactogen diverged from human growth hormone (presumably following a gene duplication) during the evolution of the primates, after their separation from the other main orders of placental mammals (41). In accordance with this it is notable that the genes for placental lactogens lie very close to those for GHs in man (12). The situation in non-primates is different. Many non-primates possess placental lactogens. Such proteins have been less fully characterized than the human placental lactogen, but in the case of the rat and mouse, for which sequences have been reported (7, 15), it is clear that the placental lactogens are more closely related to the prolactins than they are to GHs. A similar situation probably occurs in ruminants (31). In the rat and mouse there are several different forms of placental lactogen-like proteins, together with another prolactin-like protein, proliferin, first identified as expressed in proliferating fibroblasts, but now identified as produced also in the placenta (17). The physiological roles of these various placental proteins are still to be elucidated.

SPECIES SPECIFICITY OF GROWTH HORMONES: VARIATIONS IN BIOLOGICAL PROPERTIES

Species specificity was observed early in the isolation of GHs, when it was recognized that non-primate mammalian GHs had little anabolic activity in man. GH from humans is active in man (and in most other species tested) and this has formed the basis of the widely-used treatment for hypopituitary dwarfism. In accordance with their low growth-promoting activity in man, GHs from non-primates also show only low ability to bind to GH receptors derived from human cells or tissues (eg Ref. 18). These differences between the biological activities of human and non-primate GHs presumably relate to the marked structural differences between the proteins, discussed above, although the precise differences underlying them are not clear. It is notable also that human GH shows much more marked lactogenic properties, in a number of systems, than the non-primate GHs.

GHs from teleost fish appear to have little growth-promoting activity in mammals, though mammalian GHs are active in fish. Again the large differences between the amino acid sequences of GHs from teleost fish and mammals presumably underlie the differences in activity.

GHs from different species vary considerably in their ability to bind to 'receptors' prepared from different tissues. Receptors in membrane preparations prepared from rabbit liver have been most studied (4, 14, 39, 47). In this system rat, mouse and pig GHs appear to be much less active than the human, bovine or ovine hormones (4, 47). These various hormones all have very similar growth-promoting potencies when assayed in rats. It is not clear what underlies their different abilities to bind to rabbit liver receptors, but the results do suggest that potencies in radioreceptor assays do not correlate well with those determined by bioassay. The activities of GHs and prolactins from various species in radioreceptor assays from a number of tissues and species have been surveyed and discussed in detail by Nicoll et al. (28).

STRUCTURE-FUNCTION RELATIONSHIPS

A good deal of work has been reported on the investigation of structure-function relationships in GH and related proteins, based on comparison of sequences from different species and on chemical and enzymic modification studies. The recent description of the tertiary structure of pig GH will clearly influence future consideration of such relationships, and the production of variants using recombinant DNA techniques should provide a new source of well-defined derivatives. Only a few of the results obtained on structure-function relationships can be considered here. A detailed and thoughtful discussion of structure-function information that can be derived from sequence comparisons has been presented recently by Nicoll et al. (27).

Disulphide bridges

All GHs described so far have two disulphide bridges, one linking distant parts of the primary structure (in bovine GH residues Cys-53 and Cys-164), the other forming a loop near the C-terminus. The C-terminal disulphide bridge is relatively easily modified; selective modification leads to very little loss of growth-promoting activity (10, 11). In human GH reduction of both disulphide bridges can be achieved without using denaturing conditions. If the cysteine residues so produced are modified by alkylation with iodoacetic acid (with introduction of additional negative charges) the hormone loses biological activity, but alkylation with iodoacetamide gives a form of the hormone in which biological activity, and tertiary structure, is retained (6, 11). The activity of human GH in which the disulphide-bridge structure is disrupted has also been demonstrated by production of an active derivative in which Cys-165 is substituted by alanine (37) by recombinant DNA methods. Thus the tertiary structure and biological activity of the molecule is not dependent upon intact disulphide bridges, though it cannot be ruled out that such bridges stabilize the molecule in vivo.

Limited enzymic cleavage

With a number of specific enzymes, GHs can be cleaved very selectively to give discrete derivatives. Cleavage of human GH with plasmin provides a much-studied case. The enzyme cleaves after only two residues, removing a hexapeptide from the molecule and leaving a two-chain GH in which the two chains are held together by a disulphide bridge. The tertiary structure of the protein appears to be retained in the derivative, and plasmin-digested human GH appears to retain full biological activity (24); indeed, there have been some suggestions that such plasmin treatment can lead to GHs with enhanced activity, though this is controversial. After reduction and alkylation of disulphide bridges the two chains can be separated. The larger, N-terminal fragment retains significant, low biological activity. Full activity can be regained by allowing refolding of the two chains under appropriate conditions, and this has allowed some studies on the importance of the C-terminal fragment for biological activity (22).

The ability of plasmin-derived fragments of human GH to interact with antibodies has also been investigated (43). A panel of 7 monoclonal antibodies was unable to distinguish between intact GH and plasmin-cleaved GH, indicating that the residues removed by plasmin (residues 135-140) are not involved in antigenic determinants recognized by any of these antibodies. When the N- and C-terminal fragments of plasmin-digested GH were separated, ability to interact with antibodies was largely lost in the case of most of the monoclonal antibodies and a polyclonal antibody to human GH, but two of the monoclonal antibodies bound effectively to the N-terminal fragment, indicating that the epitope to which they were directed is retained in this fragment; it is presumably sequence-dependent or directed against a component of the tertiary structure that is retained in hGH_{1-134}.

Plasmin digestion of bovine GH is less specific, and gives rise to a less well-defined mixture of derivatives than is seen in the case of human GH (35).

The multiple actions of growth hormone

GH is known to have a range of actions (42). In addition to (and perhaps components of) its actions on growth promotion, it stimulates production of somatomedin C/IGF-I and a number of other proteins, and has a variety of metabolic effects including insulin-like and diabetogenic actions on carbohydrate metabolism, and lipid mobilizing effects. It has been a matter of debate for many years as to whether all of these actions stem from a single primary event (presumably interaction of GH with a single receptor) or whether they involve interaction with a number of different receptors, recognizing different regions of the tertiary structure of the hormone.

Evidence for multiple receptors for GH has come from a number of approaches, including studies using monoclonal antibodies to the hormone (36), monoclonal antibodies to the receptor (3), binding specificity (4, 14, 47) and crosslinking reagents (44, 50). However, it remains unclear whether the different receptor types identified can be associated with specific actions of the hormone. The recent cloning of the GH receptor (19) has demonstrated that at least some of the multiple receptor types are probably products of a single gene.

There have been some claims that chemical modification of GH gives derivatives with altered relative potencies in assays measuring different biological effects of the hormone (see Ref. 48 for review). Similarly there have been claims of small fragments that retain just some of the biological actions of the hormone, including insulin-like, diabetogenic and lipolytic effects. Some of this work is controversial, however, and definitive evidence for separation of the various actions of the hormone is yet to be obtained.

Modification of the N-terminal sequence of GH

A number of interesting studies has resulted in the production of GHs with altered N-terminal sequences, by recombinant DNA techniques. Methionyl human GH (9) has now been widely used for clinical studies. Methionyl bovine and porcine GH have also been reported (33). Variants of bovine GH with short N-terminal additions or deletions have also been described (45, 49) (Table 2). In most of these cases modification of the N-terminus appeared to have no effect on the biological activities of the hormone, although addition of methionine to the N-terminus of human GH did decrease its interaction with one of a panel of monoclonal antibodies (2). A more extensive deletion of sequence at the N-terminus of human GH did give a derivative with altered biological properties (8). Human GH lacking residues 1-13 had much decreased biological activity in a number of assays. It retained the ability to bind to GH receptors, however, and was in fact an inhibitor of the actions of unmodified hGH in a number of lactogenic assays.

Table 2 N-terminal sequences of some recombinant DNA-derived GHs

		Reference
Human GH	1 5 10 15 MFPTIPLSRLFDNAMLRA------------	9
	15 MLRA------------	8
Bovine GH	1 5 10 15 MFPAMSLSGLFANAVLRA------------	33, 49
	10 15 SLFANAVLRA------------	49
Ovine GH	2 5 10 (TMITNSGD)FPAMSLSGLF-----------	45
	4 10 (TMITNS)AMSLSGLF---------------	Wallis & Wallis (see page 226)

Growth-promoting activity of ovine and bovine prolactin

Although prolactins are clearly distinct from GH, they do show some growth-promoting activity in some systems (eg Ref. 13). The physiological significance of such activity is not clear, but it may provide some information about structure-function relationships in somatogenic hormones. Both ovine and bovine prolactin can displace labelled human GH from rabbit liver GH receptors. Interestingly, they differ in their ability to do this, bovine prolactin being about 3-fold more potent than the ovine hormone (4). Since the two hormones differ at only 2-3 amino acid residues (40), though different extents of glycosylation are also possible, this difference in biological activity is surprising and must be the consequence of quite a small structural alteration. We therefore investigated its basis in more detail (46).

Several different ovine and bovine prolactin preparations were compared. Their potencies were not significantly different when measured by radioimmunoassay or assays measuring lactogenic properties (radioreceptor assay based on rabbit mammary gland receptors; Nb2 cell bioassay), but bovine prolactin was about 3 times more active than the ovine hormone in assays measuring somatogenic properties (radioreceptor assay based on rabbit liver receptors; induction of somatomedin C/IGF-1 in hypophysectomized rats). For all preparations studied, contamination with GH was very low and could not explain the somatogenic activity observed. It thus seems that the small structural differences between ovine and bovine prolactin are sufficient to cause a substantial difference in somatogenic activity.

CONCLUSIONS

Despite the accumulation of a considerable amount of information about the structure and chemistry of GHs and related proteins from various species, the features of these proteins that contribute to their biological actions remain unclear. From the studies outlined above it is clear that some fairly large structural modifications can be introduced without altering biological activity, but that in other cases small changes may have a substantial effect. Perhaps the most crucial question relates to the possibility of separating the various growth-promoting and metabolic actions of the hormone. Work with recombinant DNA-produced analogues of GH presents the best hope of defining structure function relationships in more detail in the future.

REFERENCES

1 Abdel-Meguid, S.S., Shieh, H.-S., Smith, W.W., Dayringer, H.E., Violand, B.N. and Bentle, L.A. (1987). Three-dimensional structure of a genetically engineered variant of porcine growth hormone. Proceedings of the National Academy of Sciences, U.S.A. **84**, 6434-6437.

2 Aston, R., Cooper, L., Holder, A., Ivanyi, J. and Preece, M. (1985). Monoclonal antibodies to human growth hormone can distinguish between pituitary and genetically engineered forms. Molecular Immunology **22**, 271-275.

3 Barnard, R., Bundesen, P.G., Rylatt, D.B. and Waters, M.J. (1985). Evidence from the use of monoclonal antibody probes for structural heterogeneity of the growth hormone receptor. Biochemical Journal **231**, 459-468.

4 Cadman, H.F. and Wallis, M. (1981). An investigation of sites that bind human somatotropin (growth hormone) in the liver of the pregnant rabbit. Biochemical Journal **198**, 605-614.

5 Chen, H.-T., Pan, F.-M and Chang, W.-C. (1988). Purification of duck growth hormone and cloning of the complementary DNA. Biochimica et Biophysica Acta **949**, 247-251.

6 Dixon, J.S. and Li, C.H. (1966). Retention of the biological potency of human pituitary growth hormone after reduction and carbamidomethylation. Science **154**, 785-786.

7 Duckworth, M.L., Kirk, K.L. and Friesen, H.G. (1986). Isolation and identification of a cDNA clone of rat placental lactogen II. Journal of Biological Chemistry **261**, 10871-10878.

8 Gertler, A., Shamay, A., Cohen, N., Ashkenazi, A., Friesen, H.G., Levanon,A., Gorecki, M., Aviv, H., Hadary, D. and Vogel, T (1986). Inhibition of lactogenic activities of ovine prolaction and human growth hormone (hGH) by a novel form of a modified recombinant hGH. Endocrinology **118**, 720-726.

9 Goeddel, D.V., Heyneker, H.L., Hozumi, T., Arentzen, R., Itakura, K., Yansura, D.G., Ross, M.J., Miozzari, G., Crea, R and Seeburg, P.H. (1979). Direct expression in Escherichia coli of a DNA sequence coding for human growth hormone. Nature **281**, 544-548.

10 Graf, L., Li, C.H. and Bewley, T.A. (1975). Selective reduction and alkylation of the COOH-terminal disulfide bridge in bovine growth hormone. International Journal of Peptide and Protein Research **7**, 467-473.

11 Graf, L., Borvendeg, J., Barat, E., Hermann, I. and Patthy, A. (1976). Reactivity and biological importance of the disulfide bonds in human growth hormone. FEBS Letters **66**, 233-237.

12 Hirt, H., Kimelman, J., Birnbaum, M.J., Chen, E.Y., Seeburg, P.H., Eberhardt, N.L. and Barta, A. (1987). The human growth hormone gene locus: Structure, evolution and allelic variations. DNA **6**, 59-70.

13 Holder, A.T. and Wallis, M. (1977). Actions of growth hormone, prolactin and thyroxine on serum somatomedin-like activity and growth in hypopituitary dwarf mice. Journal of Endocrinology **74**, 223-229.

14 Hughes, J.P. (1979). Identification and characterization of high and low affinity binding sites for growth hormone in rabbit liver. Endocrinology **105**, 414-420.

15 Jackson, L.L., Colosi, P., Talamantes, F. and Linzer, D.I.H. (1986). Molecular cloning of mouse placental lactogen cDNA. Proceedings of the National Academy of Sciences, U.S.A. **83**, 8496-8500.

16 King, J.L. and Jukes, T.H. (1969). Non-Darwinian evolution. Science **164**, 788-798.

17 Lee, S.-J., Talamantes, F., Wilder, E., Linzer, D.I.H. and Nathans, D. (1988). Trophoblastic giant cells of the mouse placenta as the site of proliferin synthesis. Endocrinology **122**, 1761-1768.

18 Lesniak, M.A. and Roth, J. (1976). Regulation of receptor concentration by homologous hormone. Effect of human growth hormone on its receptor in IM-9 lymphocytes. Journal of Biological Chemistry **251**, 3720-3729.

19 Leung, D.W., Spencer, S.A., Cachianes, G., Hammonds, R.G., Collins, C., Henzel, W.J., Barnard, R., Waters, M.J. and Wood, W.I. (1987). Growth hormone receptor and serum binding protein: Purification, cloning and expression. Nature **330,** 537-543.

20 Lewis, U.J. (1984). Variants of growth hormone and prolactin and their posttranslational modifications. Annual Review of Physiology **46,** 33-42.

21 Lewis, U.J., Singh, R.N.P., Lindsey, T.T., Seavey, B.K. and Lambert, T.H. (1973). Enzymically modified growth hormones and the diabetogenic activity of human growth hormone. In Advances in Human Growth Hormone Research (Raiti, S., ed.), pp 349-363. U.S. Department of Health, Education and Welfare, Washington DC.

22 Li, C.H. (1978). Noncovalent interaction of the NH_2-terminal fragment of human somatotropin with the COOH-terminal fragment of human choriomammotropin to generate growth-promoting activity. Proceedings of the National Academy of Sciences, U.S.A. **75,** 1700-1702.

23 Li, C.H., Chung, D., Lahm, H. and Stein, S. (1986). The primary structure of monkey pituitary growth hormone. Archives of Biochemistry and Biophysics **245,** 287-291.

24 Li, C.H. and Graf, L. (1974). Human pituitary growth hormone: Isolation and properties of two biologically active fragments from plasmin digests. Proceedings of the National Academy of Sciences, U.S.A. **71,** 1197-1201.

25 Li, C.H., Liu, W-K. and Dixon, J.S. (1966). Human pituitary growth hormone. XII The amino acid sequence of growth hormone. Journal of the American Chemical Society **88,** 2050-2051.

26 Miller, W.L. and Eberhardt, N.L. (1983). Structure and evolution of the growth-hormone gene family. Endocrine Reviews **4,** 97-130.

27 Nicoll, C.S., Mayer, G.L. and Russell, S.M. (1986). Structural features of prolactins and growth hormones that can be related to their biological properties. Endocrine Reviews **7,** 169-203.

28 Nicoll, C.S, Tarpey, J.F., Mayer, G.L. and Russell, S.M. (1986). Similarities and differences among prolactins and growth hormones and their receptors. American Zoologist **26,** 965-983.

29 Pan, F.-M. and Chang, W.-C. (1988). Cloning and sequencing of bullfrog growth hormone complementary DNA. Biochimica et Biophysica Acta **950,** 238-242.

30 Sato, N., Watanabe, K., Murata, K., Sakaguchi, M., Kariya, Y., Kimura, S., Nonaka, M., and Kimura, A. (1988). Molecular cloning and nucleotide sequence of tuna growth hormone cDNA. Biochimica et Biophysica Acta **949,** 35-42.

31 Schuler, L.A. and Hurley, W.L. (1987). Molecular cloning of a prolactin-related mRNA expressed in bovine placenta. Proceedings of the National Academy of Sciences, U.S.A. **84,** 5650-5654.

32 Seeburg, P.H., Shine, J., Martial, J.A., Baxter, J.D. and Goodman, H.M. (1977). Nucleotide sequence and amplification in bacteria of structural gene for rat growth hormone. Nature **270,** 486-494.

33 Seeburg, P.H., Sias, S., Adelman, J., de Boer, H.A., Hayflick, J., Jhurani, P., Goeddel, D.V. and Heyneker, H.L. (1983). Efficient bacterial expression of bovine and porcine growth-hormones. DNA **2**, 37-45.

34 Souza, L.M., Boone, T.C., Murdock, D., Langley, K., Wypych, J., Fenton, D., Johnson, S., Lai, P.H., Everett, R., Hsu, R.Y. and Bosselman, R. (1984). Application of recombinant DNA technologies to studies on chicken growth hormone. Journal of Experimental Zoology **232**, 465-473.

35 Surowy, T.K. and Wallis, M. (1983). Modification of bovine somatotropin (growth hormone) with plasmin. Journal of Protein Chemistry **2**, 347-362.

36 Thomas, H.M., Green, I.C., Wallis, M. and Aston, R. (1987). Heterogeneity of growth-hormone receptors detected with monoclonal antibodies to human growth hormone. Biochemical Journal **243**, 365-372.

37 Tokunaga, T., Tanaka, T., Ikehara, M. and Ohtsuka, E. (1985). Synthesis and expression of a human growth hormone (somatotropin) gene to change cysteine-165 to alanine. European Journal of Biochemistry **153**, 445-449.

38 Tsubokawa, M. and Kawauchi, H. (1985). Complete amino acid sequence of fin whale growth hormone. International Journal of Peptide and Protein Research **25**, 297-304.

39 Tsushima, T. and Friesen, H.G. (1973). Radioreceptor assay for growth hormone. Journal of Clinical Endocrinology and Metabolism **37**, 334-337.

40 Wallis, M. (1978). The chemistry of pituitary growth hormone, prolactin and related hormones, and its relationship to biological activity. In Chemistry and Biochemistry of Amino Acids, Peptides and Proteins (Weinstein, B., ed.), vol. 5, pp 213-320. Dekker, New York.

41 Wallis, M. (1981). The molecular evolution of pituitary growth hormone, prolactin and placental lactogen: A protein family showing variable rates of evolution. Journal of Molecular Evolution **17**, 10-18.

42 Wallis, M. (1988). Mechanism of action of growth hormone. In Hormones and their Actions (Cooke, B.A., King, R.J.B. and van der Molen, H.J., eds.), part 1, pp 265-294. Elsevier, Amsterdam.

43 Wallis, M. and Daniels, M. (1983). Binding specificity of monoclonal antibodies towards fragments of human growth hormone produced by plasmin digestion. FEBS Letters **159**, 241-245.

44 Wallis, M., Daniels, M. and Webb, C.F. (1987). Heterogeneity of rabbit liver growth hormone receptors studied by cross-linking. Journal of Endocrinology **112** (Suppl.), Abstract 110.

45 Wallis, M., Webb, C.F. and Daniels, M. (1988). Ovine and bovine prolactin show markedly different potencies in somatogenic radioreceptor assays, but similar potency in lactogenic assays. Journal of Endocrinology **117** (Suppl.), Abstract 118.

46 Wallis, O.C. and Wallis, M. (1987). A recombinant DNA-derived ovine growth hormone analogue expressed in Escherichia coli. Journal of Endocrinological Investigation **10** (Suppl. 4), 8.

47 Webb, C.F., Cadman, H.F. and Wallis, M. (1986). The specificity of binding of growth hormone and prolactin to purified plasma membranes from pregnant-rabbit liver. Biochemical Journal **236,** 657-663.

48 Wilhelmi, A.E. (1982). Structure and function in growth hormone. In Hormone Drugs pp 369-381. U.S. Pharmacopeial Convention Inc., Rockville.

49 Wingfield, P.T., Graber, P., Buell, G., Rose, K., Simona, M.G. and Burleigh, B.D. (1987). Preparation and characterization of bovine growth hormones produced in recombinant Escherichia coli. Biochemical Journal **243,** 829-839.

50 Ymer, S.I. and Herington, A.C. (1987). Structural studies on membrane-bound and soluble growth-hormone-binding proteins of rabbit liver. Biochemical Journal **242,** 713-720.

GROWTH HORMONE RECEPTORS AND BINDING PROTEINS

M.J.Waters[1],S.A.Spencer[2],D.Leung[2],R.G.Hammonds[2],G.Cachianes[2],W.J.Henzel[2], W.I. Wood[2], R. Barnard[1], P. Quirk[1] and G. Hamlin[1]

[1] Department of Physiology and Pharmacology, University of Queensland, St Lucia, Queensland 4067, Australia
[2] Department of Developmental Biology, Genentech Inc, 460 Point San Bruno Blvd, San Francisco, CA 94080, U.S.A.

Growth hormone receptors are found in a wide variety of tissues and are thought to mediate the various actions of growth hormone. Recently, a growth hormone binding protein was demonstrated in serum and shown to have antigenic identity with the liver growth hormone receptor by use of a panel of monoclonal antibodies to the receptor. Here we describe the purification, part sequence and cloning of the rabbit liver growth hormone receptor. Purification and N terminal sequence analysis of the rabbit serum binding protein for GH showed it to be identical to the extracellular region of the rabbit liver GH receptor. Rabbit liver receptor and human binding protein sequences were expressed in COS-7 cells and shown to display predicted hormone specificity and antigenic characteristics. Finally, the rabbit mammary gland prolactin receptor was purified and part sequenced. This showed 34% homology with the rabbit liver GH receptor, and therefore constitutes the second member of a new class of transmembrane receptors regulating growth and lactation.

INTRODUCTION

Growth hormone (GH) acts to promote cell growth in tissues of the body through the actions of locally and systemically produced IGF-I and through direct actions on the tissues themselves (26, 28). All of these actions are believed to result in some unknown way from the initial interaction of GH with its plasma membrane bound receptor. Since the original identification of the rabbit liver GH receptor by Friesen's group in 1973 (42), a number of groups have reported studies on GH receptors in hepatocytes (2, 40, 45), adipocytes (13, 25), lymphocytes (30), macrophages (29), fibroblasts (35), preadipocytes (36), chondrocytes (28), β islet cells (10), and osteoblasts (4) from a number of species. These receptors generally appear in affinity crosslinking experiments as disulphide bonded dimers and trimers of a Mr 110,000 binding subunit, with fragmentation of the liver receptor to lower Mr forms without loss of binding ability. Where this has been examined these receptors are sialoglycoproteins with broad species specificity for their hormone (primate receptors are the exception to this). Receptor subtypes seem to occur (2, 23, 27, 47), perhaps mediating different actions of GH (eg growth versus metabolism). Where a high and low affinity receptor occur together (eg rat (8) and bovine (12) liver), physiological factors regulating receptor concentration appear to affect the high affinity form. The adipose (22) and liver (9) receptor are chronically induced by GH itself, although the liver receptor, along with the IM9 lymphocyte and mouse fibroblast receptor have been shown to be acutely down regulated by GH (34). Insulin (8) and thyroxine (21) upregulate the receptor, and fasting (32), protein malnutrition (33) and renal insufficiency (18) lower receptor levels. There are some examples of apparent genetic receptor defects, both animal ((dw mice (20) and Brattleboro rats (37)), and human (Laron dwarfs (16)).

All of the above studies on GH action and receptor regulation have been handicapped by a lack of knowledge of the structure of the receptor. We have addressed this problem by defining the structure of the liver GH receptor in the hope that it may give clues to the signal transduction mechanism. Our work has provided the additional benefit of defining the relationship between this liver receptor and the GH binding protein (BP).

RECEPTOR PURIFICATION AND CLONING

We chose to clone the rabbit liver GH receptor for a number of reasons: abundance, ease of solubilization, antigenic similarity to the tibial growth plate GH receptor (3), similarity of regulation to the better studied rat liver receptor, and because the liver is the major source of GH dependent IGF-I. In addition, we had earlier developed an affinity chromatography method for purification of this receptor (43).

Because initial attempts to clone the receptor from expressed cDNA and genomic libraries using receptor monoclonals were unsuccessful, we used an oligonucleotide probe based on a part sequence from the purified receptor for cloning (30). The receptor was purified to ~ 40% purity (59,000 fold, yield 30%, based on Scatchard analysis) from solubilized microsomes using a small controlled pore glass-hGH affinity column which was extensively washed with 0.5 M NaCl and 4 M urea before elution of the receptor with 4 M $MgCl_2$ (41). Apart from its high capacity which allows for a reduction in size and hence non-specific binding, the advantage of this column over the previous Affigel hGH column lies in its high affinity for receptor, which means that it can be washed with urea before eluting the majority of the receptor with $MgCl_2$. This further lowers non- specific binding. The low level of prolactin receptor which elutes in the $MgCl_2$ fraction is removed in the subsequent electroelution procedure. Blockade of the prolactin receptor component of [^{125}I]-hGH binding (with ovine prolactin) yields a linear Scatchard plot of the expected affinity (10^{10} M^{-1}).

A prominent Mr 130,000 band, identified as the receptor by immuno and hormone blotting, was electroeluted from reduced SDS gels and cleaved with trypsin or S. aureus V8 protease to yield peptide fragments which were purified by reverse phase HPLC. Gas phase sequencing of 10 peptides and the N terminus provided 140 residues of sequence. N-terminal analysis revealed ubiquitin sequence at 20-50% that of the receptor. Since covalently attached ubiquitin mediates the ATP dependent degradation of proteins, this finding fits with studies on the rat liver (7) and adipose tissue receptors (24) which demonstrate that these receptors are turned over rapidly (t½ ~ 45 min) independent of hormone binding. The PDGF and lymphocyte homing receptors are also ubiquitinylated (30).

Screening of λgt 10 cDNA libraries (random and oligo dT primed) of rabbit liver with a 57 mer oligonucleotide probe based on a tryptic sequence (T_4, 538-556) yielded 29 positive clones. Some of these were sequenced by the dideoxy method after cloning into PUC 118/119 vectors, and using fragment probes obtained from this initial set of clones the entire sequence of the rabbit receptor was elucidated from the cDNA library. One of the fragment probes was then used to screen a human liver cDNA library, and this yielded the full human sequence by a similar strategy. For both the rabbit and human sequences two independent clones were isolated and sequenced for all coding regions. The rabbit sequence is 3880 bp (including a 1965 bp 3′ untranslated region) and the human sequence is 4375 bp, with a 2400 bp 3′ untranslated sequence. This translates to a 638 amino acid protein with an 18 amino acid signal sequence, giving a 620 residue mature protein. Alternatively spliced 5′ untranslated sequences were found 12 bp 5′ of the start codon. In addition, two independent rabbit clones diverged from the others beginning equivalently to 5 residues C terminal to the transmembrane domain, such that 4 new amino acids were coded before encountering a stop codon. It is possible that this sequence corresponds to the binding protein precursor. Examination of the amino acid sequence translated from the full length nucleotide sequences shows an overall homology between rabbit and human of 84% (Figs. 1 and 2). A recent report of the N terminal sequence of a mouse liver receptor (39) shows 56% homology with the rabbit sequence in this poorly conserved region (rabbit/human 53%). Hydropathy analysis showed only one potential transmembrane sequence (247-271) and five potential N glycosylation sites in the

Fig. 1 Amino acid sequence of rabbit and human liver growth hormone receptors showing homology with part sequences of rabbit mammary prolactin receptor

```
                                                                    KETFTCW
RAB PrlR                   1                                         *** *
RAB
GHR     MDLWQLLLTVALAGSSDAFSGSEATPATLGRASESVQRVHPGLGTNSSGKPKFTKCRSPELETFSCH
HUMAN     SIGNAL  SEQUENCE       A I S  PW L S N   K    KE          R
GHR
(non identical)


        WRPGADGGL  PTXYTLTYHK                          KKHTXIWXIYIITVXANXQ
        *  *   **  *    * *                               * **  * *    *
50      WTDGVHHGLKSPGSVQLFYIRRNTQEWTQEWKECPDYVSAGENSCYFNSSYTSIWIPYCIKLTNNGG
           E    T NL PI    T                             F            S


                                               KWLPPTLVDVRSGWLXXQYEIRLKPEPA
                                                * **   **  **    **   *
118     MVDQKCFSVEEIVQPDPPIGLNWTLLNVSLTGIHADIQVRWEPPPNADVQKGWIVLEYELQYKEVNE
        T  E     D         A                      A R   I    M

        AEWETHFAGQQTQFKILSLYPGQK
*          *     **
184     TQWKMMDPVLSTSVPVYSLRLDKEYEVRVRSRQRSSEKYGEFSEVLYVTLPQMSPFTCEEDFR
         K      I T        KV          K  N GN               Q       Y


                                                  KPFFTHLLE
        TRANSMEMBRANE  SEQUENCE                   *     **
247     FPWFLIIIFGIFGLTVMLFVFIFSKQQRIKMLILPPVPVPKIKGIDPDLLKEGKLEEVNTILA
           L                 L


310     IQDSYKPEFYNDDSWVEFIELDIDDPDEKTEGSDTDRLLSNSHQKSLSVLAAKDDDSGRTSCY
         H       HS             E      E       SD  E  H N GV  G       C


373     EPDILENDFNASDGCDGNSEVAQPQRLKGEADLLCLDQKNQNNSPYHDVSPAAQQPEVVLAEED
              T    N IHE T                              AC  T   S IQ  KN


437     KPRPLLTGEIESTLQAAPSQLSNPNSLANIDFYAQVSDITPAGSVVLSPGQKNKAGNSQCDAHP
          Q  P EGA   H   HI     S  S                            M    M


501     EVVSLCQTNFIMDNAYFCEADAKKCIAVAPHVDVESRVEPSFNQEDIYITTESLTTTAERSGTA
         M     E  L               P    IK   HIQ  L              A G P  G


565     EDAPGSEMPVPDYTSIHLVQSPQGLVLNAATLPLPDKEFLSSCGYVSTDQLNKILP
         HV              I       I   TA                       M
```

extracellular region. Glycosylation could largely account for the discrepancy between the calculated molecular weight (70 kDa) and the Mr on SDS gels (130 kDa). Since there are 7 cysteines in the extracellular region, the receptor is a thiol protein, and this would account for its propensity to form disulphide linked polymers, and could be a factor in signal transduction mediated by microaggregation.

A full length sequence of the rabbit receptor was expressed in monkey COS-7 cells with an expression vector previously developed for factor VIII. The receptor was expressed at about 200,000 copies per cell and displayed the expected specificity characteristics (hGH > bGH > oPRL) with one class of receptor apparent by Scatchard analysis (Fig. 3). Interestingly, epitope expression analysis with two inhibitory GH receptor monoclonal antibodies showed two classes of sites for MAb 263 despite the fact that expression derived from a single gene construct. It seems the epitope recognized by this monoclonal can be modified in some way post translationally. Our recent studies indicate that this epitope does not include carbohydrate determinants, ubiquitin or phosphorylated residues. A third receptor subtype (type 3 - (4)) was not seen with the expressed receptor (the residual binding seen at maximal MAb 7 concentrations is endogenous prolactin receptor in this kidney cell line).

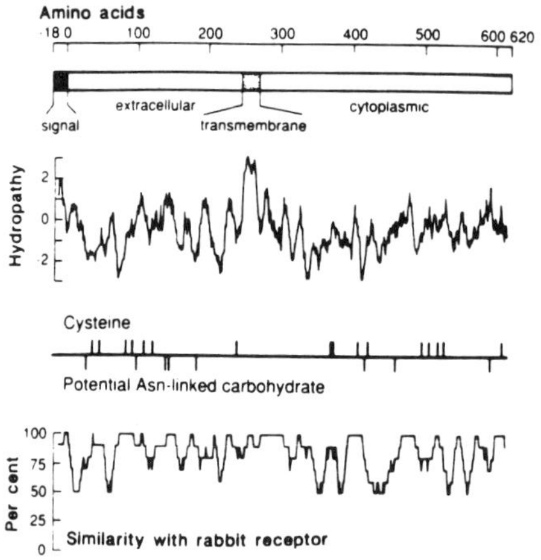

Fig. 2 Schematic of the human GH receptor. The hydropathy plot is from the method of Kyte and Doolittle with a window of 10 residues; positive values indicate increasing hydrophobicity. The potential N-linked glycosylation sites are NXS/T. The similarity with the rabbit receptor is for exact matches over a window of 10 amino acids. (Reproduced by permission of Nature, 31).

SERUM BINDING PROTEIN

This was first reported in serum by Herington's (46) and Baumann's (6) groups, and then shown to be antigenically identical to the liver receptor in the rabbit by use of a panel of monoclonal antibodies (5). Fig. 4 shows that the human serum BP can be immunoprecipitated by 4 of 7 monoclonal antibodies to the rabbit liver GH receptor. The immunoprecipitated BP shows primate GH specificity, a characteristic of all known human GH receptors.

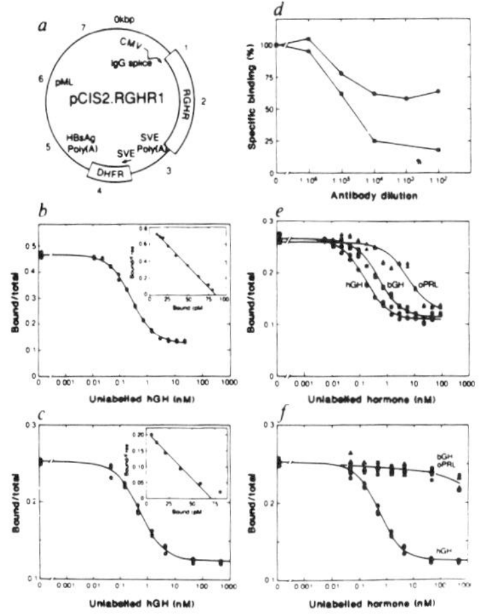

Fig. 3 Expression of rabbit and human GH receptor cDNA clones in COS-7 monkey kidney cells

(a) Expression vector. (b) Competition for expressed rabbit GHR by hGH, with Scatchard analysis. (c) Competition for expressed soluble human BP by hGH, with Scatchard analysis. (d) Competition for $[^{125}I]$-hGH binding to expressed rabbit receptor by -●- MAb 263, -■- MAb 7. (e) Competition for expressed rabbit receptor by relevant hormones. (f) Competition for expressed human receptor by relevant hormones. (Reproduced by permission of Nature, 31).

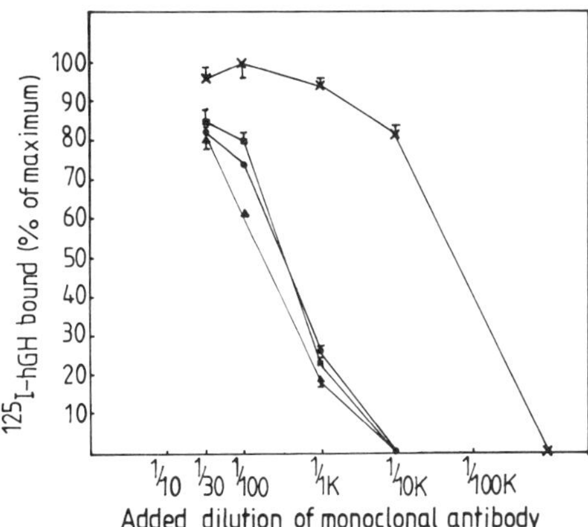

Fig. 4 Immunoprecipitation of human serum binding protein-$[^{125}I]$ hGH complex by rabbit liver GH receptor monoclonals

-x- MAb 263, -■- MAb 43, -●- MAb 1, -▲- MAb 5 250 fmoles human serum BP purified by CPG-hGH affinity chromatography (4M urea eluate) assay methodology as in (5).

The relationship between the major liver GH receptor and the serum binding protein for GH was clearly demonstrated by monoclonal antibody cross reactivity in the rabbit (5). However, in order to provide definitive proof of this relationship we purified the rabbit serum BP 396,000 fold by a two step procedure involving the CPG-hGH column followed by gelfiltration on Sephacryl S-300. The resulting material showed a major, broad band at Mr 51,000 on reduced SDS gels, and this was shown to correspond to the BP subunit by affinity crosslinking. This band was electroeluted and its N terminal sequence determined. Comparison of the 3 simultaneous N terminal sequences obtained with that of the rabbit GH receptor showed identity for 37 residues, with the N terminal heterogeneity being a result of partial proteolysis of the main sequence. Cyanogen bromide cleavage of the residue left after N terminal sequencing followed by further sequence analysis of that residue extended the sequence homology to residue 239 of the GH receptor, 8 residues before the transmembrane segment. The equivalent human sequence was expressed in COS-7 cells by use of a construct in the expression vector which made use of a 70 bp oligonucleotide to terminate the human sequence at residue 247, adjacent to the transmembrane segment. Transfected cells produced a soluble GH binding protein which bound only human (not bovine) GH with high affinity (3×10^9 M^{-1}) and which was polyethylene glycol soluble unless an anti GH receptor monoclonal was used to render it precipitable, in an identical fashion to the native human BP (Fig. 5). Availability of this BP will facilitate a number of structural, physiological and clinical studies.

It has been possible to set up a two site immunoradiometric assay based on two of these monoclonals which is independent of plasma GH concentrations, and which shows a wide range of concentrations of BP in clinical samples (Fig. 5).

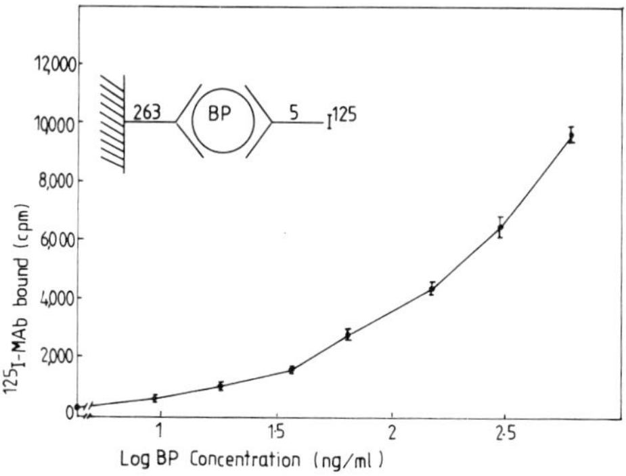

Fig. 5 Two site immunoradiometric assay for human GH binding protein

Standards are recombinant human BP. MAb 5 is radiolabelled trace. MAb 263 is solid phase adsorbed to a microtiter plate (manuscript in preparation).

HOMOLOGY WITH THE PROLACTIN RECEPTOR

A search of the protein sequence data banks for homologous proteins did not reveal any proteins homologous to our GH receptor sequence. This protein is not a tyrosine kinase and does not display any of the cysteine-rich extracellular regions seen with some other growth factor receptors.

Because of the homology between prolactins and growth hormones (25-30%) and because we had previously shown antigenic cross reactivity between GH and prolactin receptors (44), we elected to determine the sequence of the rabbit mammary gland prolactin receptor. This receptor was purified approximately 100,000 fold by a combination of hGH affinity chromatography and octyl Sepharose chromatography. The Mr 44,000 subunit was shown to correspond to the prolactin receptor by immunoblotting with a monoclonal antibody and by affinity crosslinking experiments.

Initial sequencing of this material demonstrated that it was N terminally blocked. Lysine C digestion of this material followed by reverse phase HPLC yielded six peptide sequences and 108 residues of sequence. It is clear from Fig. 3 that there is homology between these sequences and that of the rabbit liver GH receptor (34%). Subsequent to this work, Boutin et al. (11) reported the cloning and sequence of the rat liver receptor using part sequence based cDNA probes. Their sequence shows 69% identity to our rabbit mammary gland sequence, confirming our data. The rat liver prolactin receptor sequence is only 291 amino acids long, with a short (38 residue) cytoplasmic tail, which contrasts with the 349 residue tail of the GH receptor. It would seem likely that the GH receptor carries out a number of cytoplasmic functions with this cytoplasmic sequence, but that the functional role of the rat liver prolactin receptor is uncertain. Until the full mammary gland sequence is obtained by cloning we cannot be certain whether the protein that we have isolated (which is intimately involved in regulating lactogenesis) is a cleaved form of a larger receptor with a cytoplasmic region equivalent to that of the GH receptor. A higher Mr form of the rabbit and pig mammary prolactin receptor has been reported (15, 38).

PROSPECTIVE

A recent study (19) has provided evidence that the GH receptor in a number of cell types (including liver) is tyrosine phosphorylated on binding of GH, and these authors suggest that their GH receptor is a tyrosine kinase. As indicated, our sequence is not homologous to known tyrosine kinases, although there are tyrosine residues in the cytoplasmic region which are phosphorylatable. Although two other groups have been unable to demonstrate phosphorylation of the GH receptor (1, 17), we have recently been able to confirm the findings of Foster et al. (19) using 3T3-F442A preadipocytes. A small but significant proportion (1-6%) of affinity labelled 3T3 F442A preadipocyte GH receptors do bind to a phosphotyrosine antibody column, and these show a Mr of 130,000 on SDS gels. Accordingly, we must postulate either (i) the GH receptor is closely associated with a tyrosine kinase (subunit?) or (ii) there is a second GH receptor which is a tyrosine kinase. As mentioned in the introduction, there is a variety of evidence for GH receptor subtypes.

Our contention that the receptor we have cloned is intimately involved in the growth process is based on the following observations: (1) Laron dwarfs are not GH responsive, and lack both the liver GH receptor and serum GH binding protein (14, 16). (2) Restriction fragment patterns of the GH receptor sequence are altered in Laron dwarfs. (3) The 4.7 Kb mRNA for the cloned receptor is found in GH target tissues, including liver, 3T3-F442A preadipocytes and IM9 lymphocytes. (4) A monoclonal antibody able to stain GH receptors in growth plate chondrocytes of rabbit and human long bones also reacts with the expressed receptor in the predicted manner. (5) The mammary gland prolactin receptor shows 34% homology with the cloned GH receptor. Until we have expressed the receptor in a cell type able to produce a growth related response on addition of GH, it is not possible to state beyond doubt that we have cloned the definitive GH receptor. However, we believe the evidence strongly supports this conclusion.

ACKNOWLEDGEMENTS

Portions of this work were supported by NH&MRC (Australia) grants to MJW.

REFERENCES

1 Asakawa, K., Grunberger, G., McElduff, A. and Gorden, P. (1988). Polypeptide hormone receptor phosphorylation: Is there a role in receptor-mediated endocytosis of human growth hormone? Endocrinology **117**, 631-637.

2 Barnard, R., Bundesen, P.G., Rylatt, D.B. and Waters, M.J. (1985). Evidence from the use of monoclonal antibody probes for structural heterogeneity of the growth hormone receptor. Biochemical Journal **231**, 459-468.

3 Barnard, R., Haynes, K.M., Werther, G.A. and Waters, M.J. (1988). The ontogeny of growth hormone receptors in the rabbit tibia. Endocrinology **122**, 2562-2569.

4 Barnard, R., Martin, T.J., Ng, K.W. and Waters, M.J., unpublished.

5 Barnard, R. and Waters, M.J. (1986). Serum and liver cytosolic growth-hormone-binding proteins are antigenically identical with liver membrane 'receptor' types 1 and 2. Biochemical Journal **237**, 885-892.

6 Baumann, G., Stolar, M.W., Amburn, K., Barsanso, C.P. and de Vries, B.C. (1986). A specific growth hormone-binding protein in human plasma: Initial characterization. Journal of Clinical Endocrinology and Metabolism **62**, 134-141.

7 Baxter, R.C. (1985). Measurement of growth hormone and prolactin receptor turnover in rat liver. Endocrinology **117**, 650-655.

8 Baxter, R.C., Bryson, J.M. and Turtle, J.R. (1980). Somatogenic receptors of rat liver: Regulation by insulin. Endocrinology **107**, 1176-1181.

9 Baxter, R.C. and Zaltsman, Z. (1984). Induction of hepatic receptors for growth hormone (GH) and prolactin by GH infusion is sex independent. Endocrinology **115**, 2009-2014.

10 Billestrup, N. and Martin, J. (1985). Growth hormone binding to specific receptors stimulates growth and function of cloned insulin-producing rat insulinoma RIN-SAH cells. Endocrinology **116**, 1175-1181.

11 Boutin, J-M., Jollcoeur, C., Okamura, H., Gagnon, J., Edery, M., Shirota, M., Banville, D., Dusanter-Fourt, I., Djiane, J. and Kelly, P.A. (1988). Cloning and expression of the rat prolactin receptor, a member of the growth hormone/prolactin receptor gene family. Cell **53**, 69-77.

12 Breier, B.H., Gluckman, P.D. and Bass, J.J. (1988). The somatotrophic axis in young steers: Influence of nutritional status and oestradiol-17β on hepatic high- and low-affinity somatotrophic binding sites. Journal of Endocrinology **116**, 169-177.

13 Carter-Su, C., Schwartz, J. and Kukuchi, G. (1984). Identification of a high affinity growth hormone receptor in rat adipocyte membranes. Journal of Biological Chemistry **259**, 1099-1104.

14 Daughaday, W.H. and Trivedi, B. (1987). Absence of serum growth hormone binding protein in patients with growth hormone receptor deficiency (Laron dwarfism). Proceedings of the National Academy of Sciences U.S.A. **84**, 4636-4639.

15 Dusanter-Fourt, I., Kelly, P.AS. and Djiane, J. (1987). Immunological recognition of the prolactin receptor. Biochimie **69,** 639-646.

16 Eshet, R., Laron, Z., Pertzelan, A., Arnon, R. and Dintzman, M. (1984). Defect of human growth hormone receptors in the liver of two patients with Laron-type dwarfism. Israel Journal of Medical Science **20,** 8-11.

17 Fields, T.J. and Hughes, J.P. (1984). Phosphorylation of proteins in a partially purified preparation of liver receptors is regulated by human growth hormone. 7th International Congress of Endocrinology (Quebec) Abstr 664.

18 Finidori, J., Postel-Vinay, M.C. and Kleinknecht, C. (1980). Lactogenic and somatotropic binding sites in liver membranes of rats with renal insufficiency. Endocrinology **106,** 1960-1965.

19 Foster, C.M., Shafer, J.A., Rozsa, F.W., Wang, X., Lewis, S.D., Renken, D.A., Natale, J.E., Schwartz, J. and Carter-Su, C. (1988). Growth hormone promoted tyrosyl phosphorylation of growth hormone receptors in murine 3T3-F442A fibroblasts and adipocytes. Biochemistry **27,** 326-334.

20 Fouchereau-Peron, M., Broer, Y. and Rosselin, G. (1980). Growth hormone and insulin binding to isolated hepatocytes in the genetically dwarf mouse. Biochimica Biophysica Acta **631,** 451-462.

21 Fouchereau-Peron, M., Broer, Y. and Rosselin, G. (1981). Triiodothyroxine and growth hormone exert an opposite effect on the binding of growth hormone and insulin by hepatocytes from dwarf mouse. Biochimica Biophysica Acta **677,** 445-452.

22 Gause, I. and Eden, S. (1986). Induction of growth hormone (GH) receptors in adipocytes of hypophysectomized rats by GH. Endocrinology **118,** 119-124.

23 Goodman, H.M. and Kostyo, J.L. (1981). Altered profiles of biological activity of growth hormone fragments on adipocyte metabolism. Endocrinology **108,** 553-558.

24 Gorin, E. and Goodman, H.M. (1985). Turnover of growth hormone receptors in rat adipocytes. Endocrinology **116,** 1796-1805.

25 Gritching, G., Levy, L.K. and Goodman, H.M. (1983). Relationship between binding and biological effects of human growth hormone in rat adipocytes. Endocrinology **113,** 1111-1120.

26 Hughes, J.P. and Friesen, H.G. (1985). The nature and regulation of the receptors for pituitary growth hormone. Annual Review of Physiology **47,** 469-482.

27 Hughes, J.P., Tokuhiro, E., Simpson, J.S.A. and Friesen, H.G. (1983). 20K is bound with high affinity by one rat and one of two rabbit growth hormone receptors. Endocrinology **113,** 1904-1906.

28 Isaksson, O.G., Lindahl, A., Nilsson, A. and Isgaard, J. (1987). Mechanism of the stimulatory effect of growth hormone on longitudinal bone growth. Endocrine Reviews **8,** 426-438.

29 Kover, K., Hung, C.H. and Moore, W.V. (1986). The characteristics of hGH binding to the liver macrophages. Hormone and Metabolic Research **18,** 26-30.

30 Lesniak. M.A. and Roth. J. (1976). Regulation of receptor concentration by homologous hormone. Effect of human growth hormone on its receptor in IM-9 lymphocytes. Journal of Biological Chemistry **251**, 3720-3729.

31 Leung. D.W.. Spencer. S.A.. Cachianes. G.. Hammonds. R.G.. Collins. C.. Henzel. W.J.. Barnard. R.. Waters. M.J. and Wood. W.I. (1987). Growth hormone receptor and serum binding protein: purification. cloning and expression. Nature **330**, 537-543.

32 Maes. M.. Underwood. L.E. and Ketelslegers. J.M. (1982). Plasma somatomedin-C in fasted and refed rats: close relationship with changes in liver somatogenic but not lactogenic binding sites. Journal of Endocrinology **97**, 243-252.

33 Maes. M.. Underwood. L.E. and Ketelslegers. J.M. (1984). Low serum somatomedin-C in protein deficiency: relationship with changes in liver somatogenic and lactogenic binding sites. Molecular and Cellular Endocrinology **37**, 301-309.

34 Maiter. D.. Underwood. L.E.. Maes. M. and Ketelslegers. J.M. (1988). Acute down-regulation of the somatogenic receptors in rat liver by a single injection of growth hormone. Endocrinology **122**, 1291-1296.

35 Murphy. L.J.. Vrhovsek. E. and Lazarus. L. (1983). Identification and characterization of specific growth hormone receptors in cultured human fribroblasts. Journal of Clinical Endocrinology and Metabolism **57**, 1117-1124.

36 Nixon. T. and Green. H. (1983). Properties of growth hormone receptors in relation to the adipose conversion of 3T3 cells. Journal of Cellular Physiology **115**, 291-296.

37 Picard. F.B.. Wolf. B.A.. Hughes. J.N.. Duran. D.. Voirol. M.J.. Charrier. J.. Czernichow. P. and Postel Vinay. M-C. (1986). The Brattleboro rat: normal growth hormone secretion. decreased hepatic growth hormone receptors and low plasma somatomedin activity. Molecular and Cellular Endocrinology **45**, 49-56.

38 Sakai. S.. Katoh. M.. Berthon. P. and Kelly P.A. (1985). Characterization of prolactin receptors in pig mammary gland. Biochemical Journal **224**, 911-922.

39 Smith. W.C.. Colosi. P. and Talamantes. F. (1988). Isolation of two molecular weight variants of the mouse growth hormone receptor. Molecular Endocrinology **2**, 108-116.

40 Smith. W.C. and Talamantes. F. (1987). Identification and characterization of a heterogenous population of growth hormone receptors in mouse hepatic membranes. Journal of Biological Chemistry **262**, 2213-2219.

41 Spencer. S.A.. Hammonds. R.G.. Henzel. W.J.. Rodriguez. H.. Waters. M.J. and Wood. W.I. (1988). Rabbit liver growth hormone receptor and serum binding protein. Purification. characterization and sequence. Journal of Biological Chemistry **263**, 7862-7867.

42 Tsushima T. and Friesen. H.G. (1973). Radioreceptor assay for growth hormone. Journal of Clinical Endocrinological Metabolism **37**, 334-337.

43 Waters. M.J. and Friesen. H.G. (1979). Purification and partial characterization of a nonprimate growth hormone receptor. Journal of Biological Chemistry **254**, 6815-6825.

44 Waters, M.J., Lusins, S., Friesen, H.G. and Thorburn, G.D. (1980). Use of anti-receptor antibodies as probes for studying the structure of growth hormone and prolactin receptors. 7th International Congress of Endocrinology (Melbourne) Abst 844.

45 Yamada, K. and Donner, D.B. (1984). Structures of the somatotropin receptor and prolactin receptor of rat hepatocytes characterized by affinity labelling. Biochemical Journal **220,** 361-369.

46 Ymer, S.I. and Herington, A.C. (1985). Evidence for the specific binding of growth hormone to a receptor-like protein in rabbit serum. Molecular and Cellular Endocrinology **41,** 153-161.

47 Ymer, S.I. and Herington, A.C. (1987). Structural studies on membrane-bound and soluble growth-hormone-binding proteins of rabbit liver. Biochemical Journal **242,** 713-730.

THE REGULATION OF THE GROWTH HORMONE RECEPTOR

P.D. Gluckman and B.H. Breier

Developmental Physiology Laboratory, Department of Paediatrics, University of Auckland, Auckland, New Zealand

The first step in growth hormone (GH) action is binding to a cell membrane receptor. Heterogeneity of these cell surface binding sites is present at several levels. At least two affinity states exist for the somatogenic receptor and the capacity of higher affinity site correlates with growth rate in ruminants. The somatogenic receptor is absent in hepatic tissue in many species before birth, however its appearance and regulation in varying tissues may differ. Chronic GH administration increases GH binding in the liver of domestic animal species. Single doses of GH may also down regulate GH receptors. Nutrition is a dominant influence on GH receptors. In ruminants the number and affinity state of the receptor varies across the normal anticipated range of nutritional intake. The high affinity receptor is not demonstrable at maintenance intake. Oestradiol, which is a growth promotant in ruminants, has a major effect on the capacity of the GH receptor. The GH receptor is under active endocrine and nutritional regulation. In the ruminant such regulation is a major determinant of the state of the somatotropic axis.

INTRODUCTION

The interaction of growth hormone (GH) with binding sites on the cell surface is the first step in the biological action of GH; little is known of the subsequent steps in this cascade. High affinity membrane binding for GH can be demonstrated in many tissues although most studies have concentrated on the liver and adipose tissue, two major sites for GH action. GH also binds to cytosolic and circulating binding proteins; these are now known to be structurally related to the membrane GH receptor. This paper will focus on the physiological regulation of GH binding sites on cell membranes.

HETEROGENEITY OF THE GH RECEPTOR

Consideration of this field is complex because there are at least three major classes of receptor for hormones of the GH/prolactin (PRL) family. Somatogenic receptors have been defined as receptors which have higher affinity for classical somatogenic hormones such as bovine and ovine GH. Lactogenic receptors have highest affinity for hormones with strong lactogenic and weak somatogenic activity such as ovine PRL. Recently a specific receptor for placental lactogen (PL) has also been demonstrated in the sheep and its presence can be inferred in other species (17). Further, some hormones have affinity for both somatogenic and lactogenic receptors. This is notably the case for hGH in subprimate tissues: the use of hGH as ligand in many studies in rodents has probably led to confusion because of this heterogeneous binding. Conversely, primate somatogenic receptors appear to be highly specific for primate GHs.

Recently a somatogenic receptor was cloned from both the rabbit and human liver (25). There was a marked sequence homology despite the considerable evolutionary divergence in the sequence of GH itself. Further it was shown that the circulating binding protein for GH was identical to the extramembranous portion of the cloned receptor protein. Evidence that the cloned receptor is of biological significance is provided by the observation that in children with GH resistance due to Laron dwarfism both membrane binding and circulating binding protein are absent (15). However the receptor sequence does not contain any sequence consistent with known signal transduction mechanisms and the mechanism for signal transduction remains uncertain.

There is mounting evidence for heterogeneity of the GH receptor. Using a series of monoclonal antisera. Barnard and Waters demonstrated the presence of related but distinct receptor subtypes in the rabbit liver and further showed the cytosolic binding protein to be related to the membrane receptors, possibly representing either internalised or freshly synthesised receptor protein (2. 3).

Functional evidence for heterogeneity of the somatogenic receptor is provided by a number of radioligand binding studies. In reviewing such studies it is essential to consider the potential of the ligand used to also bind to lactogenic sites and whether ligand degradation has been excluded. Ideally studies should use homologous hormones to reduce the potential for erroneous interpretation. A number of studies have demonstrated that the binding of GH to membrane receptors is heterogeneous after excluding the possibility of binding to lactogenic sites. In these studies it has been possible to fit a two site model by Scatchard or similar analysis. Such studies include those by Hughes of rat or rabbit GH binding to the rabbit liver (23) and there are several similar reports in the rat (9. 34).

We have studied this hetereogeneity in cattle, sheep and pigs. In each of these species, provided the binding of GH is high, curvilinear Scatchard plots can be demonstrated. In such studies it is critical to firstly process the livers in the presence of proteolytic enzyme inhibitors and to secondly pretreat the membranes with $MgCl_2$ to remove endogenous hormone bound to high affinity receptors. In the absence of pretreatment with $MgCl_2$, Scatchard analysis of the binding of bGH to bovine hepatic membranes suggests a single binding site of low affinity (Ka 5 l/nmol) whereas after $MgCl_2$ washing a curvilinear plot is seen with the same low affinity site observed but a higher affinity site (Ka 50 l/nmol) is revealed (10). The regulation and significance of these two sites has been considered. The capacity of the high affinity site correlates highly with growth rate and plasma IGF-I levels in growing steers (10, 11). In contrast the capacity of the lower affinity site showed no relationship to growth rate. Further, in animals with detectable high affinity sites, GH induced a somatogenic response (a rise in plasma IGF-I) whereas in undernourished steers in which the high affinity receptor was not detectable, no response to intravenous bGH was seen (11). These studies strongly imply that the high affinity site is the functional somatogenic receptor. They suggest that the lower affinity site is of lesser significance or subserves an alternate function (eg clearance). Whether the two sites are alternate states of the same receptor protein remains to be determined. Preliminary evidence showed no difference in the inhibition of binding by a GH receptor monoclonal antiserum to tissues rich or deficient in the high affinity site (Gluckman, Waters, Barnard and Breier, unpublished). An alteration in the state of the receptor protein is compatable with data from monoclonal epitopic mapping (2. 3) which might suggest that the receptor protein may alter conformation. Such observations, if confirmed in other species, suggest that many studies based on linear models of GH binding may be of limited value and must be interpreted with caution.

ONTOGENY OF GH RECEPTORS

As reviewed elsewhere prenatal somatic growth is not regulated significantly by GH (19). In most species the binding of GH, at least to hepatic tissues, undergoes marked ontogenic change, binding being low or absent before birth. In general the appearance of the hepatic receptor correlates with the age at which growth has been demonstrated to develop GH dependence. In the rat binding of bGH was undetectable in the infant rat until 8 days postnatal and then rose preceding the pubertal growth spurt. Lactogenic receptors had a different ontogenic pattern (27). Other studies in the rodent have been conflicting presumably reflecting the methodological difficulties discussed previously.

In the sheep, the binding of oGH to hepatic membranes is insignificant before birth (16, 21) and is to receptors of lactogenic or placental lactogenic nature. oGH binding rapidly increases a few days after birth to adult values and this rise parallels a rise in plasma IGF-I concentration (21). Using covalent coupling to hepatic membrane proteins, bGH binding was not detectable in the calf at 2 days of age (1). bGH binding to liver membranes in calves rises gradually between 3 and 12 wk accompanied by a progressive rise in IGF-I levels (12).

In the pig we have demonstrated that the binding of both porcine and bovine GH to hepatic membranes rises gradually over the first 105 days after birth and then sharply between 105 and 165 days. Scatchard analysis suggests a 10 fold increase in the capacity of the higher affinity receptor over this period; the increase is paralleled by an increase in plasma IGF-I. Only in the older pigs are curvilinear plots seen due to the lower affinity site becoming apparent. However some binding to the liver is detectable at birth (unpublished observations). Similarly in man, binding of hGH to midgestation human liver has been reported (22). However the cell type to which binding is observed is not known and the fetal liver contains considerable hemopoietic tissue.

The ontogenesis of GH receptors is not necessarily parallel in all tissues. There is evidence that GH does have some action in the fetus of species where hepatic binding is absent prior to birth. For example in the sheep fetus, GH has been shown to be lipolytic and diabetogenic. In vitro studies have suggested that GH is essential to adipocyte differentiation. Similarly in the rat, GH appears to have a role in islet cell replication (for review see 20). Using anti-receptor antisera, GH receptors could be demonstrated in neonatal rabbit chondrocytes in the region of the secondary ossification centres although they were not detectable in the growth plate itself until somewhat later (4). Conversely in man where hepatic binding is demonstrable before birth, hGH does not bind to skeletal muscle from fetal tissues (22).

ENDOCRINE AND METABOLIC REGULATION OF THE GH RECEPTOR

Growth hormone

There is considerable evidence that chronic exposure to GH increases the number of GH binding sites in the liver and in adipose tissue. In both the rabbit and sheep, hypophysectomy reduces GH binding which is partially restored by GH treatment (30). There are similar data in the rat (6, 18) although contradictory observations are also reported. In the intact animal chronic GH administration has been demonstrated to increase the binding of GH to the liver in rats (6, 8) and pigs (14). We have studied lambs treated for 8 weeks with bGH in doses ranging from 0 to 0.55 mg/kg/day. The specific binding of oGH increased from 7.1% in saline treated lambs to 16.5% in lambs treated with 0.55 mg/kg/day bGH. Scatchard analysis showed that the increase in binding was primarily due to a 20 fold increase in the capacity of the high affinity site (Ka 5 l/nmol) with only a small increase in the capacity of the lower affinity site (Ka = 0.3 l/nmol). It is of interest that GH has the ability to increase the growth rate even under conditions of lower nutrition when high affinity GH receptors cannot normally be detected. It may be that the first step in a somatogenic response to exogenous GH under these conditions is an induction of the high affinity receptor state. The mechanistic basis of this induction cannot be discussed until the basis of differing affinity states of the GH receptor is understood. Limited data suggest that the adipocyte GH receptor is similarly maintained by GH (18).

There are in vivo data that might suggest that a single injection of GH may acutely and transiently down regulate its receptor at least in hypophysectomised rats (29). Such observations are supported by in vitro studies although it is not always possible to exclude artefacts due to receptor occupancy (31).

Nevertheless the number of functional receptors is clearly regulated by GH and at the doses of GH being used to manipulate animal growth, upregulation of the number of high affinity receptors may be a central aspect of the response to GH. Recent reports have raised the possibility of differential effects of pulsatile versus continuous GH therapy on the induction of the GH receptor and on the secretion of IGF-I (24). However the data are too limited at this time to assess the role of the pattern of GH secretion in determining tissue responsiveness.

Nutrition

Nutrition has a dominant influence on responsivity to GH. Observations in the fasted rat show that fasting leads to a rapid fall in GH binding to liver and binding is restored quickly on refeeding (26). Plasma IGF-I levels paralleled the changes in GH binding. With relative milk deprivation GH binding was depressed by 14 days postnatal, soon after GH receptors appear in this species (28). As insulin restores GH binding in diabetic rats (7) it has been suggested the nutritional effect may be mediated by insulin. There is additional evidence suggesting post receptor defects as well in malnourished rats (28); however the characterisation of the GH binding in such studies has been limited thus restricting interpretation of the data.

The nutritional influence on the GH receptor has been extensively investigated by us in the bovine (10, 11). Nutritional effects on the release of IGF-I can be demonstrated by 8 weeks of age (12) and the marked effect of weaning on IGF-I is presumably mediated by changes in GH receptors. Examination of GH secretion and plasma IGF-I levels over 24 hours in steers on high, intermediate or maintenance planes of nutrition demonstrate that in animals on high nutritional intake GH secretion was low but that at intermediate and maintenance planes of nutrition, GH secretion was higher although IGF-I secretion was maintained except at the lowest plane of intake. The progressive change in the GH:IGF-I ratio across the range of intakes suggested a change in GH responsitivity. When hepatic binding was examined on the highest and lowest planes of intake, we found that at maintenance no high affinity GH binding sites were demonstrable, whereas at the high plane they were always present although they were almost fully saturated at basal GH concentrations (10). Despite being fully saturated in vitro, the highly dynamic state of the GH receptor system (5) means that in vivo regulation of the number of high affinity receptors can have considerable functional significance. GH is secreted in a highly pulsatile fashion and has a rapid half life. The GH receptor has rapid turnover - estimates vary from 30 to 90 min (5, 32). Further there is evidence to suggest that turnover is increased by receptor occupancy (33).

The marked change in high affinity GH receptors across the anticipated range of intake in the ruminant (10) suggests that changing receptor number may be a major regulatory mechanism for determining activity within the somatotropic axis. Thus nutritional considerations may be critical in determining the utility of exogenous GH therapy. It is noteworthy that the high affinity site which correlates with growth rate (10) is essentially fully occupied by the normally low circulating GH concentration and it may be the upregulating effect of GH to increase the number of the high affinity receptor which may be a dominant effect of exogenous GH therapy.

Other hormones

Oestrogens are potent growth promotants in the ruminant, provided nutrition is maintained at high plane. We have demonstrated in the steer that chronic oestradiol therapy elevates GH secretion only slightly (10). However on the high plane of nutrition it increases the capacity of the high affinity GH receptor by 350% and has only a small effect on the lower affinity site. On a low plane of nutrition, oestradiol had only a marginal effect on weight gain and no high affinity receptors were detected (10). Thus the effect of oestradiol to promote growth correlated well with a massive induction of high affinity GH binding sites.

Little is known of other potential endocrine regulators of the GH receptor or of its regulation in tissues other than the liver. There are limited data regarding thyroid hormones and cortisol suggesting a possible role (18). The possible role of regulation by the IGFs or somatostatin has not been reported.

CONCLUDING REMARKS

Study of the growth hormone receptor has been limited by methodological difficulties and inadequate consideration of the heterogeneity of binding sites. In the ruminant it is clear that nutritional and endocrine regulation of the high affinity hepatic GH receptor is a major site of regulation of the GH axis. The optimal use of GH to manipulate animal growth will depend on utilising these data. The recent cloning of a GH receptor may allow a rapid expansion of knowledge. Key questions which remain include: what is the role of the GH receptor in different tissues; how is it differentially regulated in different tissues?; what is the relationship between high and lower affinity GH binding sites?; and what are the post receptor mechanisms involved in GH action?

ACKNOWLEDGEMENTS

The authors' research is supported by grants by the Medical Research Council of New Zealand, the Ministry of Agriculture and Fisheries, the National Childrens' Health Research Foundation and the Auckland Medical Research Foundation.

REFERENCES

1 Badinga, L., Collier, R.J., Thatcher, W.W., Quintana, S.J. and Bazer, F.W. (1987). Covalent coupling of bovine growth hormone to its receptor in bovine liver membranes. Molecular and Cellular Endocrinology **52,** 85-89.

2 Barnard, R. and Waters, M.J. (1986). Evidence for differential binding of growth hormones to membrane and cytosolic GH binding proteins of rabbit liver. Journal of Receptor Research **6,** 209-225.

3 Barnard, R., Bundesen, P.G., Rylatt, D.B. and Waters, M.J. (1985). Evidence from the use of monoclonal antibody probes for structural heterogeneity of the growth hormone receptor. Biochemical Journal **231,** 459-468.

4 Barnard, R., Haynes, K.M., Werther, G.A. and Waters, M.J. (1988). The ontogeny of growth hormone receptors in the rabbit tibia. Endocrinology **122,** 2562-2569.

5 Baxter, R.C. (1985). Measurement of growth hormone and prolactin receptor turnover in rat liver. Endocrinology **117,** 650-655.

6 Baxter, R.C. and Zaltsman, Z. (1984). Induction of hepatic receptors for growth hormone (GH) and prolactin by GH infusion is sex independent. Endocrinology **115,** 2009-2014.

7 Baxter, R.C., Bryson, J.M. and Turtle, J.R. (1980). Somatogenic receptors of rat liver: regulation by insulin. Endocrinology **107,** 1176-1181.

8 Baxter, R.C. Zaltsman, Z., and Turtle, J.R. (1982). Induction of somatogenic receptors in livers of hypersomatotropic rats. Endocrinology **111,** 1020-1022.

9 Baxter, R.C., Zaltsman, Z. and Turtle, J.R. (1984). Rat growth hormone (GH) but not prolactin (PRL) induces both GH and PRL receptors in female rat liver. Endocrinology **114,** 1892-1901.

10 Breier, B.H., Gluckman, P.D. and Bass, J.J. (1988a). The somatotrophic axis in young steers: influence of nutritional status and oestradiol-17β on hepatic high- and low-affinity somatotrophic binding sites. Journal of Endocrinology **116**, 169-177.

11 Breier, B.H., Gluckman, P.D. and Bass, J.J. (1988). Influence of nutritional status and estradiol 17β on plasma growth hormone, insulin-like growth factors 1 and 2 and on the response to exogenous growth hormone in young steers. Journal of Endocrinology **118**, 243-250.

12 Breier, B.H., Gluckman, P.D. and Bass, J.J. (1988). Plasma concentrations of insulin-like growth factor-I and insulin in the infant calf: ontogeny and influence of altered nutrition. Journal of Endocrinology **119**, 43-50.

13 Breier, B.H., Bass, J.J., Butler, J.H. and Gluckman, P.D. (1986). The somatotropic axis in young steers: influence of nutritional status on pulsatile growth hormone (GH) release and circulating insulin like growth factor-I (IGF-I) concentrations. Journal of Endocrinology **111**, 209-215.

14 Chung, C.S. and Etherton, T.D. (1986). Characterization of porcine growth hormone (pGH) binding to porcine liver microsomes: Chronic administration of pGH induces pGH binding. Endocrinology **119**, 780-786.

15 Daughaday, W.H. and Trivedi, B. (1987). Absence of serum growth hormone binding protein in patients with growth hormone receptor deficiency (Laron dwarfism). Proceedings of the National Academy of Sciences of the United States of America **84**, 4636-4640.

16 Freemark, M., Comer, M. and Handwerger, S. (1986). Placental lactogen and GH receptors in sheep liver: striking differences in ontogeny and function. American Journal of Physiology **251**, E328-E333.

17 Freemark, M., Comer, M. and Korner, G. (1988). Differential solubilization of placental lactogen (PL)- and growth hormone-binding sites: Further evidence for a unique PL receptor in fetal and maternal liver. Endocrinology **122**, 2771-2779.

18 Gause, I. and Eden, S. (1986). Induction of growth hormone (GH) receptors in adipocytes of hypophysectomized rats by GH. Endocrinology **118**, 119-124.

19 Gluckman, P.D. (1986). Hormones and fetal growth. In Oxford Reviews in Reproductive Biology **8**, pp 1-60. Oxford University Press, Oxford.

20 Gluckman, P.D., Bassett, N.S. and Butler, J.H. (1985). The somatotrophic axis in the fetus and newborn. In Perinatal Endocrinology (Albrecht, E., ed.), pp 1-20. Perinatology Press, Ithaca.

21 Gluckman, P.D., Butler, J.H. and Elliott, T.B. (1983). The ontogeny of somatotrophic binding sites in ovine hepatic membranes. Endocrinology **112**, 1607-1612.

22 Hill, D.J., Freemark, M., Strain, A.J., Handwerger, S. and Milner, R.D.G. (1988). Placental lactogen and growth hormone receptors in human fetal tissues: Relationship to fetal plasma human placental lactogen concentrations and fetal growth. Journal of Clinical Endocrinology and Metabolism **66**, 1283-1291.

23 Hughes, J.P. (1979). Identification and characterization of high and low affinity binding sites for growth hormone in rabbit liver. Endocrinology **105**, 414-420.

24 Ketelslegers, J.M., Underwood, L.E., Maes, M. and Maiter, D. (1988). Intermittent and continuous growth hormone administration produce different effect on serum IGF-I and liver GH receptors. Endocrinology **122,** 984A (abstract).

25 Leung, D.W., Spencer, S.A., Cachianes, G., Hammonds, R.G., Collins, C., Henzel, W.J., Barnard, R., Waters, M.J. and Wood, W.I. (1987). Growth hormone receptor and serum binding protein: purification, cloning and expression. Nature **330,** 537-543.

26 Maes, M., Underwood, L.E. and Ketelslegers, J.M. (1983). Plasma somatomedin-C in fasted and refed rats: close relationship with changes in liver somatogenic but not lactogenic binding sites. Journal of Endocrinology **97,** 243-252.

27 Maes, M., De Hertogh, R., Watrin-Granger, P. and Ketelslegers, J.M. (1983). Ontogeny of liver somatotrophic and lactogenic binding sites in male and female rats. Endocrinology **113,** 1325-1333.

28 Maes, M., Underwood, L.E., Gerard, G. and Ketelslegers, J.M. (1984). Relationship between plasma somatomedin-C and liver somatogenic binding sites in neonatal rats during malnutrition and after short and long term refeeding. Endocrinology **115,** 786-792.

29 Maiter, D., Underwood, L.E., Maes, M. and Ketelslegers, J.M. (1988). Acute down-regulation of the somatogenic receptors in rat liver by a single injection of growth hormone. Endocrinology **122,** 1291-1296.

30 Posner, B.I., Patel, B., Vezinhet, A. and Charrier, J. (1980). Pituitary-dependent growth hormone receptors in rabbit and sheep liver. Endocrinology **107,** 1954-1958.

31 Roupas, P. and Herington, A.C. (1986). Growth hormone receptors in cultured adipocytes: a model to study receptor regulation. Molecular and Cellular Endocrinology **47,** 81-90.

32 Roupas, P. and Herington, A.C. (1987). Receptor-mediated endocytosis and degradative processing of growth hormone by rat adipocytes in primary culture. Endocrinology **120,** 2158-2165.

33 Roupas, P. and Herington, A.C. (1988). Intracellular processing of growth hormone receptors by adipocytes in primary culture. Molecular and Cellular Endocrinology **57,** 93-99.

34 Takahashi, S. and Meites, J. (1987). GH binding to liver in young and old female rates: Relation to somatomedin-C secretion. Proceedings of the Society for Experimental Biology and Medicine **186,** 229-233.

MODULATION OF GROWTH HORMONE RELEASE: FROM CNS TO THE SECRETORY EVENT

S.R. Rawlings and W.T. Mason

Department of Neuroendocrinology, AFRC Institute of Animal Physiology and Genetics Research, Cambridge Research Station, Babraham, Cambridge CB2 4AT, U.K.

INTRODUCTION

Growth hormone (GH) release is regulated by neurohormones released from secretory neurones terminating in the median eminence (ME) of the hypothalamus, which are transported by the hypophysial portal system to the adenohypophysis (Fig. 1). Other pathways of the central nervous system (CNS) act to influence the release of these hypothalamic peptides, and so in this way the CNS can exert its influence over GH secretion. Physiological stimuli that influence GH release such as rises in blood glucagon and insulin, stress, the onset of sleep, in most cases act via receptors in the CNS to influence this complex regulatory pathway. Other factors such as oestrogen or somatomedins act more directly on the GH-containing cell (somatotroph) itself. A number of previous reviews have considered a number of the different aspects of GH secretion control (31, 39). This short chapter will outline some of the more recent work on the major pathways controlling GH release at both the CNS and the single cell level.

There appear to be two neurosecretory peptide systems in the hypothalamus that act as the major common pathways controlling the release of GH. The first contains a 44-amino acid peptide, Growth Hormone Releasing Hormone (GHRH), which, as the name suggests, stimulates GH release, and the second contains a complementary GH-inhibitory peptide, Somatostatin (SS).

GHRH AND SS IN THE HYPOTHALAMUS

Extensive immunocytochemical studies have localised the main area of GHRH-containing cell bodies to the arcuate nucleus. From here, the axons project to the median eminence (ME) where they terminate as specialised neurosecretory terminals in close apposition to the blood vessels of the portal system. GHRH released from these terminals is thought to enter the portal blood to be carried to the adenohypophysis. GHRH-cell bodies have been found in other areas of the hypothalamus, but has the most restricted distribution in the CNS of any of the neuropeptides influencing pituitary function.

In contrast, SS neurones have a very wide distribution and are found outside of the confines of the hypothalamus in addition to the role this peptide plays in gut function. SS neurones projecting to the ME can be found in the periventricular hypothalamic (PV) and medial-basal amygdaloid nuclei. In the external layer of the ME, axonal terminals of GHRH and SS show the same pattern of distribution and close proximity to one another (8).

OTHER HYPOTHALAMIC FACTORS INFLUENCING GH RELEASE

There are a number of other factors present in the ME, which are known to influence GH release at the pituitary level (Table 2). These include thyrotropin-releasing hormone (TRH), bombesin, motilin, cholecystokinin (CCK), vasoactive intestinal peptide (VIP), vasopressin, and the neurotransmitters dopamine (DA), adrenaline (ADR) and acetylcholine (ACh). The importance and exact role of these factors is still not fully resolved.

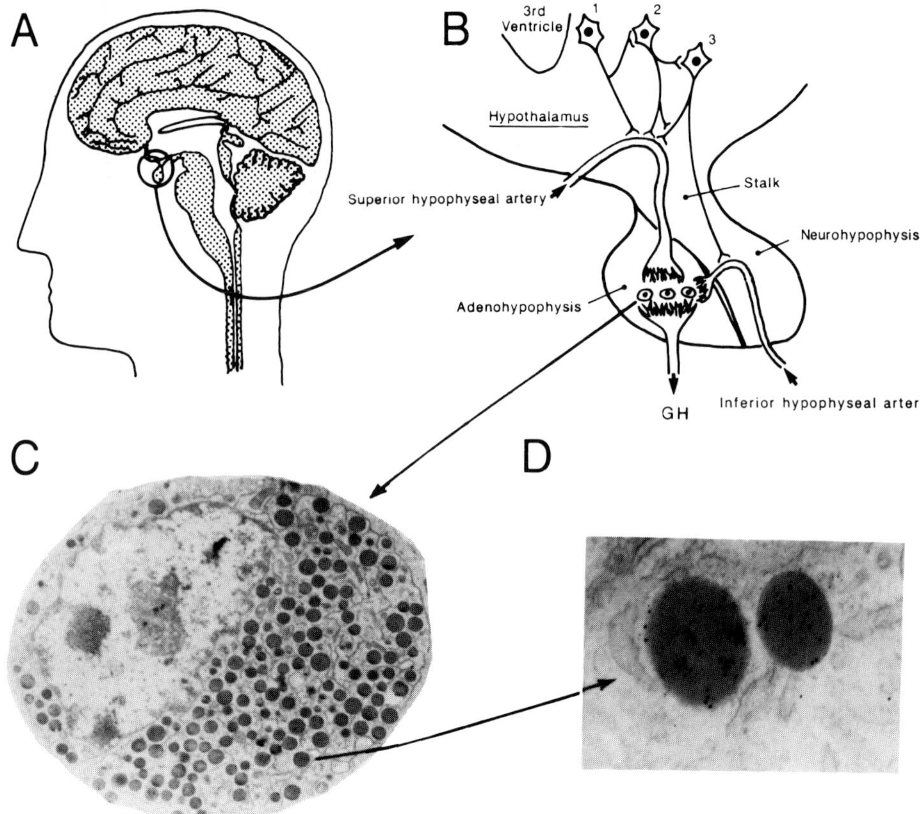

Fig. 1 The GH-secreting cell - the somatotroph

(A) The position of the pituitary gland in the human. (B) The anterior portion of this gland (the adenohypophysis) contains cells secreting a number of important hormones, including GH. This hormone secretion is under the control of factors released from neurones terminating in the ME of the hypothalamus (1, 2, 3), and these inputs may interact as shown to modulate neuropeptide release (after 32). (C) A picture of a bovine somatotroph using electron microscopy stained with an antiserum to bovine GH, and marked with gold particles. Magnification x 6206. (D) A higher power electron micrograph showing that the gold particles (and thus the GH) is contained exclusively within secretory granules. Magnification x 45,000

CONTROL OF HYPOTHALAMIC PEPTIDES

It has long been established that a monoamine system projects to the hypothalamus and can influence hypothalamic peptide, and thus GH, release (38). Pharmacological studies show that α_2-adrenoreceptor activation stimulates GHRH release in the ME (29, 6), whereas α_1-receptors exclusively stimulate SS systems to reduce GH release (6). Thus the central adrenaline system can act to both increase and decrease GH release in the rat.

The adrenergic system is not alone, and there are a number of other neurotransmitter systems that modulate GH secretion through their effects on GHRH and SS. Both DA and ACh increase GHRH release by direct, or indirect, action on rat GHRH-containing neurones (21). The control of GH release by central opioid systems is more complex, although they have stimulatory effects, via μ- or δ-receptors, on GHRH release (29, 41), proposed to be mediated via the adrenaline system (38), and/or GABA neurones (32). Both leu-enkephalin and β-endorphin, (and VIP) significantly inhibit K^+-induced SS release from the medial basal hypothalamus (MBH) (10). Substance P (sP) will stimulate GHRH release in the rat (21), and GABA has been shown in different studies to stimulate GH release via the stimulation of GHRH (32), or exert a tonic inhibition of GH release by inhibiting the secretion of a GH-releasing factor (12).

Recent work employing double staining techniques show that GHRH-cell bodies in the arcuate nucleus, and axons in the ME, are surrounded by dense networks of fibre varicosities containing SS, sP and enkephalin (Enk) (8). This suggests that the central regulation of GH release may occur both at the level of the ME and the cell bodies.

There is evidence to suggest that negative feedback occurs at a number of levels to control GH release. Both GHRH and SS apparently feedback to inhibit their own release, as well as sending fibres to mutually stimulate secretion of their opposing peptide (33). GH may itself act centrally to inhibit GHRH or stimulate SS release (15), or may do so via its action to increase Insulin-like Growth Factors (IGF) levels in the plasma, which feedback on both the hypothalamus and the pituitary gland (36).

Table 1 Factors influencing GH release; CNS mediated effects

Factors ↑ GH release	Peptide mediating the response	Factors ↓ GH release	Peptide mediating the response
Catecholamines:-		Catecholamines:-	
NA, ADR		β_2-receptor	↑ SS
α_2-receptor	↑ GHRH, ↓ SS	α_1-receptor	↑ SS
Dopamine		Dopamine	↑ SS
GABA	↑ GHRH, ↓ SS	GABA	↓ GHRH
ACh (m)	↓ SS (GABA)	ACh (m)	↑ SS
5-HT		5-HT	
Histamine (H_1)	↑ GHRH	Histamine (H_1)	
Neuropeptide Y	↑ GHRH, ↓ SS	Neuropeptide Y	↓ GHRH, ↑ SS
Opioids (μ/δ)	↑ GHRH	Opioids (K)	
TRH		TRH	↑ SS
SS	↑ GHRH, ↓ SS	GHRH	↓ GHRH, ↑ SS
Galanin	↓ SS (ADR)	IGF-I, IGF-II	↑ SS
Bombesin		Growth Hormone	↓ GHRH, ↑ SS
Substance P	↑ GHRH	Substance P	↑ SS
Neuropeptide Y		Motilin	↓ Motilin
Neurotensin		Calcitonin	↑ SS
Cholecystokinin		GRP	↑ SS
Melatonin	↓ SS	GH	↑ SS
ANP	↓ SS		

ADR, effect via adrenergic system; ANP, atrial natriuretic polypeptide; GRP, gastrin-releasing peptide; SS, somatostatin; GABA, γ-aminobutyric acid; ↑ increased, ↓ decreased release.

Table 2 Factors influencing GH; adenohypophysial effects

Factors	Effect on GH Release	Effect on GH Synthesis	General comments
GHRH	↑	↑	via cAMP system
Adrenaline	↑		via cAMP system
β_2/β_1-receptors	↑		via cAMP system
α_1-receptors	↑		
ACh	↑		via phosphatidylinositol system synergism with GHRH
TRH	↑		via phosphatidylinositol system synergism with GHRH
Bombesin	↑		synergism with GHRH
Opioids (μ/δ)	↑		
Motilin	↑		
CCK	↑		
VIP	↑		via cAMP system
PHI-27	↑		synergism with GHRH
β-estradiol	↑	↑	augments GHRH effect
Interleukin-1	↑		
Angiotensin II	↑		from rats up to 14 days of age
Dopamine	↑↓		↑ basal, ↓ GHRH x GH release inhibits rise in cAMP levels
T_3		↑↓	↑ in rat, ↓ in bovine
Somatostatin		↓	stimulates Gi protein
IGF-I, IGF-II	↓	↓	↓ GHRH x GH release

Peptide histidine isoleucine (PHI-27). x = stimulated, ↑ increased, ↓ decreased release.

EPISODIC GH RELEASE

In a number of animal species GH is released in an episodic pattern throughout the day. This may be due in turn to the episodic release of GHRH and/or SS, and be controlled by catecholaminergic neurones projecting from higher areas of the CNS (38). Pituitary cells are possibly stimulated more effectively by an intermittent exposure to releasing factors in high concentration, rather than a continuous exposure which can lead to down regulation and desensitisation. Significantly, GHRH receptors on rat somatotrophs desensitise to prolonged exposure to GHRH in vitro (5), and continuous infusion of SS is surprisingly ineffective in reducing growth, in view of its powerful inhibitory effects on GH secretion. However, there is disagreement as to which of the two hypothalamic peptides is ultimately responsible for the episodic GH release pattern. One suggestion is that GHRH and SS are released tonically into the portal blood, and superimposed on this is a 3-4 hour surge of both peptides providing for the integration of the ultradian rhythm of the GH secretion (37). A number of studies have suggested that SS withdrawl sets the timing of the episodic burst of GH release, while GHRH sets the magnitude (23, 35). However destroying the PVN in the rat decreases SS levels in the ME by 85% with no effect on the secretory pattern of GH release (40). Differences between male and female GH secretory patterns in the rat are also apparent (35).

Whatever the method of regulation, the GH secretory pattern is probably very important in its action. Thus, for example, liver GH receptors, steroid metabolising enzymes, carbonic anhydrase isozymes and other proteins are regulated differentially by the pattern of GH concentration they are exposed to, rather than the dose of GH used.

SOMATOTROPHS, SECOND MESSENGERS AND ION CHANNELS

GH in the adenohypophysis is contained solely in an acidophilic cell type called a somatotroph, although there appears to be a certain heterogeneity in this cell population (13). The GH is stored within the cell in discrete vesicles called secretory granules and can be released from the cell by fusion of the granule with the plasma membrane (a process called exocytosis) in response to an appropriate stimulus. The intracellular pathways linking receptor activation to the modulation of exocytosis is still poorly understood. However the requirement for a rise in intracellular free calcium ion concentration ($[Ca]_i$) for the secretion of a hormone seems to be the case in most endocrine cells.

The somatotroph cell, like its neighbours in the adenohypophysis, possesses a variety of ion channels in its cell membrane, whose opening is modulated not only by chemical factors, but also by the voltage across the membrane (26; Fig. 2B). These ion channels, in most cases specific for a particular ionic species, allow the movement of charged ions such as sodium (Na^+), potassium (K^+), calcium (Ca^{2+}) and chloride (Cl^-) between the cytoplasm and the extracellular environment. Modulation of such channel opening, and thus the movement of ionic charge, probably plays an important role in cell signalling (see 26 for review).

GHRH CONTROL OF GH RELEASE

Initial experiments with natural and synthetic GHRH peptides demonstrated a high biological potency in the picomolar range to stimulate GH release using rat pituitary cell monolayer cultures. GHRH also stimulates synthesis of GH (14), presumably via the transcription of the GH gene (2), and the production of GH mRNA (17).

GHRH appears to exert its influence on the somatotroph cell by activating adenylate cyclase (AC) to produce the intracellular messenger cyclic adenosine-3′,5′-monophosphate (cAMP) (3). The link between GHRH receptor binding and AC activation is a cholera toxin-sensitive G-protein termed (G_s) (7). The rise in cAMP levels that result influences cell activity through the phosphorylation (and thus activation or inactivation) of specific active peptides eg cAMP-dependent protein kinase. It seems likely that the secretory response to a rise in cAMP is Ca^{2+}-dependent, since verapamil (a blocker of voltage-dependent Ca^{2+} channels in the plasma membrane) will block cAMP-stimulated GH release (2, 3). GHRH will stimulate a $[Ca]_i$ rise in both rat and bovine somatotrophs within 60 s (18; Fig. 2A), possibly through the activation of voltage-dependent calcium channels in the cell membrane (Fig. 2B).

This system is however not as simple as it may appear. GHRH, in addition to its cAMP stimulating effects, can stimulte phosphatidylinositol labelling (4), and arachidonate and PGE_2 production (20, 11) in cultured adenohypophysial cells. The involvement of both protein kinase C (PK-C) (36) and calmodulin (28) has also been implicated in the GHRH stimulation of GH release. Interestingly, GHRH apparently stimulates Ca^{2+} mobilisation at a concentration about ten times lower than that observed to cause a measurable increase in biochemical messengers such as cAMP, phosphatidylinositol, or arachidonate (25)!

Although GHRH stimulated GH release is probably mediated by both cAMP and Ca^{2+}, a rise in $[Ca]_i$ is not a sufficient stimulus for switching on GH synthesis (2). However, a

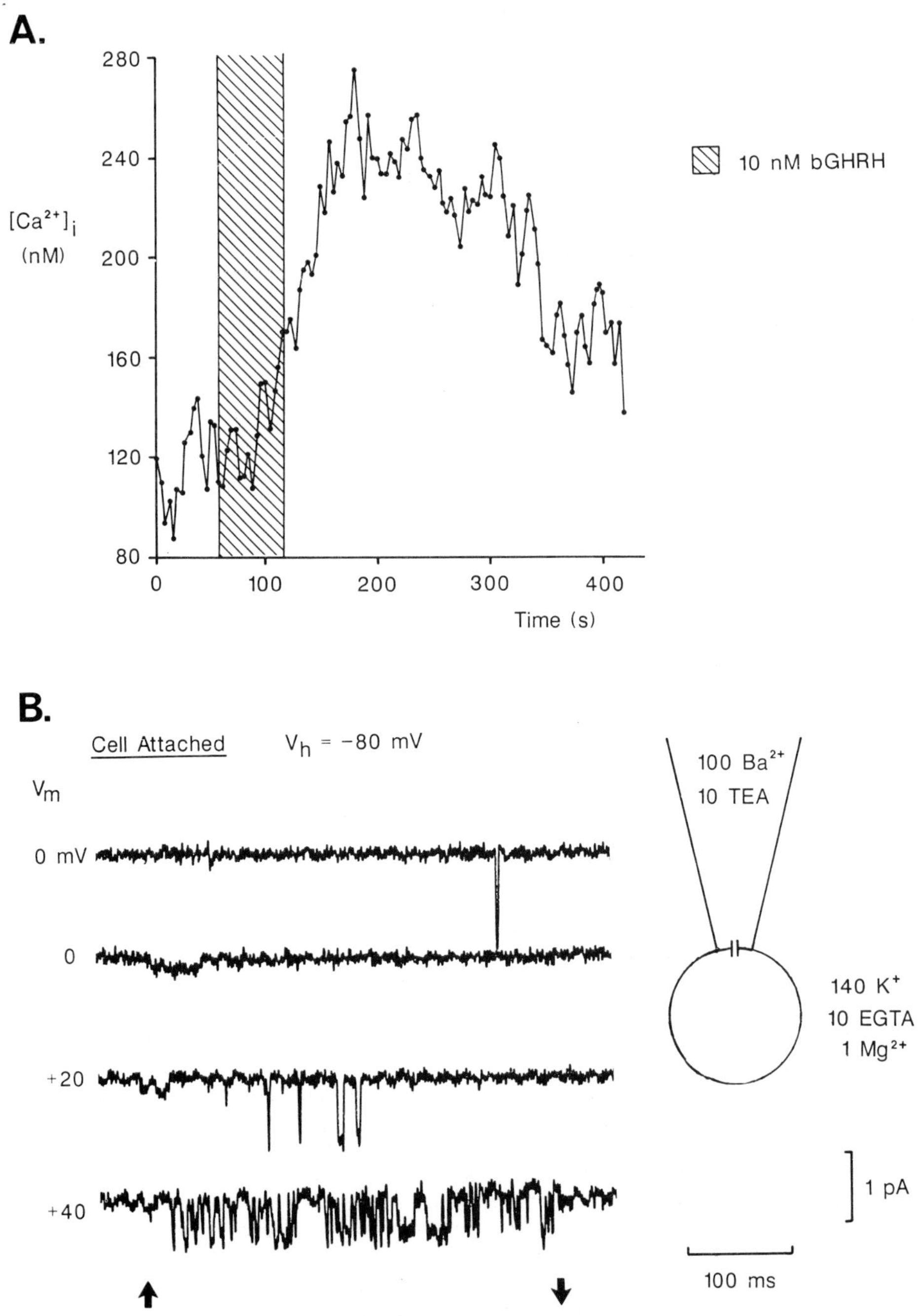

A.
280
240
200
160
120
80
[Ca²⁺]i (nM)
0
100
200
300
400
Time (s)
10 nM bGHRH
B.
Cell Attached
Vh = -80 mV
Vm
0 mV
0
+20
+40
100 Ba²⁺
10 TEA
140 K⁺
10 EGTA
1 Mg²⁺
1 pA
100 ms

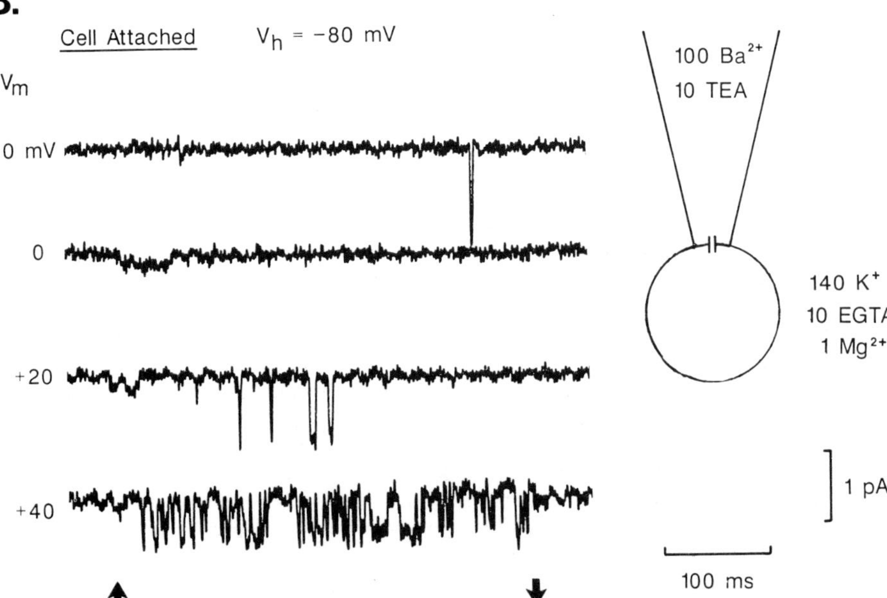

rise in cAMP levels appears to be a sufficient signal to switch on GH gene transcription in rat somatotrophs, and probably exerts its effect via the phosphorylation of putative regulatory nuclear proteins (2).

SS CONTROL OF GH RELEASE

Unlike GHRH, SS has an influence on the release of a number of pituitary hormones. Activation of the SS receptor on adenohypophysial cells stimulates a G-protein (G_i) that acts to inhibit AC production of cAMP (7, 24). However, although it is clear that SS can reduce $[Ca]_i$ in rat somatotrophs (18), whether its major effect is by reducing cAMP levels in the cell (7) or by an action at a later stage in the pathway (22) is still unclear. In fact electrophysiological studies show that SS can inhibit voltage-activated Ca^{2+} currents possibly via the stimulation of a K^+ conductance in the cell membrane (30).

INTERRELATIONSHIPS BETWEEN GH SECRETING FACTORS

Table 2 lists other factors that appear to influence GH release by an action at the pituitary level. However the physiological significance of many of these responses is still in some doubt.

There is however a good deal of evidence to implicate a role for catecholamines in controlling GH release. Both adrenaline (ADR) and DA stimulate basal GH release from cultured anterior pituitary cells, and, using a push-pull perfusion technique both these amines (and noradrenaline) have been detected in the rat anterior pituitary gland (9). DA however inhibits the GHRH stimulated GH rise. The ADR effect is probably mediated through the β_2-receptor (β_1 in pigs) coupled to the generation of cAMP (34). Adrenaline (but not NA or DA) is released into the hypophysial portal system in response to heat stress (16). Interestingly, Baes and Denef have suggested that the ADR stimulated GH release is influenced by cell-cell communication (with lactotrophs) within the rat adenohypophysis (1).

The IGFs produced by a rise in plasma GH levels provide a good example of factors

---

Fig. 2 (A) By loading isolated bovine somatotroph cells with a chemical probe (fura-2) that alters its fluorescent properties when it binds to Ca^{2+}, we are able to record these changes in fluorescence at two specific wavelengths and relate them to changes in intracellular free Ca^{2+} concentration ($[Ca]_i$) within the cell. Application of bGHRH (a factor known to stimulate GH release from these cells) to the medium surrounding this somatotroph cell resulted in a delayed rise in $[Ca]_i$. This rise is probably due to influx of Ca^{2+} from the extracellular medium. (B) Using the technique of the cell attached patch clamp it is possible to isolate a small portion (or "patch") of a cell membrane and record the flow of electrical current through single ion channels in that membrane. In (B) is shown such a recording demonstrating single Ca^{2+} channels in an identified bovine somatotroph. These channels are activated by voltage (potentials are shown on the left-hand side, arrows show the on and off of the voltage pulse), and their opening may be modulated by intracellular chemical messengers (not shown here). The diagram in inset shows the recording configuration employed. Barium (Ba^{2+}) was used since it maximises the current flow through Ca^{2+} channels. The opening of such channels may allow the influx of Ca^{2+}, and thus a rise in $[Ca]_i$ in response to bGHRH (A), which is postulated to result ultimately in the release of GH.

feeding back to inhibit both GH secretion and synthesis (43).

A number of factors, such as ACh and TRH, cause GH release through the stimulation of phosphoinositidase C to produce the putative intracellular messengers D-myo-inositol-1, 4,5-trisphosphate (IP_3) and diacylglycerol (DAG). IP_3 is thought to act on intracellular stores to release Ca^{2+} into the cytoplasm, while DAG probably exerts its action via PK-C to increase influx of Ca^{2+} from the extracellular space. There is a marked synergism of these GH releasing factors with factors (such as GHRH) that act via the cAMP-generating system (19).

CODA

It has been possible in this short text to give only the briefest of overviews of the series of complex pathways that interact to control GH release. We have concentrated on the more up-to-date information in this review, but, for further reference, we recommend the papers cited in the opening paragraph (31, 39).

ACKNOWLEDGEMENTS

S.R.R. is supported by the Meat and Livestock Commission, and we wish to thank Dr P Wooding for the electron microscopy pictures, and Drs. M. Berridge, T. Check and R. Moreton for their collaboration in producing the data for Fig. 2A.

REFERENCES

1 Baes, M. and Denef, C. (1987). Evidence that stimulation of growth hormone release by epinephrine and vasoactive intestinal peptide is based on cell-to-cell communication in the pituitary. Endocrinology **120,** 280-290.

2 Barinaga, M., Bilezikjian, L.M., Vale, W.W., Rosenfeld, M.G. and Evans, R.M. (1985). Independent effects of growth hormone releasing factor on growth hormone release and gene transcription. Nature **314,** 279-281.

3 Bilezikjian, L.M. and Vale, W.W. (1983). Stimulation of adenosine 3′,5′-monophosphate production by growth hormone-releasing factor and its inhibition by somatostatin in anterior pituitary cells in vitro. Endocrinology **113,** 1726-1731.

4 Canonico, P.L., Cronin, M.J., Thorner, M.O. and MacLeod, R.M. (1983). Human pancreatic GRF stimulates phosphatidylinositol labelling in cultured anterior pituitary cells. American Journal of Physiology **245,** E587-E590.

5 Ceda, G.P. and Hoffman, A.R. (1985). Growth hormone-releasing factor desensitization in rat anterior pituitary cells in vitro. Endocrinology **116,** 1334-1340.

6 Cella, S.G., Locatelli, V., de Gennaro, V., Wehrenberg, W.B. and Müller, E.E. (1987). Pharmacological manipulations of α-adrenoreceptors in the infant rat and effects on growth hormone secretion. Study of the underlying mechanisms of action. Endocrinology **120,** 1639-1643.

7 Cronin, M.J., Hewlett, E.L., Evans, W.S., Thorner, M.O. and Rogol, A.D. (1984). Human pancreatic tumor growth hormone (GH)-releasing factor and cyclic adenosine 3′,5′-monophosphate evoke GH release from anterior pituitary cells; the effects of pertussis toxin, cholera toxin, forskolin and cycloheximide. Endocrinology **114,** 904-913.

8 Daikoku, S., Hisano, S., Hitoshi, K., Chikamori-Aoyama, M., Kagotani, Y., Zhang, R. and Chihara, K. (1988). Ultrastructural evidence for neuronal regulation of growth hormone secretion. Neuroendocrinology **47**, 405-415.

9 Dluzen, D.E. and Ramirez, V.D. (1988). In vivo neurotransmitter levels in the anterior pituitary of freely behaving intact and castrated male rats determined with push-pull perfusion and high pressure liquid chromatography coupled with electrochemical detection. Endocrinology **122**, 2861-2864.

10 Drouva, S.V., Epelbaum, J., Tapia-Arancibia, L., Laplante, E. and Kordon, C. (1981). Opiate receptors modulate LHRH and SRIF release from mediobasal hypothalamic neurons. Neuroendocrinology **32**, 163-167.

11 Fafeur, V., Gouin, E. and Dray, F. (1985). Growth hormone-releasing factor (GRF) stimulates PGE2 production in rat anterior pituitary. Evidence for a PGE2 involvement in GRF-induced GH release. Biochemical and Biophysical Research Communications **126**, 725-733.

12 Fiók, J., Acs, Z., Makara, G.B. and Erdo, S.L. (1984). Site of γ-aminobutyric acid (GABA)-mediated inhibition of growth hormone secretion in the rat. Neuroendocrinology **39**, 510-515.

13 Frawley, L.S. and Neill, J.D. (1984). A reverse hemolytic plaque assay for microscopic visualisation of growth hormone release from individual cells: evidence for somatotrope heterogeneity. Neuroendocrinology **39**, 484-487.

14 Fukata, J., Diamond, D.J. and Martin, J.B. (1985). Effects of rat growth hormone (rGH)-releasing factor and somatostatin on the release and synthesis of rGH in dispersed pituitary cells. Endocrinology **117**, 457-467.

15 Ganzetti, I., de Gennaro, V., Redaelli, M., Müller, E.E. and Cocchi, D. (1986). Effect of hypophysectomy and growth hormone replacement on hypothalamic GHRH. Peptides **7**, 1011-1014.

16 Gibbs, D.M. (1985). Hypothalamic epinephrine is released into hypophysial portal blood during stress. Brain Research **335**, 360-364.

17 Gick, G.G., Zeytin, F.N., Brazeau, P., Ling, N.C., Esch, F.S. and Bancroft, C. (1984). Growth hormone-releasing factor regulates growth hormone mRNA in primary cultures of rat pituitary cells. Proceedings of the National Academy of Sciences, U.S.A. **81**, 1553-1555.

18 Holl, R.W., Thorner, M.O. and Leong, D.A. (1988). Intracellular calcium concentration and growth hormone secretion in individual somatotrophs: effects of growth hormone-releasing factor and somatostatin. Endocrinology **122**, 2927-2932.

19 Ingram, C.D. and Bicknell, R.J. (1986). Synergistic interaction in bovine pituitary cultures between growth hormone-releasing factor and other hypophysiotrophic factors. Journal of Endocrinology **109**, 67-74.

20 Judd, A.M., Koji, K. and MacLeod, R.M. (1985). GRF increases release of growth hormone and arachidonate from anterior pituitary cells. American Journal of Physiology **248**, E438-E442.

21 Kakucska, I. and Makara, G.B. (1983). Various putative neurotransmitters affect growth hormone (GH) release in rats with anterolateral hypothalamic

deafferentation of the medial basal hypothalamus: evidence for mediation by a GH-releasing factor. Endocrinology **113**, 318-323.

22 Kraicer. J. and Spence, J.W. (1981). Release of growth hormone from purified somatotrophs: use of high K^+ and the ionophore A23187 to elucidate inter-relationships among Ca^{2+}, adenosine 3′,5′-monophosphate, and somatostatin. Endocrinology **108**, 651-657.

23 Kraicer. J., Sheppard. M.S.. Luke. J.. Lussier. B.. Moor. B.C. and Cowan. J.S. (1988). Effect of withdrawl of somatostatin and growth hormone (GH)-releasing factor on GH release in vitro. Endocrinology **122**, 1810-1815.

24 Lewis. D.L.. Weight. F.F. and Luini. A. (1986). A guanine nucleotide-binding protein mediates the inhibition of voltage-dependent calcium current by somatostatin in a pituitary cell line. Proceedings of the National Academy of Sciences. U.S.A. **83**, 9035-9039.

25 Login. I.S.. Judd. A.M. and MacLeod. R.M. (1986). Association of $^{45}Ca^{2+}$ mobilisation with stimulation of growth hormone (GH) release by GH-releasing factor in dispersed normal rat pituitary cells. Endocrinology **118**, 239-243.

26 Mason. W.T.. Rawlings. S.R.. Cobbett. P.. Sikdar. S.K.. Zorec. R., Akerman. S.N.. Benham. C.D.. Berridge. M.J.. Cheek. T. and Moreton. R.B. (1988). Control of secretion in anterior pituitary cells - linking ion channels. messengers and exocytosis. Journal of Experimental Biology **405**, 577-593.

27 Mason, W.T. and Rawlings. S.R. (1988). Whole cell recordings of ionic currents in bovine somatotrophs and their involvement in growth hormone secretion. Journal of Physiology **405**, 577-593.

28 Merritt. J.E.. Dobson. P.R.M.. Wojcikiewicz. R.J.H.. Baird. J.G. and Brown. B.L. (1984). Studies in the involvement of calcium and calmodulin in the action of growth-hormone-releasing factor. Bioscience Reports **4**, 995-1000.

29 Miki. N.. Ono. M. and Shizume. K. (1984). Evidence that opiatergic and α-adrenergic mechanisms stimulate rat growth hormone release via growth hormone-releasing factor (GRF). Endocrinology **114**, 1950-1952.

30 Mollard. P.. Vacher. P.. Dufy. B. and Barker. J.L. (1988). Somatostatin blocks Ca^{2+} action potential activity in prolactin-secreting pituitary tumor cells through coordinate actions on K^+ and Ca^{2+} conductances. Endocrinology **123**, 721-732.

31 Muller. E.E. (1987). Neural control of somatotropic function. Physiological Reviews **67**, 962-1053.

32 Murakami. Y.. Kato. Y., Kabayama. Y., Inoue, T., Koshiyama. H. and Imura. H. (1987). Involvement of hypothalamic growth hormone (GH)-releasing factor in GH secretion induced by intracerebroventricular injection of somatostatin in rats. Endocrinology **120**, 311-316.

33 Murakami. Y.. Kato. Y.. Kabayama. Y.. Tojo. K.. Inoue. T. and Imura. H. (1985). Involvement of growth hormone-releasing factor in growth hormone secretion induced by gamma-aminobutyric acid in conscious rats. Endocrinology **117**, 787-789.

34 Perkins, S.N., Evans, W.S., Thorner, M.O., Gibbs, D.M. and Cronin, M.J. (1985). β-Adrenergic binding and secretory responses of the anterior pituitary. Endocrinology **117**, 1818-1825.

35 Robinson, I.C.A.F. and Clark, R.G. (1987). The secretory pattern of GH and its significance for growth in the rat. In Growth hormone - basic and clinical aspects (Isaksson, O., Binder, C., Hall, K. and Hökfelt, B., eds.), pp 109-127. Elsevier, Amsterdam.

36 Shibasaki, T., Yamauchi, N., Hotta, M., Masuda, A., Imaki, T., Demura, H., Ling, N. and Shizume, K. (1986). In vitro release of growth hormone-releasing factor from rat hypothalamus: effect of insulin-like growth factor-1. Regulatory Peptides **15**, 47-53.

37 Summers, S.T., Canonico, P.L., MacLeod, R.M., Rogol, A.D. and Cronin, M.J. (1985). Phorbol esters affect pituitary growth hormone (GH) and prolactin release: the interaction with GH releasing factor, somatostatin and bromocriptine. European Journal of Pharmacology **111**, 371-376.

38 Tannenbaum, G.S. and Ling, N. (1985). The interrelationship of growth hormone (GH)-releasing factor and somatostatin in generation of the ultradian rhythm of GH secretion. Endocrinology **115**, 1952-1957.

39 Terry, L.C., Crowley, W.R. and Johnson, M.D. (1982). Regulation of episodic growth hormone secretion by the central epinephrine system. Journal of Clinical Investigation **69**, 104-112.

40 Thorner, M.O. and Cronin, M.J. (1985). Growth hormone-releasing factor: clinical and basic studies. In Neuroendocrine Perspectives (Müller, E.E., MacLeod, R.M. and Frohman, L.A., eds.) vol. 4, pp 95-144. Elsevier, Amsterdam.

41 Urman, S., Kaler, L. and Critchlow, V. (1985). Effects of hypothalamic periventricular lesions on pulsatile growth hormone secretion. Neuroendocrinology **41**, 357-362.

42 Wehrenberg, W.B., Bloch, B. and Ling, N. (1985). Pituitary secretion of growth hormone in response to opioid peptides and opiates is mediated through growth hormone-releasing factor. Neuroendocrinology **41**, 13-16.

43 Yamashita, S. and Melmed, S. (1987). Insulin-like growth factor I regulation of growth hormone gene transcription in primary rat pituitary cells. Journal of Clinical Investigation **79**, 449-452.

NEUROREGULATION OF GROWTH HORMONE SECRETION

M.D. Page, C. Dieguez and M.F. Scanlon

Neuroendocrine Group, Department of Medicine, University of Wales College of Medicine, Heath Park, Cardiff CF4 4XN, Wales, U.K.

GHRH and somatostatin are the important final common hypothalamic pathways in the control of somatotroph function and they act together in a concerted fashion to mediate this effect. Data accrued largely from in vivo studies suggest that many of the factors which modulate GH secretion do so through control of the somatostatinergic neurones in the hypothalamus against the background of an important priming action of GHRH. Acute inhibitory feedback control can occur through GH itself whereas chronic feedback control may well operate through IGF-I. There is no evidence that down-regulation or desensitisation of the somatotroph is an important in vivo event and variability in GH responsiveness to exogenously administered GHRH can be explained via a GH-mediated feedback pathway. The idea has emerged of a "hypothalamic cholinergic-somatostatinergic unit" by means of which acetyl choline tonically inhibits somatostatin release. This unit is sensitive to peripheral signals such as those mediated by glucose and GH itself. Further studies should now be directed towards understanding the nature of these modulatory interactions at the cellular level.

INTRODUCTION

Tissue growth is dependent on adequate release of pituitary growth hormone (GH) prior to epiphysial fusion which occurs post-pubertally under the influence of sex steriods. For many years extracted human pituitary GH was used in the treatment of short stature due to GH deficiency. However the occurrence of a "Jacob Creutzfeldt-like illness" in several individuals who had previously been treated with such preparations has led to the abandonment of the use of this agent. Fortunately, technological advances have led to the availability of both recombinant human GH and synthetic human hypothalamic growth hormone releasing hormone (GHRH), each of which is being evaluated at present in the treatment of GH deficiency states. The purpose of this brief review is to outline recent developments in understanding of the major peptidergic pathways in the control of GH secretion and the manner of their interaction with conventional neurotransmitter pathways. It is also necessary to consider the question of desensitisation and the operation of feedback signals in the control of the somatotroph.

GH itself has two separate pathways of action (Fig. 1). It stimulates general body growth through the mediation largely of IGF-I which is produced mainly by the liver under the direct influence of GH. However, IGF-I, and probably several other growth factors, also play important paracrine roles in the control of the growth and function of many tissues. These locally produced growth mediators may also be, to some extent, under the control of GH. GH also plays an important role in metabolism, stimulating lipolysis, increased production of non-esterified and free fatty acids and the production of ketones. In excess, GH causes carbohydrate intolerance which may lead on to frank diabetes mellitus. Various products of each of these separate pathways of GH action feed back on both the hypothalamus and the anterior pituitary to control the function of the somatotroph cells. In consequence, the secretion of GH is determined, at any given point in time, by the interplay of a variety of different factors.

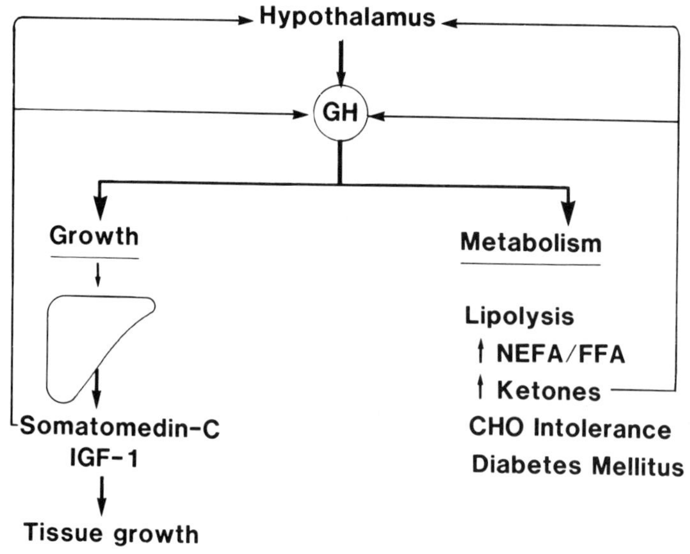

Fig. 1 Schematic outline of separate pathways of GH action.

PEPTIDERGIC CONTROL OF GH RELEASE

It is now clearly established that hypothalamic control of somatotroph function is mediated by two peptides, somatostatin which is inhibitory and GHRH which is stimulatory. It has been demonstrated by passive immunisation techniques in free-living rats and by direct measurement, albeit in anaesthetised animals, that these two peptides probably work in concert such that pulses of GH release are a consequence of a pulse of GHRH release accompanied by a reduction in somatostatin release into hypophysial portal blood (20, 28). When GHRH is administered to normal human subjects or GH deficient children, the majority of whom have hypothalamic GHRH deficiency, GH responses can be very variable. When repeated doses of GHRH are administered to normal subjects at two-hourly intervals, the GH response to the second bolus of GHRH is abolished (24). Furthermore, continuous infusion of GHRH causes GH release initially but GH secretion falls to baseline levels within 6 hours (27). Is this reduction in responsiveness of the somatotroph due to desensitisation, depletion of intracellular stores of GH, or the operation of various inhibitory feedback pathways?

A variety of studies have indicated that although desensitisation of the somatotroph does occur if GHRH is administered at a high enough dose over a long enough period of time in vitro (2, 6, 9), this is in no way comparable in terms of degree to the marked desensitisation of the gonadotroph which follows repeated GnRH administration (8). Further studies have shown that loss of responsiveness of the somatotroph to GHRH in vitro has at least two components: firstly there is a reduction in the Bmax response to GHRH which is probably due to depletion of intracellular stores since this can be reversed by treating with somatostatin along with GHRH (Fig. 2). Secondly, however, the cyclic AMP response to repeated administration of GHRH is also reduced (6, 9) and this cannot be reversed by co-incubation with somatostatin (9). These data suggest that there is uncoupling of the GHRH receptor from adenylate cyclase, a process which is

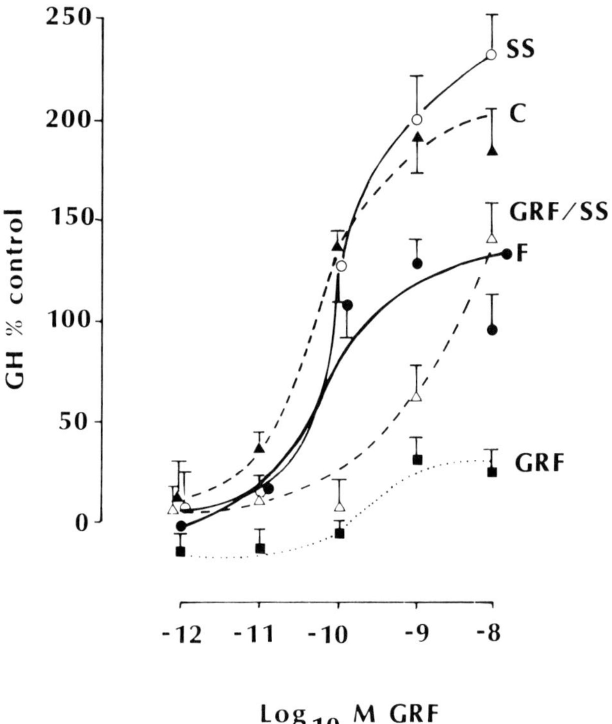

Fig. 2 Effect of pretreatment of cultured rat AP cells with either saline (C), 1 μmol/l somatostatin (SS), 10 nmol/l GRF plus 1 μmol/l SS (GRF/SS) or 10 μmol/l forskolin (F) on the subsequent GH responses to different concentrations of GRF.

---

known to occur in other systems following repeated agonist administration. Although it has been demonstrated that repeated administration of GHRH leads to about a 50% reduction in the numbers of GHRH receptors, this does not, in itself, reduce GH responsiveness to GHRH and probably only 30-40% occupancy of such receptors is necessary for maximal responsiveness (3). Despite these in vitro observations however, there are no data which demonstrate that desensitisation of the somatotroph is important in vivo.

There is accumulating evidence that depletion of intracellular stores of GH does not explain the loss of GH responsiveness to GHRH in vivo. The GH response to α-adrenergic agonism is preserved 2 hours after pretreatment with GHRH (25) and the GH response to hypoglycaemia and arginine infusion are not only preserved but enhanced (18, 24). Furthermore the simultaneous administration of maximal doses of GHRH and insulin leads to increased GH release compared with either agent alone (16) suggesting the operation of separate pathways. The GH response to L-dopa administration is abolished by prior GHRH administration (18). These data indicate that the somatotroph is not depleted of GH and does not become refractory to all stimuli following GHRH pretreatment. Insulin-induced hypoglycaemia, arginine, clonidine and alpha-adrenergic activation do not require the release of endogenous hypothalamic GHRH in order to cause GH secretion and could be influencing somatostatinergic tone. In contrast, L-dopa may well act via the release of endogenous GHRH, though an effect on SS remains possible since dopamine agonism with bromocriptine potentiates the GH response to GHRH (26). The results of further studies have suggested that the reduction in GH response to GHRH after prior GHRH treatment is probably mediated by acute inhibitory feedback pathways operating through hypothalamic somatostatinergic pathways.

FEEDBACK PATHWAYS AND CHOLINERGIC CONTROL

Melmed and colleagues in the States (14) have shown in a variety of studies that IGF-I itself has an inhibitory feedback role directly at the level of the somatotroph, to limit GH responsiveness to GHRH, forskolin, cyclic AMP, TPA, glucocorticoids and T_3. This is in addition to the known effects of IGF-I to stimulate somatostatin release from the hypothalamus in vitro (1). An important point about such effects of IGF-I is that they are slow, occurring over several hours, and may therefore constitute a system for "setting" the background level of responsiveness of the somatotroph to other signals. In the more acute setting there are convincing data suggesting that GH itself can act as a negative feedback inhibitor of somatotroph responsiveness. It has been known for some years that GH causes stimulation of somatostatin release from the rat hypothalamus in vitro (23). More recently it has been demonstrated in human in vivo studies that GH administration leads to abolition of subsequent GH responsiveness to GHRH over a short time period before there is any detectable rise in circulating IGF-I levels in response to the administered GH (22). This may constitute an acute inhibitory feedback pathway for the control of somatotroph function.

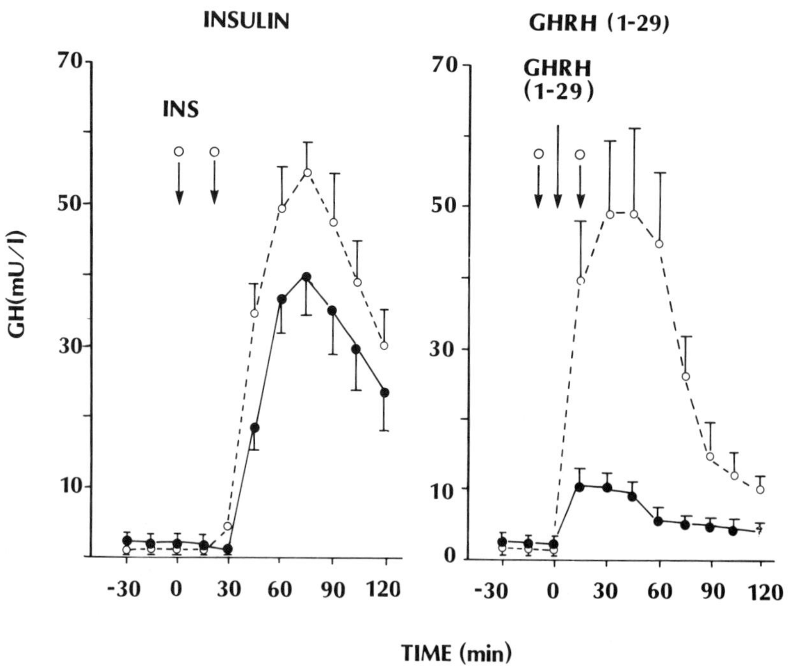

Fig. 3 GH responses to GHRH 1-29 (100 μg iv) or insulin (0.1 units/kg iv) administered at 0 minutes, with ●---● and without o---o atropine pretreatment (600 μg at time 0).

Recent evidence has highlighted the importance of cholinergic pathways in the control of GH secretion and has indicated probable interaction of a "cholinergic-somatostatinergic hypothalamic unit" with peripheral feedback pathways mediated by glucose and GH. Cholinergic activation with pyridostigmine leads to enhanced GH responsiveness to GHRH whereas cholinergic muscarinic blockade with drugs such as atropine or pirenzepine (Fig. 3) leads to abolition of this response (4, 11, 12, 13). In addition cholinergic muscarinic blockade abolishes physiological nocturnal slow wave sleep-related GH release (Fig. 4) and also GH responses to exercise and food (5, 10, 15, 17, 19). Finally cholinergic muscarinic blockade abolishes the GH responses to all known GH secretagogues (7) with the exception of insulin-induced hypoglycaemia (7). Although such evidence does not exclude a direct pituitary action of acetylcholine to stimulate GH release, such a pathway cannot account for the abolition of the GH response to exogenously administered GHRH by cholinergic blockade. In the in vivo setting it is possible that

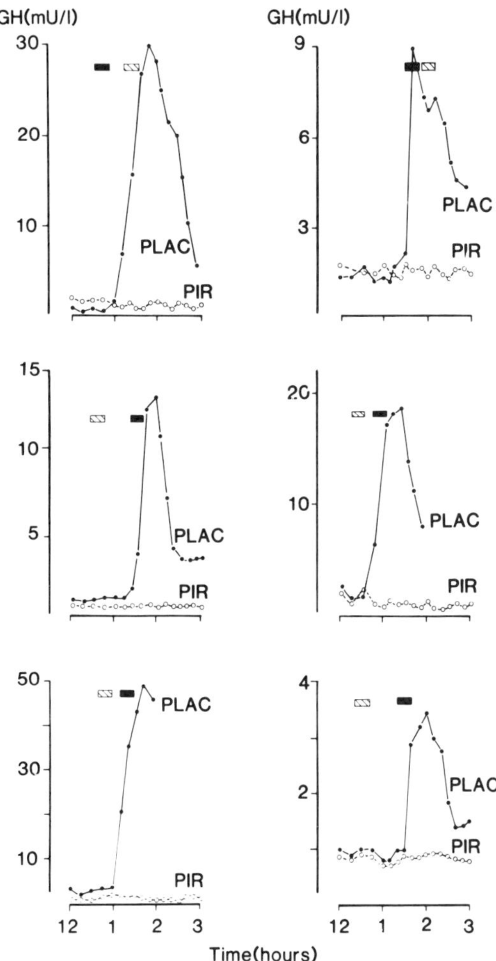

Fig. 4 GH release in 6 normal males from midnight to 3 a.m. with (o---o) and without (●---●) pirenzepine pretreatment (100 mg at 10 p.m. and 12 m.n.). Episodes of slow wave sleep on placebo (■) and pirenzepine (▧) study nights.

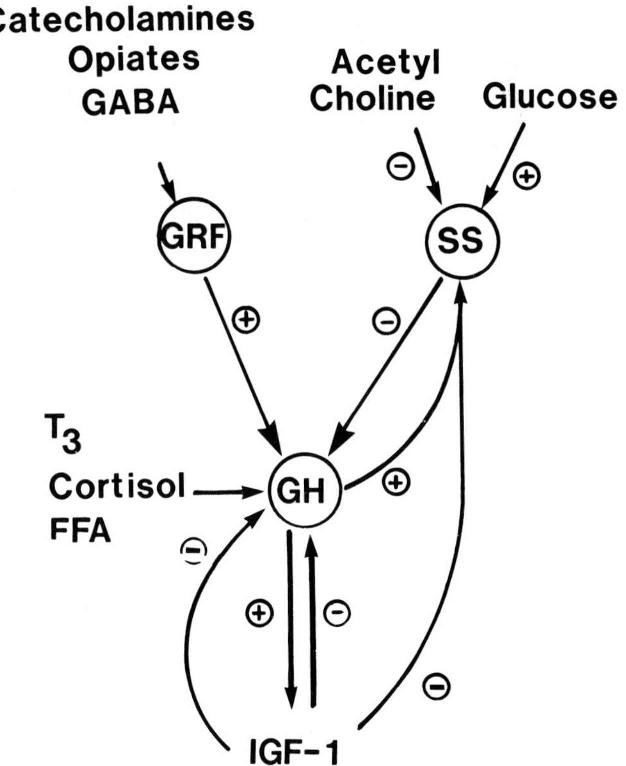

Fig. 5 Schematic representation of major pathways in GH control. Interaction between acetylcholine, glucose, GH, IGF-I and the somatostatin (SS)-producing cell is depicted although the cellular nature of such interactions is unknown.

---

cholinergic mechanisms are operating at the hypothalamic level through modulation of the somatostatinergic neurone such that there is a tonic inhibitory cholinergic control of somatostatin release (Fig. 5). The relative sparing of the GH response to insulin-induced hypoglycaemia by cholinergic blockade is consistent with the hypothesis that cholinergic pathways and glucose modulate somatostatin-producing neurons in a hierarchical fashion in that hypoglycaemia can override the cholinergic influence when appropriate physiological demands are made. Investigation of this cholinergic system in relation to inhibitory feedback control by GH itself indicates that the GH response to GHRH which is abolished by prior GH or GHRH administration, can be restored by activation of cholinergic pathways with pyridostigmine (12, 21). These data suggest that the inhibitory GH-mediated pathway may well act via modulation of the release of endogenous somatostatin. Pharmacological utilisation of this pathway may lead to a more predictable response to repeated administration of GHRH in the in vivo situation.

REFERENCES

1 Berelowitz, M., Szabo, M. and Frohman, L.A. (1981). Somatomedin-C mediates growth hormone negative feedback by effects on both hypothalamus and the pituitary. Science **212,** 1279-1281.

2 Bilezikjian, L.M. and Vale, W.W. (1984). Chronic exposure of cultured rat anterior pituitary cells to GRF causes partial loss of responsiveness to GRF. Endocrinology **115,** 2032-2034.

3 Bilezikjian, L.M., Seifert, H. and Vale, W (1986). Desensitization to growth hormone-releasing factor (GRF) is associated with down-regulation of GRF-binding sites. Endocrinology **118,** 2045-2052.

4 Casanueva, F.F., Betti, R., Cella, S.G., Müller, E.E. and Mantegazza, P. (1983). Effects of agonists of cholinergic neurotransmission on growth hormone release in the dog. Acta Endocrinologica **103,** 15-20.

5 Casanueva, F.F., Villanueva, L., Cabranes, J.A., Cabezas-Cervato, J. and Fernandez-Cruz, A. (1984). Cholinergic mediation of growth hormone secretion elicited by arginine, clomidine and physical exercise in man. Journal of Clinical Endocrinology and Metabolism **59,** 526-530.

6 Ceda, G.P. and Hoffman, A.R. (1985). Growth hormone-releasing factor desensitisation in rat anterior pituitary cells in vitro. Endocrinology **116,** 1334-1340.

7 Dieguez, C., Page, M.D. and Scanlon, M.F. (1988). Growth hormone neuroregulation and its alteration in disease states. Clinical Endocrinology **28,** 109-143.

8 Dieguez, C., Foord, S.M., Shewring, G., Edwards, C.A., Heyburn, P.J., Peters, J.R., Hall, R. and Scanlon, M.F. (1984). The effects of long term growth hormone releasing factor (GRF 1-40) administration on growth hormone secretion and synthesis in vitro. Biochemical and Biophysical Research Communications **121,** 111-117.

9 Edwards, C.A., Dieguez, C., Ham, J., Peters, J.R. and Scanlon, M.F. (1988). Evidence that growth hormone depletion and uncoupling of the regulatory protein of adenylate cyclase (N_s) both contribute to the desensitisation of growth hormone responses to growth hormone releasing factor. Journal of Endocrinology **116,** 185-190.

10 Leveston, S.A. and Cryer, P.E. (1980). Endogenous cholinergic modulation of growth-hormone secretion in normal and acromegalic humans. Metabolism **29,** 703-706.

11 Massara, F., Ghigo, E., Goffi, S., Molinatti, G.M., Müller, E.E. and Camanni, F. (1984). Blockage of hp-GRF-40-induced GH release in normal men by a cholinergic muscarinic antagonist. Journal of Clinical Endocrinology and Metabolism **59,** 1025-1026.

12 Massara, F., Ghigo, E., Molinatti, P., Mazza, E., Locatelli, V., Müller, E.E. and Camanni, F. (1986). Potentiation of cholinergic tone by pyridostigmine bromide re-instates and potentiates the growth hormone responsiveness to intermittent administration of growth hormone-releasing factor in man. Acta Endocrinologica **113,** 12-16.

13 Massara, F., Ghigo, E., Demislis, K., Tangolo, D., Mazza, E., Locatelli, V., Müller, E.E., Molinatti, G.M. and Camanni, F. (1986). Cholinergic involvement in the growth hormone releasing hormone-induced growth hormone release: Studies in normal and acromegalic subjects. Neuroendocrinology **43,** 670-675.

14 Melmed, S. (1988). Pituitary Growth Factors. Neuroendocrine Perspective **6,** 27-46. Springer-Verlag, New York.

15 Page, M.D., Bevan, J.S., Dieguez, C., Peters, J.R. and Scanlon, M.F. (1989). Cholinergic blockade with pirenzepine improves carbohydrate tolerance and abolishes the GH response to meals in normal subjects. Clinical Endocrinology (in press).

16 Page, M.D., Koppeschaar, H.P.F., Edwards, C.A., Dieguez, C. and Scanlon, M.F. (1987). Additive effects of growth hormone releasing factor and insulin hypoglycaemia on growth hormone release in man. Clinical Endocrinology **26,** 589-595.

17 Page, M.D., Koppeschaar, H.P.F., Dieguez, C., Gibbs, J.T., Hall, R., Peters, J.R. and Scanlon, M.F. (1987). Cholinergic muscarinic receptor blockade with pirenzepine abolishes slow wave sleep-related growth hormone release in young patients with insulin-dependent-diabetes mellitus. Clinical Endocrinology **26,** 355-359.

18 Page, M.D., Dieguez, C., Valcavi, R., Edwards, C., Hall, R. and Scanlon, M.F. (1988). Growth hormone (GH) responses to arginine and L-dopa alone and after GHRH pretreatment. Clinical Endocrinology **28,** 551-558.

19 Peters, J.R., Evans, P.J., Page, M.D., Hall, R., Gibbs, J.T., Dieguez, C. and Scanlon, M.F. (1986). Cholinergic muscarinic receptor blockade with pirenzepine abolishes slow wave sleep related growth hormone release in normal adult males. Clinical Endocrinology **25,** 213-217.

20 Plotsky, P.M. and Vale, W. (1985). Patterns of growth hormone-releasing factor and somatostatin secretion into the hypophysial-portal circulation of the rat. Science **230,** 461-463.

21 Ross, R.J.M., Tsagarakis, S., Grossman, A., Nhagafoong, L., Touzel, R.J., Rees, L.M. and Besser, G.M. (1987). GH feedback occurs through modulation of hypothalamic somatostatin under cholinergic control: studies with pyridostigmine and GHRH. Clinical Endocrinology **27,** 727-734.

22 Ross, R.J.M., Borges, F., Grossman, A., Smith, R., Nhagafoong, L., Rees, L.H., Savage, M.O. and Besser, G.M. (1987). Growth hormone pretreatment in man blocks the response to growth hormone-releasing hormone; evidence for a direct effect of growth hormone. Clinical Endocrinology **26,** 117-123.

23 Sheppard,, M.C., Kronheim, S. and Pimstone, B.L. (1978). Stimulation by growth hormone of somatostatin release from the rat-hypothalamus in vitro. Clinical Endocrinology **9,** 583-586.

24 Shibasaki, T., Hotta, M., Masuda, A., Imaki, T., Obara, N., Demura, H., Ling, N. and Shizume, K. (1985). Plasma GH responses to GHRH and insulin-induced hypoglycemia in man. Journal of Clinical Endocrinology and Metabolism **60,** 1265-1267.

25 Valcavi, R., Dieguez, C., Page, M.D., Zini, M., Casoli, P., Portioli, I. and Scanlon, M.F. (1988). Alpha-2-adrenergic pathways release growth hormone via a

non-GRF-dependent mechanism in normal human subjects. Clinical Endocrinology **29**, 309-316.

26 Vance, M.L., Kaiser, D.L., Frohman, L.A., Rivier, J., Vale, W.W. and Thorner, M.O. (1987). Role of dopamine in the regulation of growth hormone secretion: dopamine and bromocriptine augment growth hormone (GH)-releasing hormone-stimulated GH secretion in normal man. Journal of Clinical Endocrinology and Metabolism **64**, 1136-1141.

27 Vance, M.L., Kaiser, D.L., Evans, W.S., Thorner, M.O., Furlanetto, R., Rivier, J., Vale, W., Perisatti, G. and Frohman, L.A. (1985). Evidence for a limited growth hormone (GH)-releasing hormone (GHRH)-releasable quantity of GH: effects of 6-hour infusions of GHRH on GH secretion in normal man. Journal of Clinical Endocrinology and Metabolism **60**, 370-375.

28 Wehrenberg, W.B., Baird, A. and Ling, N. (1983). Potent interaction between glucocorticoids and growth-hormone-releasing factor in vivo. Science **221**, 556-558.

ROLE OF GROWTH HORMONE IN THE REGULATION OF ADIPOCYTE GROWTH AND FUNCTION

R.G. Vernon and D.J. Flint

Hannah Research Institute, Ayr, Scotland

INTRODUCTION

It has been known for many years, that prolonged treatment of rats with preparations rich in growth hormone (GH) decreased the amount of body fat (36, 69). More recently with the production of recombinant GH and the greater availability of pituitary-derived GH it has been shown that chronic administration of GH decreases adiposity in farm animals (7, 44). Very recently transgenic animals have been produced carrying and expressing alien GH genes, eg pigs with rat (21) or bovine (102) GH, which again exhibit a remarkably diminished degree of adiposity. These effects of GH on adiposity in farm animals are achieved without any loss, and in fact there can be a gain, in feed-conversion efficiency (7, 44). Thus manipulation of GH either by administration of the hormone or by production of transgenic animals offers an effective and efficient means of producing leaner animals.

The mechanisms whereby GH decreases adiposity have not been fully elucidated but it appears that changes in both lipid synthesis and mobilisation may be involved; the mechanism may be indirect as GH appears to alter the ability of fat cells to respond to both insulin and catecholamines. In addition GH may have a direct lipolytic effect itself. The picture is complicated, however, by observations that GH can have insulin-like effects under certain conditions, while studies with cell lines which develop into adipocyte-like cells, suggest that GH and also IGF-I promote adipocyte development.

In the following sections we document and attempt to assess the physiological relevance of the various effects listed above.

RECEPTORS FOR GH AND IGFs

GH receptors have been demonstrated on adipocytes from rats (24) and man (16) confirming the conclusion drawn from in vitro studies that the hormone can act directly on the tissue. There are about 2 x 10^4 receptors per cell in rats (24, 41, 46); number varies with depot (51), sex (46) and age (46). Hypophysectomy decreases the number of receptors and this can be reversed by GH treatment (30, 31, 42), while activation of protein kinase A leads to a decrease in GH binding, possibly due to receptor phosphorylation on serine residues (38). The ability of the receptor to bind GH variants differs, eg the 20 kDa variant of hGH is bound very poorly relative to the 22 kDa form by rat adipocytes (82). Adipocytes take up GH by receptor-mediated endocytosis and process it primarily by a degradative pathway which appears to involve lysosomal proteolysis; there is also some evidence for a non-degradative pathway which ends in release of intact hormone (37, 70). The chain of events involved in GH signal tranduction in adipocytes has not been elucidated. It is not known if the receptor has tyrosine kinase activity itself, but there is evidence for the binding of GH to its receptor leading to phosphorylation of receptor tyrosine residues in a fibroblast derived adipocyte cell line (28). The receptor is unlikely to interact with a GTP-binding protein as we have found that pertussis toxin which ADP-ribosylates and inactivates several GTP-binding proteins did not impair GH action in sheep adipose tissue (95).

Receptors for IGF-II have been demonstrated on rat adipocytes (57) but are absent from human adipocytes (16). Insulin promotes a re-distribution of IGF-II receptors from the interior of the cell to the plasma membrane (63, 99); there is a concomitant decrease in phosphorylation status of the IGF-II receptor due to insulin inhibiting a plasma membrane tyrosine kinase (13). Hypophysectomy leads to an increased number of IGF-II receptors on adipocytes and decreased ability of insulin to promote re-distribution towards the plasma membrane; these effects are reversed by chronic treatment with GH (54). There do not, however, appear to be any IGF-I receptors on rat (57) or human (16) adipocytes at least.

CHRONIC EFFECTS OF GH ON INSULIN ACTION AND LIPID SYNTHESIS

Effects of GH treatment in vivo on adipose tissue metabolism depend on the state of the animal. Studies with intact rats showed that chronic elevation of GH decreased rates of glucose oxidation and fatty acid synthesis in adipose tissue (14, 36) while in intact growing pigs GH treatment reduced the rate of lipogenesis and the activity of several lipogenic enzymes (55). However, chronic treatment of steers with GH had no effects on the rate of lipogenesis in adipose tissue (66). Conversely, administration of an antiserum to rat GH (ArGH) for 2 days increased the activity of acetyl CoA carboxylase in rat adipose tissue (95). Treatment of rats with ArGH for 1 to 6 h increased glucose metabolism in adipocytes but had no effect on the response or sensitivity to insulin (76) but after 6 days of treatment there was a marked increase in sensitivity to insulin (32). There have also been a number of studies examining the effects of chronic GH treatment of hypophysectomized (hypox) rats . The whole endocrine system is in disarray in such animals, for not only do they lack all pituitary hormones but serum insulin may decrease (86) with additional consequences for adipose tissue metabolism; interpretation of findings is thus complex. Hypophysectomy leads to a greatly enhanced rate of glucose transport into the adipocyte (71, 73) and a substantial increase in the activity of the low Km phosphodiesterase (72), neither of which can be increased further by insulin; GH therapy restores normal rates and insulin responsiveness of these two functions (71, 72, 73). In contrast the ability of adipocytes to metabolise glucose is normally reduced, probably largely due to hypoinsulinaemia; this can be minimised by feeding on very high carbohydrate diets, including glucose in the drinking water or insulin therapy (33). GH treatment of hypox rats maintained on high carbohydrate diets (but not other diets) decreased the rates of glucose oxidation and lipogenesis (36). These observations from intact and hypox rats show that GH can have a chronic inhibitory effect on glucose transport and utilisation, and on lipogenesis in adipose tissue. The effect of GH on glucose metabolism appears to be dependent on an adequate insulin level in the animal, probably arising from a requirement for insulin to maintain the activity (amount) of key enzymes, and for this aspect GH can be viewed as having an insulin-antagonistic role (a view supported by in vitro data, below). In contrast, maintenance of a high capacity for glucose transport is independent of insulin in the absence of GH; however, GH, by suppressing the activity, renders it susceptible to regulation by insulin.

Incubation of adipose tissue in vitro for 24 to 72 h with GH results in a diminished rate of lipogenesis in adipose tissue from man (62), sheep (88, 93), cattle (23) and pigs (97, 98). This inhibitory effect was achieved with physiological concentrations of GH in the studies with farm animals (ED_{50} 1-2 ng/ml) (23, 93). GH had little or no effect by itself in these studies, rather it antagonised the ability of insulin to maintain or increase lipogenesis, suggesting that it is acting as an insulin-antagonist. Curiously, both the ability of insulin to maintain or enhance lipogenesis during tissue culture and the antagonism of this by GH were enhanced by glucocorticoids (23, 92, 93, 98). Tissue culture of adipose tissue from lactating sheep (which has a very low rate of lipogenesis) in the presence of insulin and dexamethasone restored the rate of lipogenesis (93) and increased the maximum activity of several lipogenic enzymes, effects which were again antagonised by GH (Table 1). These various observations from studies in vitro also suggest that GH impairs the ability of insulin to maintain the activities (amounts?) of key lipogenic enzymes and hence decreases the capacity of the

Table 1 Effect of tissue culture for 48 h in presence or absence of insulin (Ins), dexamethasone (dex) and growth hormone (GH) on enzyme activities and rates of lipogenesis and protein synthesis in sheep adipose tissue.

	Additions to culture medium			
	None	Ins + dex	Ins + dex + GH	SED
Non-lactating sheep				
Lipogenesis	624[a]	1808[b]	409[a]	125
Acetyl CoA carboxylase % in active state	36.6[a]	82.4[b]	30.9[a]	8.4
Lactating sheep				
Lipogenesis	34[a]	347[b]	29[a]	38
Glucose 6-phosphate dehydrogenase	33.2[a]	54.7[b]	39.7[a]	5.7
Acetyl CoA carboxylase				
Total activity	1.04[a]	2.33[b]	1.70[ab]	0.46
Active state	0.03[a]	0.85[b]	0.18[a]	0.22
Protein synthesis	5.7[a]	7.9[b]	8.1[b]	0.9

Results are means of 4 to 9 observations; enzyme activities are expressed as nmol/min per 10^6 cell units; rates of lipogenesis and protein synthesis are expressed as nmol acetate or leucine, respectively, incorporated/h per 10^6 cell units (93). [a,b] Values in a row with a different superscript differ significantly, $P < 0.05$.

---

tissue for lipogenesis. GH does not, however, inhibit the ability of insulin to stimulate total protein synthesis in sheep adipose tissue in vitro (Table 1), but this does not exclude an effect on the synthesis of a few specific enzymes.

Changes in enzyme activity alter the lipogenic capacity (ie maximum possible rate) of a cell; lipogenesis is also subject to acute control by changes in the activation status of key enzymes, in particular acetyl CoA carboxylase. The initial decrease in lipogenesis in response to GH seen in sheep adipose tissue in culture is due to GH antagonising the ability of insulin to activate this enzyme (Table 1). Furthermore, during early lactation in sheep (a state associated with a high serum GH concentration in this species), the activation status of acetyl CoA carboxylase is decreased in adipose tissue and the ability of insulin to activate the enzyme is lost, but can be restored by maintaining the tissue in culture with insulin and dexamethasone for 48 h; restoration of responsiveness is prevented by GH (Table 1) and also by actinomycin D, an inhibitor of gene transcription (90). These observations suggest that GH impedes the production of some component of the signal transduction pathway whereby insulin activates acetyl CoA carboxylase.

In contrast to the above in another study with rat adipose tissue, culture with GH for 20 h decreased basal glucose transport and as a consequence glucose metabolism but had no effect on the response to insulin or on insulin binding (56). The reason for these apparent discrepancies is not clear.

The mechanism by which GH achieves its insulin-antagonistic effects is not known. The effects of GH can be mimicked by actinomycin D and also by agents (eg 3-methylisobutylxanthine, cholera toxin) (95) which increase adenylate cyclase activity and hence lipolysis. However, insulin antagonistic effects of GH can occur in the absence of any change in lipolytic rate (23, 88, 93, 98) excluding an obligatory involvement of the adenylate cyclase - protein kinase A axis.

ACUTE, INSULIN-LIKE EFFECTS OF GH ON ADIPOCYTE METABOLISM

In addition to the insulin-antagonistic effects of GH described above, under some conditions GH can also have insulin-like effects on adipose tissue metabolism. This aspect has been studied in greatest detail in the rat. In this species the insulin-like effects of GH are acute, transient (there is a subsequent period of refractoriness to GH) and appear to be dependent on a prior period of GH deficiency (34, 35). Thus in adipose tissue from hypox rats, GH increased glucose uptake, glucose oxidation, glycogen synthesis, pyruvate decarboxylation, leucine oxidation, glucose conversion to fatty acids and total lipids and inhibited adrenaline-stimulated lipolysis (34), while recent additions to this list include decreased phosphorylation of hormone-sensitive lipase (4) and even an increased number of IGF-II receptors in the plasma membrane (54)! Insulin-like effects of GH have also been observed in intact rats after treatment with antiserum to GH, after thyroidectomy (which decreases serum GH concentration), in intact weanling male rats (where endogenous GH is naturally very low), and in Dwarf mice (which are GH deficient) (14, 34). Insulin-like effects of GH are found in adipose tissue from intact rats subjected to surgical stress about 3 h prior to slaughter (12). Adrenalectomy also results in appearance of the acute insulin-like effects of GH and there is some evidence to suggest that a factor released from the pituitary gland along with ACTH mediates the development of sensitivity to GH in stressed rats (11). In contrast, insulin-like effects of GH are not normally seen with adipose tissue from intact non-stressed adult rats, but can be induced by pre-incubating adipose tissue for 3-4 h in the absence of GH (34, 35). Induction of GH responsiveness by preincubation is prevented by inhibitors of protein synthesis and is associated with an increase in GH binding to adipocytes (34, 35). The suggestion that the insulin-like effects are produced by a contaminant of the GH preparations has been refuted by studies using several biosynthetic human and bovine GHs (35).

The mechanism involved in this insulin-like effect of GH in rats is unknown. It is clear that tissue need only be exposed to GH for 5-10 min in vitro for the effect to be manifest (34). The effect appears to be dependent on insulin and is lost if all traces of insulin are removed from the tissue with anti-insulin serum (27). This suggests that GH may in fact be having an 'insulin enhancing' rather than an 'insulin-like' effect. Whatever, GH must be triggering the insulin signal transduction system at some early stage; GH does not, however, interact with the insulin receptor (35). GH promotes phosphorylation of a 45 kDa protein in the plasma membrane (19) and there is some evidence for the involvement of protein kinase C (81). The refractoriness to GH which leads to the loss of this insulin-like effect is dependent on protein synthesis, but does not involve any change in GH binding to adipocytes (34, 35).

The physiological relevance of the insulin-like effect remains unclear for under normal physiological conditions animals are never GH-deficient, except, arguably the young male rat with its extreme pulsatile GH release, hence the effect should not occur in vivo; resolution of this paradox remains a challenge. An insulin-enhancing effect during recovery from stress should be advantageous as it would facilitate replenishment of lipid reserves.

A requirement for a period of GH deficiency does not appear to be essential for insulin-like effects of GH in some species. Thus GH showed a transient inhibition of basal (but not catecholamine-induced) lipolysis in reindeer adipocytes without any apparent preincubation (52) and inhibited glucagon-induced lipolysis in chicken adipocytes after only an hour preincubation (it is not clear if even this is essential) (9, 10). Conflicting reports pertain to acute effects of GH in sheep adipose tissue:

one preliminary report showed that recombinant GH at a concentration of 1 μg/ml stimulated lipogenesis (65) while another group show that ovine GH at concentrations of 0.1 or 1 μg/ml inhibited lipogenesis with no effect at lower (physiological concentrations) (85). With pig adipose tissue physiological concentrations of GH showed an acute insulin-antagonistic effect on lipogenesis while at a concentration of 1 μg/ml GH increased lipogenesis (97). Explanation of the findings with reindeer and chicken are not clear (acute stress before slaughter?) while the physiological relevance of the insulin-like effects in sheep and pigs is doubtful.

EFFECTS OF GH ON LIPOLYSIS

GH has been reported to have direct lipolytic effects on adipose tissue and also to influence lipolysis indirectly by altering the ability of adipocytes to respond to lipolytic factors such as catecholamines.

Many studies in a variety of species including ruminants have shown that injecting animals with preparations of GH leads to an increase in plasma fatty acid and glycerol concentrations in vivo (36, 87); in contrast to the immediate effect of exogenous catecholamines, lipolytic effects of GH in vivo usually required an hour or more to become manifest. More rapid effects of GH on lipolysis found in some early in vitro studies with adipose tissue from laboratory species were due to contaminants such as TSH (36, 69). However, the presence of contaminants such as TSH would not account for the slow, dexamethasone and protein synthesis-dependent lipolytic effect of GH observed in studies with adipose tissue from both intact and hypox rats (34, 36, 69) an effect which has a similar time-course to that observed in the in vivo studies described above. In tissue from hypox rats the lipolytic effect increases as the acute insulin-like effect disappears (35). Suggestions that this slow effect of GH on lipolysis is due to contaminants would appear to be refuted by the demonstration that recombinant GH preparations can increase lipolysis in sheep (43) in vivo and in rat (35), chicken (9) and sheep (53) adipose tissue in vitro. It appears that this slow lipolytic effect of GH can be prevented by proteinase inhibitors, suggesting that a cleavage product of GH is the active factor (49); fragments of GH are known to be lipolytic (64).

In contrast to the above, most recent studies with highly purified pituitary GH or recombinant GH have failed to reveal lipolytic effects of GH. Pituitary GH and, or, recombinant GH failed to stimulate lipolysis in vivo in cattle (7, 60) and rabbits (6, 50) and in adipose tissue from rats (25, 26), rabbits (2, 6), sheep (20, 65, 85), reindeer (52) and pigs (98) in vitro during incubations of a few hours; most of these studies were carried out in the presence and absence of glucocorticoid. Prolonged incubation of adipose tissue for several days also failed to show an effect on lipolysis in adipose tissue from man (62), rats (56), sheep (88, 93), pigs (98) and cattle (23). The addition of a pituitary GH preparation which had no intrinsic lipolytic activity by itself did however prevent the decrease in lipolysis caused by dexamethasone during prolonged incubation of sheep adipose tissue (93).

Trying to reconcile the various conflicting observations described above is not easy. Even highly purified preparations of pituitary GH are heterogenous due to the existence of different forms of the hormone (7) which could account for some of the differences. Fragments of GH are also lipolytic (64) hence chemical changes during solubilization or storage or perhaps as a result of proteolytic cleavage during incubation in vitro (49) or in vivo (43) could also account for the discrepancies. For the present we cannot say categorically that GH itself is, or is not, lipolytic.

In addition to these questionable direct lipolytic actions of GH there is evidence that GH acts to modify the responsiveness of adipose tissue to other hormones. GH treatment of steers fed ad libitum or feed-restricted (66), lactating cows (59, 78, 79) and growing pigs (103) increased the response to adrenaline in vivo in terms of increased glycerol and fatty acid concentrations in blood. GH increased the response but not sensitivity (ED_{50}) to the lipolytic effect of adrenaline in cattle in vivo (78). However, despite the enhanced response to catecholamines in vivo, Peters (66) found that chronic treatment of steers with GH resulted in a diminished response of subcutaneous adipose tissue to catecholamines in vitro! Litter removal from lactating rats resulted in a decreased lipoytic response of adipocytes to noradrenaline in vitro; this adaptation could be prevented by injecting such rats with GH for 2 days and could be mimicked by giving lactating rats an antiserum to rat GH for 2 days (94): preliminary studies suggest that these changes in response are paralleled by changes in the number of β-receptors of adipocyte plasma membranes (95). We have also found that maintenance of sheep adipose tissue in culture for 48 h in the presence of GH increased the number of β-adrenergic receptors of adipocytes (100). The number of β-adrenergic receptors of adiocytes is increased during early lactation in cattle (48) and sheep (100), a time when serum GH is also elevated.

Lipolysis is also subject to regulation by several antilipolytic factors, in particular insulin, adenosine and even catecholamines themselves acting through the α_2-adrenergic receptor. Injecting rats with GH over two days diminished the antilipolytic response of adipocytes to adenosine (94), while maintenance of sheep adipose tissue in culture with GH decreased the degree of α_2-mediated inhibition of lipolysis by noradrenaline (91). In contrast, prolonged treatment with GH enhanced the antilipolytic effect of insulin in lactating cows (78, 79) and growing pigs (103) in vivo. These latter results are surprising in view of the insulin-antagonistic effects of GH on lipogenesis; the effect was observed after prolonged treatment with GH and so is distinct from the acute 'insulin-like' effect of GH described above.

MEDIATION BY IGFs

For some tissues (eg mammary gland) it would appear that effects of GH are mediated by IGF-I. In contrast, the presence of GH receptors and the absence of IGF-I receptors (in rat and human adipocytes at least) would suggest that in adipose tissue effects of GH are direct and are not mediated by IGF-I. Further evidence for this comes from studies such as those of Schoenle et al. (74) who found that effects of hypophysectomy in rats on glucose metabolism of adipose tissue could be prevented by infusion of GH but not IGF-I or IGF-II. Studies in vitro also suggest that effects of IGFs on adipose tissue from rats (29), man (5) and ruminants (22, 93) are mediated via the insulin receptor (ie effects are insulin-like and require high concentrations of IGF-I). In contrast, a very recent study by Lewis et al. (53) suggests that human IGF-I and IGF-II at concentrations of 0.4-4.0 ng/ml and 0.4-4.0 pg/ml respectively (but not insulin) are lipolytic in sheep adipose tissue in vitro. The authors suggest that the lower sensitivity to IGF-I probably reflects it acting via the IGF-II receptor. In contrast human IGF-I and IGF-II at a concentration of about 1 ng/ml had no lipolytic effect on human adipocytes and were in fact slightly antilipolytic (5), but the human adipocyte appears to lack IGF-II as well as IGF-I receptors (16). These potentially exciting findings of Lewis et al. need to be treated with some caution however. If IGFs were intermediates in the lipolytic effect of GH one would expect to see an effect of GH on lipolysis in vivo perhaps after a lag phase, as was observed in some earlier studies whereas many recent studies suggest that GH does not stimulate lipolysis in vivo or in vitro (see above). Nevertheless their observation is consistent with the lipolytic effects of GH in vitro which are slow to develop and require protein synthesis. Also, despite the absence of IGF-I receptors, adipose tissue contains IGF-I (15) while adipocytes produce IGF-I mRNA (17), suggesting that adipocytes themselves produce IGF-I. Furthermore, the amount of IGF-I falls on hypophysectomy (15). We have, however, failed to detect any lipolytic effects of human IGF-I or rat MSA over the concentration range of 0.04-4 ng/ml in sheep adipose tissue under apparently the same conditions used by Lewis et al. (95).

ADIPOGENESIS

A role for both GH and IGF-I in adipocyte development has been proposed by Green in his 'dual effector' theory of adipogenesis (40). Fat-filled adipocytes do not divide themselves; rather adipocyte development is thought to involve proliferation and differentiation of adipocyte precursor cells, located within the stromovascular fraction of the tissue, followed by lipid accretion by the differentiated adipocytes (89). During differentiation or immediately after there may be further proliferation (termed 'clonal expansion') (40). Study of this process has been facilitated by the discovery that certain fibroblast-derived cell lines will develop into adipocyte-like cells in culture (89). Using one such cell line (3T3-3442) Green and colleagues showed that a serum factor which stimulated differentiation of precursor cells was GH (61). GH, as part of the differentiation process, increases the responsiveness of 3T3-3442 cells to IGF-I (104), and in one cell line at least promotes IGF-I production (17); IGF-I then promotes a phase of 'clonal expansion' (40). This sequence is the basis of the 'dual effector' theory. GH also stimulates differentiation in another distinct cell line, the OB1771 line (18). In contrast GH had no effect on proliferation or differentiation of 3T3-L1 cells in a defined medium, whereas IGF-I stimulated differentation of these cells (83). The relevance of these exciting findings to events in vivo is, however, uncertain. Such cell lines can differentiate in vivo (39) as well as in vitro into cells which are remarkably like adipocytes. Differentiated 3T3-3442 cells for example even exhibit an acute, transient, insulin-like response to GH (75). There are nevertheless differences; 3T3-3442 adipocytes differ from true adipocytes in that they bind and respond equally well to both 22k and 20 kDa hGH variants (77) and more importantly they have IGF-I receptors (57, 80) (IGF-I is a potent fibroblast mitogen) (40). Furthermore there is a disconcerting variation in the factors required to promote proliferation and differentiation of the different cell lines as noted above (83, 89).

Evidence in support of a role for GH in adipogenesis in vivo is slight. GH stimulated DNA synthesis in adipose tissue in rats (47) but had no effect in sheep (95). GH also had no effect on the proliferation or differentiation in culture of precursor cells derived from sheep (8) or rat (101) adipose tissue. As described in preceding sections GH administration results in diminished adiposity which is contrary to what one might expect from its postulated role in adipogenesis. Changes in serum GH during fetal development do not accord with a role in adipocyte development (89) while hypophysectomy of fetal lambs increases the amount of subcutaneous fat (1) and this can be reversed by GH therapy (84). Hypophysectomy of lambs at either 25 or 50 days of age decreased muscle growth but had no effect on adipose tissue development, and when performed at 100 days of age, again resulted in enhanced adiposity (96). These results appear inconsistent with a role of GH in adipogenesis, but changes in the size of pre-existing adipocytes could have obscured changes in total numbers of cells or numbers of differentiated but unfilled adipocytes. This caution is emphasised by the finding that while decapitation of fetal pigs increased the amount of adipose tissue, morphological examination of the tissue revealed that mature adipocytes were larger in decapitated fetuses, and also had a higher lipogenic capacity, but there were fewer clusters of cells suggesting a decreased cell number (45, 68). These observations are consistent with GH having a role in adipogenesis but again fetal decapitation removes a variety of factors and an agent other than GH would appear to be responsible for impaired adipogenesis in such pigs (67). Administration of GH to weanling rats for 15 days (adipogenesis would be very active in these animals) decreased fat cell size but had no apparent effect on cell number (3). Similarly, implantation of a GH-secreting tumour into either 1 or 17 wk old rats again reduced adipocyte size but had no effect on number (58). Again in these two studies only the number of lipid containing adipocytes was measured hence changes in the number of differentiated but unfilled adipocytes in response to GH cannot be excluded. Available evidence thus does not support a role of GH in adipogenesis in vivo but neither does it rigorously exclude such a role, although it would seem unlikely. The absence of IGF-I receptors would also appear to exclude a role for this factor, yet, as described above, adipocytes can produce IGF-I and the possibility that receptors are present on precursor cells remains to be tested.

CONCLUSIONS

Whilst the physiological significance of the acute insulin-like effects of GH on adipocytes remains uncertain and the claims of direct lipolytic effects of GH and a role for GH and IGF-I in adipogenesis remain unproven, it appears that the hormone has important chronic effects on adipose tissue function. A picture is emerging in which prolonged elevation of serum GH concentration either naturally, as in early lactation in ruminants, or artificially, results in a fall in the lipogenic capacity of adipose tissue and a failure of insulin to activate key enzymes of lipid synthesis. In addition the responsiveness of adipocytes to the lipolytic effects of catecholamines (and perhaps other hormones) which stimulate lipolysis through the adenylate cyclase-protein kinase-A cascade is enhanced. Even here, there are a few reports which appear inconsistent with this view. In addition there is the paradoxical chronic effect of GH increasing the acute antilipolytic effect of insulin while the role of GH in modulating the antilipolytic effects of adenosine and α_2-adrenergic response requires clarification. Nevertheless the effects described above provide a plausible mechanism for the decrease in adiposity associated with an elevated serum GH. Thus we envisage the chronic effects of GH on adipose tissue metabolism as being homeorhetic, altering the responsiveness of the tissue to acutely active hormones. The importance of decreased lipid synthesis and enhanced lipolysis will depend on the energy balance of the animal, decreased lipogenesis being important in positive energy balance while enhanced lipolysis coming into play if the animal is in negative energy balance. These adaptations are a necesary adjunct to the productive effects of GH promoting carcass growth and increasing milk yield. Such effects, which appear to be mediated by IGFs, require additional nutrients for the responsive tissues. Effects of GH on adipose tissue and also liver metabolism increase the availability of nutrients for these tissues and so are thus an integral part of the homeorhetic effect of GH within the animal as a whole.

REFERENCES

1 Alexander, G. (1978). Quantitative development of adipose tissue in foetal sheep. Australian Journal of Biological Science **31**, 489-503.

2 Barenton, B., Batifol, V., Combarnous Y., Dulor, J.P., Durand, P. and Vezinhet, A. (1984). Reevaluation of lipolytic activity of growth hormone in rabbit adipocytes. Biochemical and Biophysical Research Communications **122**, 197-203.

3 Batchelor, B.R. and Stern, J.S. (1973). The effect of growth hormone upon glucose metabolism and cellularity in rat adipose tissue. Hormone and Metabolic Research **5**, 37-41.

4 Björgell, P., Rosberg, S., Isaksson, O. and Belfrage, P (1984). The antilipolytic, insulin-like effect of growth hormone is caused by a net decrease of hormone-sensitive lipase phosphorylation. Endocrinology **115**, 1151-1156.

5 Bolinder, J., Lindblad, A., Engfeldt, P. and Arner, P. (1987). Studies of acute effects of insulin-like growth factors I and II in human fat cells. Journal of Clinical Endocrinology and Metabolism **65**, 732-737.

6 Bowden, C.R., White, K.D., Lewis, U.J. and Tutwiler, G.F. (1985). Highly purified human growth hormone fails to stimulate lipolysis in rabbit adipocytes in vitro or in rabbits in vivo. Metabolism **4**, 237-243.

7 Boyd, R.D. and Bauman, D.E. (1988). Mechanism of action of somatotropin in growth. In Current Concepts of Animal Growth (Campion, D.R., Hausman, G.J. and Martin, R.J., eds.). New York (in press).

8 Broad, T.E. and Ham, R.G. (1983). Growth and adipose differentiation of sheep preadipocyte fibroblasts in serum-free medium. European Journal of Biochemistry **135**, 33-39.

9 Campbell, R.M. and Scanes, C.G. (1985). Lipolytic activity of purified pituitary and bacterially-derived growth hormone on chicken adipose tissue in vitro. Proceedings of the Society for Experimental Biology and Medicine **180**, 513-517.

10 Campbell, R.M. and Scanes, C.G. (1987). Growth hormone inhibition of glucagon- and cAMP-induced lipolysis by chicken adipose tissue in vitro. Proceedings of the Society for Experimental Biology and Medicine **184**, 456-460.

11 Coiro, V., Goodman, H.M. (1987). Pituitary secretions related to adrenocorticotrophic hormone induce sensitivity of adipose tissue to the insulin-like actions of growth hormone. Neuroendocrinology **45**, 165-171.

12 Coiro, V., Grichting, G. and Goodman, H.M. (1981). Induction of insulin-like responses to growth hormone by stress. Endocrinology **109**, 2213-2219.

13 Corvera, S. Yagaloff, K.A., Whithead, R.E. and Czech, M.P. (1988). Tyrosine phosphorylation of the receptor for insulin-like growth factor II is inhibited in plasma membranes from insulin-treated rat adipocytes. Biochemical Journal **250**, 47-52.

14 Davidson, M.B. (1987). Effect of growth hormone on carbohydrate and lipid metabolism. Endocrinological Reviews **8**, 115-131.

15 D'Ercole, A.J., Stiles, A.D. and Underwood, L.E. (1984). Tissue concentration of somatomedin C: Further evidence for multiple sites of synthesis and paracrine or autocrine mechanisms of action. Proceedings of the National Academy of Sciences, U.S.A. **81**, 935-939.

16 DiGirolamo, M., Edén, S., Enberg, G., Isaksson, O., Lönnroth, P., Hall, K. and Smith, U. (1986). Specific binding of human growth hormone but not insulin-like growth factors by human adipocytes. FEBS Letters **205**, 15-19.

17 Doglio, A., Dani, C., Fredrikson, G., Grimaldi, P. and Ailhaud, G. (1987). Acute regulation of insulin-like growth factor-I gene expression by growth hormone during adipose cell differentiation. EMBO Journal **6**, 4011-4016.

18 Doglio, A., Dani, C., Grimaldi, P. and Ailhaud, G. (1986). Growth hormone regulation of the expression of differentiation-dependent genes in preadipocyte Ob1771 cells. Biochemical Journal **238**, 123-129.

19 Donnér, J., Eriksson, H. and Belfrage, P. (1986). The acute GH action in rat adipocytes is associated with enhanced phosphorylation of a 46 kDa plasma membrane protein enriched by GH-sepharose. FEBS Letters **208**, 269-272.

20 Duquette, P.F., Scanes, C.G. and Muir, L.A. (1984). Effects of ovine growth hormone and other anterior pituitary hormones on lipolysis of rat and ovine adipose tissue in vitro. Journal of Animal Science **58**, 1191-1197.

21 Ebert, K.M., Low, M.J., Overstrom, E.W., Buonomo, F.C., Baile, C.A., Roberts, T.M., Lee, A., Mandel, G. and Goodman, R.H. (1988). A moloney MLV-rat somatotropin fusion gene produces biologically active somatotropin in a transgenic pig. Molecular Endocrinology **2**, 277-283.

22 Etherton, T.D. and Evock, C.M. (1986). Stimulation of lipogenesis in bovine adipose tissue by insulin and insulin-like growth factor. Journal of Animal Science **62**, 357-362.

23 Etherton, T.D., Evock, C.M. and Kensinger, R.S. (1987). Native and recombinant bovine growth hormone antagonize insulin action in cultured bovine adipose tissue. Endocrinology **121**, 699-703.

24 Fagin, K.D., Lackey, S.L., Reagan, C.R. and DiGirolamo, M. (1980). Specific binding of growth hormone by rat adipocytes. Endocrinology **107**, 608-615.

25 Frigeri, L.G. (1980). Absence of in vitro dexamethasone-dependent lipolytic activity from highly purified growth hormone. Endocrinology **107**, 738-743.

26 Frigeri, L.G., Robel, G. and Stebbing, N. (1982). Bacteria-derived human growth hormone lacks lipolytic activity in rat adipose tissue. Biochemical and Biophysical Research Communications **104**, 1041-1046.

27 Frigeri, L.G., Teguh, K. and Lewis, U.J. (1987). In vitro insulin-like actions of human growth hormone: A study with an insulin antibody. Hormone and Metabolic Research **19**, 464-469.

28 Foster, C.M., Shafer, J.A., Rozsa, F.W., Wang, X., Lewis, S.D., Renken, D.A., Natale, J.E., Schwartz, J. and Carter-Su, C. (1988). Growth hormone promoted tyrosyl phosphorylation of growth hormone receptors in murine 3T3-F442A fibroblasts and adipocytes. Biochemistry **27**, 326-334.

29 Froesch, E.R., Schmid, C., Schwander, J. and Zapf, J. (1985). Actions of insulin-like growth factors. Annual Review of Physiology **47**, 443-467.

30 Gause, I. and Edén, S. (1986). Induction of growth hormone (GH) receptors in adipocytes of hypophysectomized rats by GH. Endocrinology **118**, 119-124.

31 Gause, I., Edén, S., Isaksson, O., DiGirolamo, M. and Smith, U. (1985). Changes in growth hormone binding and metabolic effects of growth hormone in rat adipocytes following hypophysectomy. Acta Physiologica Scandinavica **124**, 229-238.

32 Gause, I., Edén, S., Jansson, J.O. and Isaksson, O. (1983). Effects of in vivo administration of antiserum to rat growth hormone on body growth and insulin responsiveness in adipose tissue. Endocrinology **112**, 1559-1566.

33 Gause, I., Isaksson, O., Lindahl, A. and Edén, S. (1985). Effect of insulin treatment of hypophysectomized rats on adipose tissue responsiveness to insulin and growth hormone. Endocrinology **116**, 945-951.

34 Goodman, H.M., Grichting, G., Coiro, V. (1986). Growth hormone action on adipocytes. In Human Growth Hormone (Raiti, S. and Tolman, R.A., eds.), pp 499-512. Plenum Medical Book Company, New York.

35 Goodman, H.M., Coiro, V., Frick, G.P., Gorin, E., Grichting, G., Honeyman, T.W. and Szecowka, J. (1987). Effects of growth hormone on the rat adipocyte: a model for studying direct actions of growth hormone. Endocrinologica Japonica **34** (Suppl. 1), 59-72.

36 Goodman, H.M. and Schwartz, J. (1974). Growth hormone and lipid metabolism. In Handbook of Physiology section 7, vol, 4 part II (Knobil, E and Sawyer, W., eds.), pp 211-231. American Physiological Society, Washington DC.

37 Gorin, E., Grichting, G. and Goodman, H.M. (1984). Binding and degradation of [^{125}I]human growth hormone in rat adipocytes. Endocrinology **115,** 467-475.

38 Gorin, E., Honeyman, T.W., Tai, L.R. and Goodman, H.M. (1988). Adenosine 3,5-monophosphate-dependent loss of growth hormone binding in rat adipocytes. Endocrinology **123,** 328-334.

39 Green, H. and Kehinde, O. (1979). Formation of normally differentiated subcutaneous fat pads by an established preadipose cell line. Journal of Cellular Physiology **101,** 169-172.

40 Green, H., Morikawa, M. and Nixon, T. (1985). A dual effector theory of growth hormone action. Differentiation **29,** 195-198.

41 Grichting, G., Levy, L.K. and Goodman, H.M. (1983). Relationship between binding and biological effects of human growth hormone in rat adipocytes. Endocrinology **113,** 1111-1120.

42 Grichting, G. and Goodman,. H.M. (1986). Growth hormone maintains its own receptors in rat adipocytes. Endocrinology **119,** 847-854.

43 Hart, I.C., Chadwick, P.M.E., Boone, T.C., Langley, K.E., Rudman, C. and Souza, L.M. (1984). A comparison of the growth-promoting lipolytic, diabetogenic and immunological properties of pituitary and recombinant-DNA-derived bovine growth hormone (somatotropin). Biochemical Journal **224,** 93-100.

44 Hart, I.C. and Johnsson, I.D. (1986). Growth hormone and growth in meat producing animals. In Control and Manipulation of Animal Growth (Buttery, P.J., Lindsay, D.B. and Haynes, N.B., eds.), pp 135-159. Butterworths, London.

45 Hausman, G.J., Campion, D.R., McNamara, J.P., Richardson, R.L. and Martin, R.J. (1981). Adipose tissue development in the fetal pig after decapitation. Journal of Animal Science **53,** 1634-1644.

46 Herington, A.C. (1981). Identification and characterization of growth hormone receptors on isolated rat adipocytes. Journal of Receptor Research **2,** 299-316.

47 Hollenberg, C.H., Vost, A. and Patten, R.L. (1970). Regulation of adipose tissue mass: control of fat cell development and lipid content. Recent Progress in Hormone Research **26,** 463-503.

48 Jaster, E.H. and Wegner, T.N. (1981). Beta-adrenergic receptor involvement in lipolysis of dairy cattle subcutaneous adipose tissue during dry and lactating state. Journal of Dairy Science **64,** 1655-1663.

49 Keda, Y.M. and Pankov, Y.A. (1987). Role of proteolytic processes in the stimulation of the lipolysis in adipose tissue by somatotropin, adrenocorticotropin and β-lipotropin. Biokhimiya **52,** 1107-1115.

50 Knudtzon, J. and Edminson, P.D. (1985). Different acute in vivo effects of bacterial derived and pituitary growth hormone preparations on plasma levels of glucagon, insulin and free fatty acids in rabbits. Hormone Research **21,** 10-18.

51 LaFranchi, S., Hanna, C.E., Torresani, T., Schoenle, E. and Illig, R. (1985). Comparison of growth hormone binding and metabolic response in rat adipocytes of epididymal, subcutaneous, and retroperitoneal origin. Acta Endocrinologica **110,** 50-55.

52 Larsen, T.S. and Nilssen, K.J. (1985). On the hormonal regulation of lipolysis in isolated reindeer adipocytes. Acta Physiologica Scandinavica **125,** 547-552.

53 Lewis, K.J., Molan, P.C., Bass, J.J. and Gluckman, P.D. (1988). The lipolytic activity of low concentrations of insulin-like growth factors in ovine adipose tissue. Endocrinology **122,** 2554-2557.

54 Lönnroth, P., Assmundsson, K., Edén, S., Enberg, G., Gause, I., Hall, K. and Smith, U. (1987). Regulation of insulin-like growth factor II receptors by growth hormone and insulin in rat adipocytes. Proceedings of the National Academy of Sciences, U.S.A. **84,** 3619-3622.

55 Magri, K.A., Gopinath, R. and Etherton, T.D. (1987). Inhibition of lipogenic enzyme activities by porcine growth hormone (pGH). Journal of Dairy Science **65,** Suppl. 1, 258.

56 Maloff, B.L., Levine, J.H. and Lockwood, D.H. (1980). Direct effects of growth hormone on insulin action in rat adipose tissue maintained in vitro. Endocrinology **107,** 538-544.

57 Massagué, J. and Czech, M.P. (1982). The subunit structures of two distinct receptors for insulin-like growth factors I and II and their relationship to the insulin receptor. Journal of Biological Chemistry **257,** 5038-5045.

58 McCusker, R.H., Campion, D.R. and Cartwright, A.L. (1986). Effect of growth hormone-secreting tumors on adipose tissue cellularity in young and mature rats. Growth **50,** 128-137.

59 McCutcheon, S.N. and Bauman, D.E. (1986). Effect of chronic growth hormone treatment on response to epinephrine and thyrotropin-release hormone in lactating cows. Journal of Dairy Science **69,** 44-51.

60 McDowell, G.H., Hart, I.C., Bines, J.A., Lindsay, D.B. and Kirby, A.C. (1987). Effects of pituitary-derived bovine growth hormone on production parameters and biokinetics of key metabolites in lactating dairy cows at peak and mid-lactation. Australian Journal of Biological Science **40,** 191-202.

61 Morikawa, M., Nixon, T. and Green, H. (1982). Growth hormone and the adipose conversion of 3T3 cells. Cell **29,** 783-789.

62 Nyberg, G. and Smith, U. (1977). Human adipose tissue in culture. Hormone and Metabolic Research **9,** 22-27.

63 Oka, Y., Mottola, C., Oppenheimer, C.L. and Czech, M.P. (1984). Insulin activates the appearance of insulin-like growth factor II receptors on the adipocyte cell surface. Proceedings of the National Academy of Sciences, U.S.A. **81,** 4028-4032.

64 Paladini, A.C., Pena, C. and Poskus, E. (1983). Molecular biology of growth hormone. CRC Critical Review of Biochemistry **15,** 25-56.

65 Peterla, T.A., Ricks, C.A. and Scanes, C.G. (1987). Comparison of effects of β-agonists and recombinant bovine somatotropin on lipolysis and lipogenesis in ovine adipose tissue. Journal of Animal Science **65** (Suppl. 1), 278.

66 Peters, J.P. (1986). Consequences of accelerated gain and growth hormone administration for lipid metabolism in growing beef steers. Journal of Nutrition **116,** 2490-2503.

67 Ramsay, T.G., Hausman, G.J. and Martin, R.J. (1987). Pre-adipocyte proliferation and differentiation in response to hormone supplementation of decapitated fetal pig sera. Journal of Animal Science **64,** 735-744.

68 Ramsay, T.G. Hausman, G.J. and Martin, R.J. (1987). Central endocrine regulation of the development of hormone responses in porcine fetal adipose tissue. Journal of Animal Science **64,** 745-751.

69 Rao, A.J. and Ramachandran, J. (1977). Growth hormone and the regulation of lipolysis. In Hormonal Proteins and Peptides vol. IV (Li, C.H., ed.), pp 43-61. Academic Press, London.

70 Roupas, P. and Herington, A.C. (1987). Receptor-mediated endocytosis and degradative processing of growth hormone by rat adipocytes in primary culture. Endocrinology **120,** 2158-2165.

71 Schoenle, E., Zapf, J. and Froesch, E.R. (1979). Effects of insulin on glucose metabolism and glucose transport in fat cells of hormone-treated hypophysectomized rats: evidence that growth hormone restricts glucose transport. Endocrinology **105,** 1237-1242.

72 Schoenle, E., Zapf, J. and Froesch, E.R. (1981). In vivo control of insulin-sensitive phosphodiesterase in rat adipocytes by growth hormone and its parallelism to glucose transport. Endocrinology **109,** 561-566.

73 Schoenle, E., Zapf, J. and Froesch, E.R. (1982). Glucose transport in adipocytes and its control by growth hormone in vivo. American Journal of Physiology **242,** E368-E372.

74 Schoenle, E., Zapf, J. and Froesch, E.R. (1983). Regulation of rat adipocyte glucose transport by growth hormone: no mediation by insulin-like growth factors. Endocrinology **112,** 384-386.

75 Schwartz, J. and Carter-Su, C. (1988). Effects of growth hormone on glucose metabolism and glucose transport in 3T3-F442A cells: Dependence on cell differentiation. Endocrinology **122,** 2247-2256.

76 Schwartz, J. and Edén, S. (1985). Acute growth hormone deficiency rapidly alters glucose metabolism in rat adipocytes. Relation to insulin responses and binding. Endocrinology **116,** 1806-1812.

77 Schwartz, J. and Foster, C.M. (1986). Pituitary and recombinant deoxyribonucleic acid-derived human growth hormones alter glucose metabolism in 3T3 adipocytes. Journal of Clinical Endocrinology and Metabolism **62,** 791-794.

78 Sechen, S.J., Dunshea, F.R. and Bauman, D.E. (1988). Mechanism of bovine somatotropin (bST) in lactating cows: effect on response to homeostatic signals (epinephrine and insulin). Journal of Dairy Science **71** (Suppl. 1), 168.

79 Sechen, S.J., McCutcheon, S.N. and Bauman, D.E. (1985). Response to metabolic challenges in lactating dairy cows during short-term bovine growth hormone treatment. Journal of Dairy Science **68** (Suppl. 1), 170.

80 Shimizu, M., Torti, F. and Roth, R.A. (1986). Characterization of the insulin and insulin-like growth factor receptors and responsitivity of a fibroblast/ adipocyte cell line before and after differentiation. Biochemical and Biophysical Research Communications **137**, 552-558.

81 Smal, J. and De Meyts, P. (1987). Role of kinase C in the insulin-like effects of human growth hormone in rat adipocytes. Biochemical and Biophysical Research Communications **147**, 1232-1240.

82 Smal, J., Closset, J., Hennen, G. and De Meyts, P. (1987). Receptor binding properties and insulin-like effects of human growth hormone and its 20 kDa-variant in rat adipocytes. Journal of Biological Chemistry **262**, 11071-11079.

83 Smith, P.J., Wise, L.S., Berkowitz, R., Wan, C. and Rubin, C.S. (1988). Insulin-like growth factor-I is an essential regulator of the differentiation of 3T3-L1 adipocytes. Journal of Biological Chemistry **263**, 9402-9408.

84 Stevens, D. and Alexander, G. (1986). Lipid deposition after hypophysectomy and growth hormone treatment in the sheep fetus. Journal of Developmental Physiology **8**, 139-145.

85 Thornton, R.F., Tume, R.K., Larsen, T.W., Johnson, G.W. and Wynn, P.C. (1986). The effects of ovine growth hormone on lipid metabolism of isolated ovine subcutaneous adipocytes. Proceedings of the Nutrition Society of Australia **11**, 152.

86 Van Lan, V., Yamaguchi, N., Garcia, M.J., Ramey, E.R. and Penhos, J.C. (1974). Effect of hypophysectomy and adrenalectomy on glucagon and insulin concentration. Endocrinology **94**, 671-675.

87 Vernon, R.G. (1980). Lipid metabolism in the adipose tissue of ruminant animals. Progress in Lipid Research **19**, 23-106.

88 Vernon, R.G. (1982). Effects of growth hormone on fatty acid synthesis in sheep adipose tissue. International Journal of Biochemistry **14**, 255-258.

89 Vernon, R.G. (1986). The growth and metabolism of adipocytes. In Control and Manipulation of Animal Growth (Buttery, P.J., Haynes, N.B. and Lindsay, D.B., eds.), pp 67-83. Butterworths, London.

90 Vernon, R.G., Barber, M., Finley, E. and Grigor, M.R. (1988). Endocrine control of lipogenic enzyme activity in adipose tissue from lactating ewes. Proceedings of the Nutrition Society **47**, 100A.

91 Vernon, R.G., Cork, S. and Finley, E. (1988). Effect of somatotropin on adrenergic responsiveness of ovine adipose tissue. Journal of Animal Science **66** (Suppl. 1), 250.

92 Vernon, R.G. and Finley, E. (1986). Endocrine control of lipogenesis in adipose tissue from lactating sheep. Biochemical Society Transactions **14**, 635-636.

93 Vernon, R.G. and Finley, E. (1988). Roles of insulin and growth hormone in the adaptations of fatty acid synthesis in white adipose tissue during the lactation cycle in sheep. Biochemical Journal **256**, 873-878.

94 Vernon, R.G., Finley, E. and Flint, D.J. (1987). Role of growth hormone in the adaptations of lipolysis in rat adipocytes during recovery from lactation. Biochemical Journal **242**, 931-934.

95 Vernon, R.G. and Flint, D.J. Unpublished observations.

96 Vézinhet, A. (1973). Influence de l'hypophysectomie et de traitements a l'hormone somatotrope bovine sur la croissance relative de l'agneau. Annales Biologique Animale Biochimie et Biophysica **13**, 51-73.

97 Walton, P.E. and Etherton, T.D. (1986). Stimulation of lipogenesis by insulin in swine adipose tissue: antagonism by porcine growth hormone. Journal of Animal Science **62**, 1584-1595.

98 Walton, P.E., Etherton, T.D. and Evock, C.M. (1986). Antagonism of insulin action in cultured pig adipose tissue by pituitary and recombinant porcine growth hormone: potentiation by hydrocortisone. Endocrinology **118**, 2577-2581.

99 Wardzala, L.J., Simpson, I.A., Rechler, M.M. and Cushman, S.W. (1984). Potential mechanism of the stimulatory action of insulin on insulin-like growth factor II binding to isolated rat adipose cell. Journal of Biological Chemistry **259**, 8378-8383.

100 Watt, P.W., Clegg, R.A., Flint, D.J. and Vernon, R.G. (1989). Increases in sheep β-receptor number on exposure to growth hormone in vitro. In Biotechnology in Growth Regulation (Heap, R.B., Prosser, C.G. and Lamming, G.E., eds.). Butterworths, London (in press).

101 Wiederer, O. and Löffler, G. (1987). Hormonal regulation of the differentiation of rat adipocyte precursor cells in primary culture. Journal of Lipid Research **28**, 649-658.

102 Wieghart, M., Hoover, J., Choe, S.H., McGrane, M.M., Rottman, F.M., Hanson, R.W. and Wagner, T.E. (1988). Genetic engineering of livestock - transgenic pigs containing a chimeric bovine growth hormone (PEPCK/bGH) gene. Journal of Animal Science **66** (Suppl. 1), 266.

103 Wray-Cahen, D., Boyd, R.D., Bauman, D.E., Ross, D.A. and Fagin, K. (1987). Metabolic effects of porcine somatotropin (pST) in growing swine. Journal of Animal Science **65** (Suppl. 1), 261.

104 Zezulak, K.M. and Green, H. (1986). The generation of insulin-like growth factor-I-sensitive cells by growth hormone action. Science **233**, 551-553.

A COMPARISON OF THE MECHANISMS OF ACTION OF BOVINE PITUITARY-DERIVED AND RECOMBINANT SOMATOTROPIN (ST) IN INDUCING GALACTOPOIESIS IN THE COW DURING LATE LACTATION

R.B. Heap[1], I.R. Fleet[1], F.M. Fullerton[2], A.J. Davis[1], J.A. Goode[1], I.C. Hart[3], J.W. Pendleton[1], C.G. Prosser[1], L.M. Silvester[4] and T.B. Mepham[4]

[1] AFRC Institute of Animal Physiology and Genetics Research, Babraham, Cambridge CB2 4AT, U.K.
[2] Present address: Department of Environmental and Preventive Medicine, St. Bartholomew's Hospital Medical College, London EC1M 6BQ, U.K.
[3] Present address: American Cyanamid Company, Agricultural Research Division, PO Box 400, Princeton NJ 08540, U.S.A.
[4] Department of Physiology and Environmental Science, Faculty of Agricultural and Food Sciences, University of Nottingham, Sutton Bonington, Loughborough, Leics LE12 5RD, U.K.

INTRODUCTION

The somatotropic effects of growth hormone (ST) have been known since the classical work of Evans and Simpson in 1931 (10), which demonstrated an increased weight gain in rats during chronic administration of a crude bovine ST extract. Subsequently, Lee and Schaffer (21) described an associated increase in protein anabolism and a decrease in fat deposition during treatment. Current interest in somatotropin is focused on its ability to enhance the efficiency of milk production. Asimov and Krouze in 1937 (1) were among the first to report that crude pituitary extracts were galactopoietic in cows and subsequent studies showed that the response was due to somatotropin (2, 5, 6, 14, 24, 29). The development of techniques for the production of recombinant bST (re.bST) by genetic engineering has stimulated research into the nature of bST-induced galactopoiesis and the mechanisms by which this is achieved. For reviews of related studies the reader is referred to earlier publications from several laboratories (4, 15, 17, 18, 26).

This paper reports studies designed to investigate the mechanisms by which highly purified re.bST brings about a galactopoietic response during late lactation. The experiments concentrate on the early response in groups of animals studied in detail and were designed to identify differences between responses obtained with re.bST and those with pituitary-derived ST (pit.bST) which consists of multiple forms of ST.

Similar studies reported previously (17, 26) have shown that the galactopoietic response in cows treated with pit.bST gave increased milk yields ranging from about 10 to 30% with changes in gross efficiency varying from about 10 to 50%. Long-term treatment with pit.bST or re.bST, however, gave comparable increases in milk yield but a lower improvement in gross efficiency associated with a compensatory increase in food intake (15). In the present studies, special attention has been given to the question of whether re.bST-induced galactopoiesis is associated with acute cardiovascular responses and changes in mammary substrate uptake.

EXPERIMENTAL

Experiments were performed on lactating non-pregnant Jersey cows, each with a carotid artery and milk (caudal superficial epigastric) vein surgically exteriorised as skin-covered loops. During surgical preparation, blood vessels crossing the median suspensory ligament of the udder were sectioned between double ligatures to ensure that samples taken from the mammary vein represented blood draining from one-half of the udder. Cows were housed in individual stalls, fed a daily diet of dairy concentrates at milking, and given access to best quality hay in amounts calculated to exceed the maintenance and milk yield requirements. The animals were milked twice daily, the yields of the left and right udder halves recorded separately and a regular routine adopted so that all cows were well-accustomed to the procedures.

Two groups of cows were used: the first group of five animals (6 experiments) received pit.bST (31 to 55 weeks lactation, 3rd to 6th lactation; 0.07 ± 0.01 mg/kg, 1.4 units/mg (16)); the second group of six animals were given re.bST (40 to 67 weeks lactation, 2nd to 5th lactation; 0.07 ± 0.01 mg/kg, 1.34 units/mg). Subcutaneous injections were given about 1 h after morning milking from day 1 to 7. Blood samples were taken from the carotid artery (A) via an indwelling carotid arterial catheter and mammary vein (V) on days -1, 1, 7, 10 and 21 (five pairs on each day between 10.00 and 18.00 h), and measurements made of mammary blood flow (MBF), heart rate and blood pressure (five to seven determinations) throughout the same period. For MBF measurements by thermodilution (22) and mammary vein sampling, the external pudic vein was manually occluded to ensure that total mammary venous drainage was obtained via the milk vein, uncontaminated by venous blood derived from tissues other than the mammary gland. Substantial errors would be experienced in the absence of this procedure examples of which are given in Table 1. The analytical methods and statistical treatment of data are described elsewhere (12).

Table 1 Measurement of mammary blood flow (MBF) in the cow by thermodilution (ml/min ± SEM)

Animal no.	Apparent MBF (A)	Actual MBF	Non-mammary blood flow (B)	B/A x 100 (%)
A104	1747 ± 209	1385 ± 36	362	20.7
	1906 ± 110	1516 ± 96	390	20.5
	2421 ± 39	1908 ± 125	513	21.2
C42	3008 ± 230	2417 ± 156	591	19.6
D46	2306	1769 ± 3	537	23.3
	2592	2148 ± 97	444	17.1
	2838	2085 ± 160	753	26.5
E30	2709	2149	560	20.6
	3513 ± 109	2834 ± 51	679	19.3
	2999 ± 196	2443 ± 1	556	18.5

Mammary blood flow through one-half udder was determined in four surgically prepared cows (see text). Single or repeated (three or more) measurements were made while the pudic vein was clamped manually (actual MBF) or not clamped (apparent MBF). Each mean figure (± SEM) represents the average MBF on four or more days for each cow at different stages of lactation. The results emphasize the importance of excluding blood flow from non-mammary sources in order to obtain accurate values for MBF.

RESULTS AND DISCUSSION

Milk secretion, mammary blood flow and efficiency

Fig. 1 shows that the pattern and magnitude of increase in mean milk secretion was similar in pit.bST- and re.bST-treated cows. The increase in MBF also showed a similar time-course for the two treatments. The relationship between milk secretion rate and mammary blood flow is emphasized in Fig. 2 confirming the wider applicability of observations made in other species regarding the relationship between milk secretion rate and mammary blood flow. Insufficient data are available from the present study to decide whether increased secretion rate was the effect or cause of increased mammary blood flow but it is notable that MBF had frequently increased by 24 h after the first injection. Increases in milk secretion rate in this study were similar to those of previous reports for cows treated with pit.bST (9, 27), and MBF changes confirm those reported by Davis et al. (8).

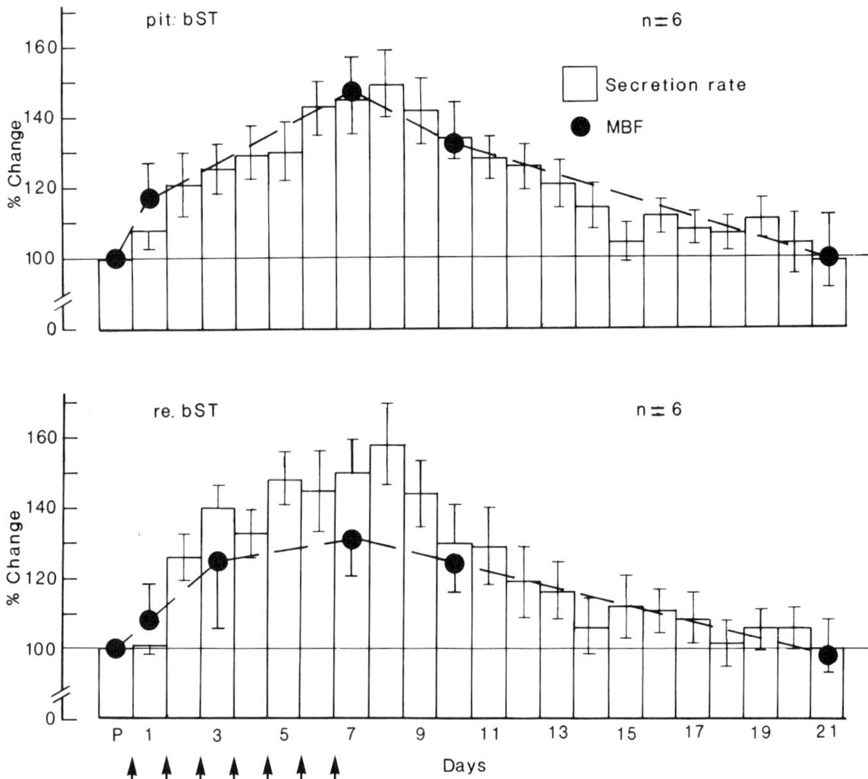

Fig. 1 Milk secretion rate (ml/h, histogram) and mammary blood flow (ml/min, dotted line) in one-half udder of Jersey cows in late lactation treated with pituitary-derived (pit.) or recombinant (re) bST. Animals were treated daily with bST from day 1 to day 7. Results are expressed as a percentage of pre-treatment values (P). Milk secretion rate was significantly increased on days 1 to 7 ($P < 0.01$) and days 8 to 14 ($P < 0.01$) compared to pre-treatment values in both treatment groups. Mammary plasma flow (pit.bST) and mammary blood flow (re.bST) were significantly greater on days 7 and 10 ($P < 0.01$ to $P < 0.05$) in both treatment groups.

Calculations of MY:MBF at specific times after treatment show that before treatment approximately 1 ml of milk was produced for 700 ml blood delivered to the gland (ratio, 1:700) in both groups, a typical figure for cows in late lactation (23). The ratio did not change significantly in cows treated with pit.bST whereas in those treated with re.bST it decreased. After 7 days treatment with re.bST the ratio MY:MBF was 1:415 ($P < 0.01$; Fig. 2), a value similar to that found at peak lactation (23, 25). These data indicate that the galactopoietic effect produced by pit.bST was associated primarily with an increase in MBF whereas re.bST induced an additional effect which resulted in enhanced mammary efficiency.

A possible explanation of the difference between treatments may lie in the circulating concentration of hormones. Pretreatment concentrations of plasma growth hormone in pit.bST and re.bST groups (0.8 ± 0.04 and 2.2 ± 0.3 ng/ml, respectively), which may differ according to energy balance in late lactation, increased significantly by day 7 of treatment (11.0 ± 0.9 and 33.0 ± 4.3 ng/ml, $P < 0.001$). This difference between the two groups was also reflected in insulin values (pretreatment, 13.5 ± 2.8 and 24.8 ± 6.4 μU/ml; day 7 of treatment, 33.3 ± 14.4 and 82.1 ± 22.4 μU/ml, for pit.bST and re.bST-treated groups, respectively). Further investigations showed that whereas the radioimmunoassay of standard amounts of pit.bST and re.bST gave parallel responses, there was a significant difference in immunoreactive potency of the two preparations (1:0.58, re.bST: pit.bST), which partially accounted for the different circulating concentrations during treatment. Thus, the difference in MY:MBF response may have been due to the occurrence of variants (13, 19, 20) or fragments (3) in the pituitary-derived preparation during treatment. We should therefore compare the mechanisms by which these two preparations of bST act in producing a galactopoietic response.

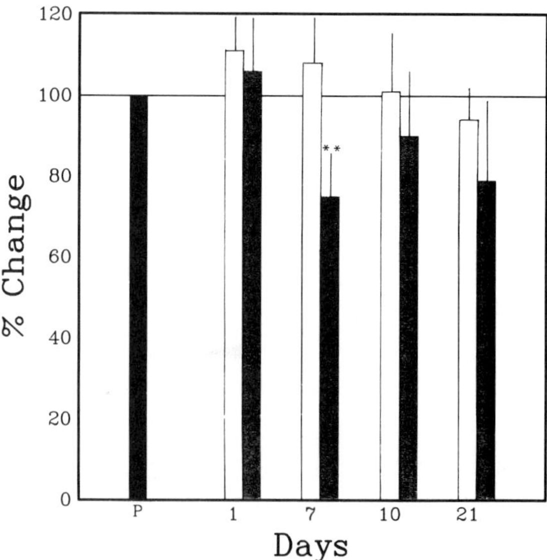

Fig. 2 Relationship between milk yield (MY) and mammary blood flow (MBF) in one-half udder of Jersey cows in late lactation treated for 7 days with pit.bST (open columns) or re.bST (shaded columns). Ratio MY:MBF expressed as a percentage of pre-treatment values (P). There was no significant difference in the pre-treatment values of MY:MBF (1:687 $\pm$ 98 ml/min for pit.bSTY and re.bST, respectively). **$P < 0.01$; treatment compared to pre-treatment.

Heart rate, blood pressure and haematocrit

There were neglible effects of the two preparations on arterial blood pressure, but a significant effect on heart rate at day 7 in animals treated with re.bST (Fig. 3). In re.bST-treated cows heart rate had returned to normal 3 days after treatment. Both preparations produced a significant reduction in packed cell volume but the effect was not significant until after the 7-day treatment (Fig. 3). A similar finding has been reported in lactating Friesland sheep treated for 7 days with pit.bST (11); it may be associated with an increase in plasma volume.

Blood gases and mammary respiratory quotient

A different measure of mammary efficiency is provided by determination of the mammary respiratory quotient. A comparison of the effects of re.bST with those of pit.bST shows that the former preparation had only a slight effect on the arteriovenous difference in 0_2 across the mammary gland, the venoarterial difference in CO_2 or the mammary RQ. In contrast, pit.bST caused a progressive reduction in mammary RQ due to a fall in the venoarterial difference in CO_2 possibly associated with increased oxidation of fatty acids (Figs. 4 and 5 (12)).

Glucose and acetate uptake

Mammary uptake of glucose increased significantly during treatment with both pit.bST and re.bST (Fig. 6). The increase resulted from raised mammary blood flow and mammary A-V difference. Mammary uptake of glucose increased within one day after injection with pit.bST but not with re.bST. The metabolic adjustments in glucose uptake contributed quantitatively to the extra glucose required for the increase in milk yield associated with both treatments. There were only small changes in milk concentrations of lactose, protein, fat, potassium, sodium and chloride, and it follows that secretion of these constituents increased proportionally to raised milk production.

Acetate uptake by the mammary gland was increased significantly during and after treatment with both pit.bST and re.bST. The increase with pit.bST was related to raised mammary blood flow but there was no change in A-V difference. With re.bST there was an increase in both A-V difference and mammary uptake at days 1 and 10 (Fig. 7). The difference between the two treatment responses may indicate that there was decreased acetate oxidation in re.bST-treated cows and that a greater quantity of precursor was available for milk fat synthesis in the face of changes in mammary glucose metabolism (7).

COMMENT

The present study shows a galactopoietic response that was quantitatively indistinguishable in terms of milk secretion rate for similar doses of pit.bST or re.bST. The magnitude of the response over 7 days of treatment exceeded the pretreatment values by about 30% and was accompanied by an increase in mammary blood flow in both instances. When examined in greater detail by relating the absolute values of milk yield to MBF in one udder-half, re.bST not only enhanced milk yield but also decreased MY:MBF.

The reasons for this difference between pit.bST and re.bST were not associated with heart rate, blood pressure, haematocrit, or glucose uptake responses which were similar for both treatments. Neither can they be explained solely in terms of mammary RQ which declined for pit.bST, but remained steady for re.bST. The galactopoietic efficacy of pit.bST was similar to that of re.bST despite a difference in immunoreactivity of 0.58:1, respectively. This implies that, unlike the highly purified and well characterised re.bST, the pit.bST preparation probably contained an active variant of

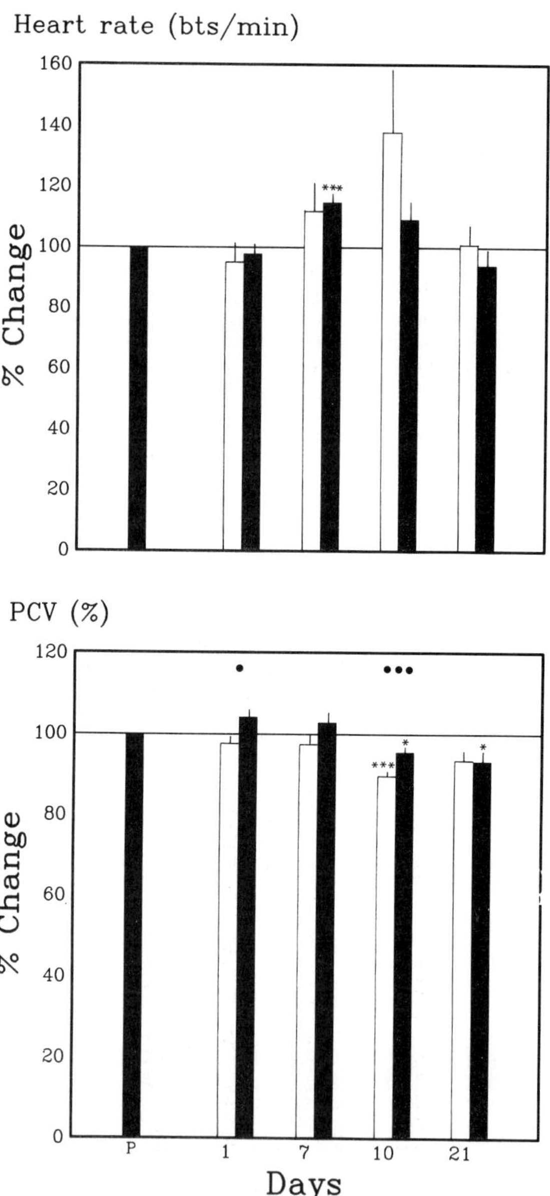

Fig. 3 Heart rate and packed cell volume (PCV) in Jersey cows in late lactation treated for 7 days with pit.bST (open columns) or re.bST (shaded columns). Results expressed as a percentage of pretreatment values (P). * $P < 0.05$; *** $P < 0.001$; treatment compared to pretreatment. • $P < 0.05$, ••• $P < 0.001$; pit.bST compared to re.bST.

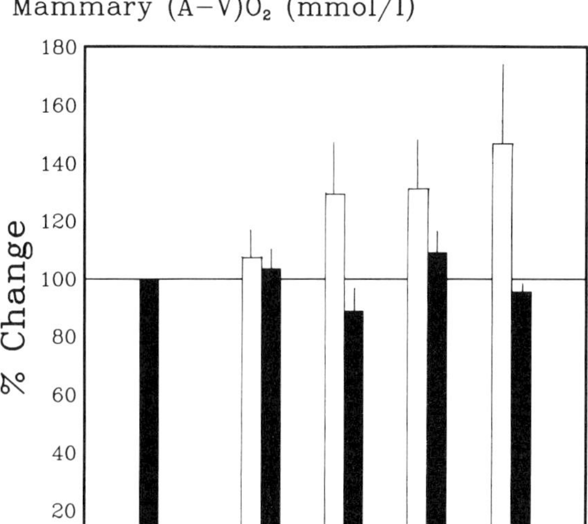

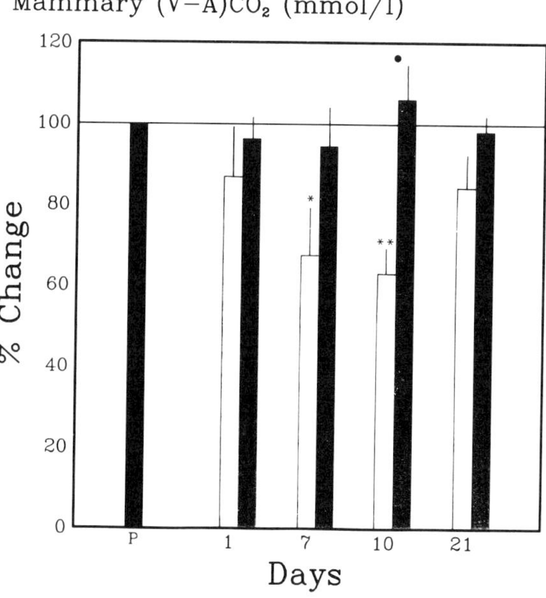

Fig. 4 Arteriovenous difference (A-V) in blood gases across the mammary gland in Jersey cows in late lactation treated for 7 days with pit.bST (open columns) or re.bST (shaded columns). Results expressed as a percentage of pretreatment values (P). * $P < 0.05$, ** $P < 0.01$; treatment compared to pretreatment. • $P < 0.05$; pit.bST compared to re.bST.

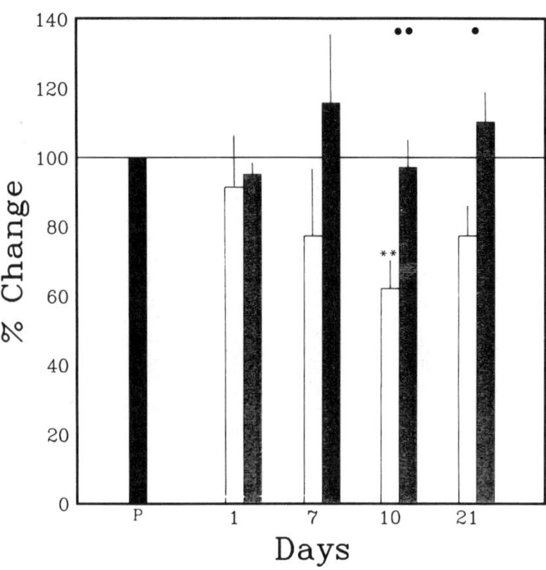

Fig. 5 Mammary respiratory quotient in Jersey cows in late lactation treated for 7 days with pit.bST (open columns) or re.bST (shaded columns). Results expressed as a percentage of pretreatment values (P). ** $P < 0.01$; treatment compared to pretreatment. • $P < 0.05$, •• $P < 0.01$; pit.bST compared to re.bST.

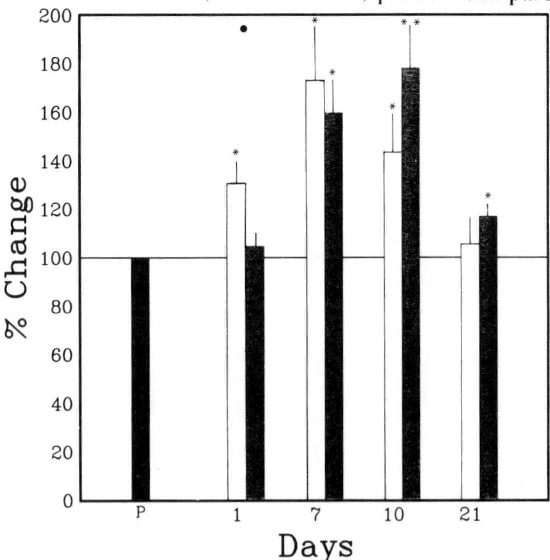

Fig. 6 Mammary uptake (mammary arteriovenous difference multiplied by mammary blood flow) of glucose in one-half udder of Jersey cows in late lactation treated for 7 days with pit.bST (open columns) or re.bST (shaded columns). Results expressed as a percentage of pretreatment values (P). * $P < 0.05$, ** $P < 0.01$; treatment compared to pretreatment. • $P < 0.05$; pit.bST compared to re.bST. Acetate uptake (mmol/min).

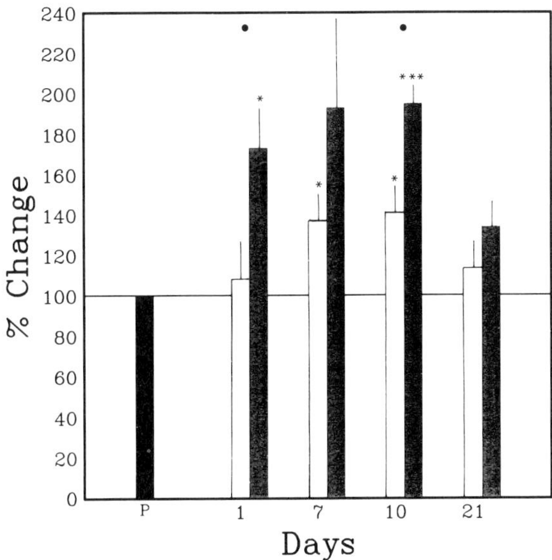

Fig. 7 Mammary uptake (mammary arteriovenous difference multiplied by mammary blood flow) of acetate in one-half udder of Jersey cows in late lactation treated for 7 days with pit.bST (open columns) or re.bST (shaded columns). Results expressed as a percentage of pretreatment values (P). * $P < 0.05$, *** $P < 0.001$; treatment compared to pretreatment. ● $P < 0.05$; pit.bST compared to re.bST.

---

bST which was not detected by radioimmunoassay. Among the metabolic responses examined, only the mammary A-V difference and mammary uptake of acetate were substantially greater in re.bST-treated animals than in pit.bST-treated cows.

Thus, a feature of the present findings is the ability of a short-term treatment with bST for 7 days to induce a substantial yet similar galactopoiesis whether ST is derived from bovine pituitaries by extraction and purification or by genetic engineering of the cloned gene. Both preparations had marked effects on increasing mammary blood flow but the metabolic responses underlying galactopoiesis showed subtle differences probably related more to the pressure of multiple forms of growth hormone in the pit.bST preparation than to any inherent difference in the action of highly purified re.bST.

ACKNOWLEDGEMENTS

We are indebted to Mr A. Bucke, Mr T. Pearce, Mr J. Herbert and Mr R. Mead for their care of the Jersey cows, Mrs J. Hood, Mrs J. Tickner and Mr C. Greenwood for help in preparation of the manuscript and Mr R.W. Ash and Dr F.A. Harrison for veterinary care. The project was partially funded by an AFRC Link Grant to Dr T.B. Mepham and Dr R.B.Heap. Recombinant bST was kindly supplied by Monsanto Company, U.S.A.

REFERENCES

1 Asimov, G.J. and Krouze, N.K. (1937). The lactogenic preparations from the anterior pituitary, and the increase of milk yield in cows. Journal of Dairy Science **20**, 289-306.

2 Bauman, D.E., Eppard, P.J., DeGeeter, M.J. and Lanza, G.M. (1985). Responses of high producing dairy cows to long-term treatment with pituitary somatotropin and recombinant somatotropin. Journal of Dairy Science **68**, 1352-1362.

3 Bauman, G., Stolar, M.W. and Buchanan, T.A. (1986). The metabolic clearance, distribution and degradation of dimeric and monomeric growth hormone (GH): Implications for the pattern of circulating GH forms. Endocrinology **119**, 1497-1501.

4 Boyd, R.D. and Bauman, D.E. (198). Mechanisms of action for somatotropin in growth. In Current Concepts of Animal Growth Regulation (Campion, D.R., Hausman, G.J. and Martin, R.J., eds.). Plenum, New York (in press).

5 Brumby, P.J. and Hancock, J. (1955). The galactopoietic role of growth hormone in dairy cattle. New Zealand Journal of Science and Technology **36**, 419-436.

6 Cowie, A.T., Forsyth, I.A. and Hart, I.C. (1980) Hormonal control of lactation. Springer, Berlin.

7 Davis, S.R., Collier, R.J., McNamara, J.P., Head, H.H., Croon, W.J. and Wilcox, C.J. (1988). Effects of thyroxine and growth hormone treatment of dairy cows on mammary uptake of glucose, oxygen and other milk fat precursors. Journal of Animal Science **66**, 80-89.

8 Davis, S.R., Collier, R.J., McNamara, J.P., Head, H.H. and Sussman, W. (1988). Effects of thyroxine and growth hormone treatment of dairy cows on milk yield, cardiac output and mammary blood flow. Journal of Animal Science **69**, 70-79.

9 Eppard, P.J., Bauman, D.E. and McCutcheon, S.N. (1985a). Effect of dose of bovine growth hormone on lactation of dairy cows. Journal of Dairy Science **68**, 1109-1115.

10 Evans, H.M. and Simpson, M.E. (1931). Hormones of the anterior hypophysis. American Journal of Physiology **98**, 511-546.

11 Fleet, I.R., Fullerton, F.M., Heap, R.B., Mepham, T.B., Gluckman, P.D. and Hart, I.C. (1988). Cardiovascular and metabolic responses during growth hormone treatment of lactating sheep. Journal of Dairy Research **55**, 479-485.

12 Fullerton, F.M., Fleet, I.R., Heap, R.B., Hart, I.C. and Mepham, T.B. (1989). Cardiovascular responses and mammary substrate uptake in Jersey cows treated with pituitary-derived growth hormone during late lactation. Journal of Dairy Research (in press).

13 Hampson, R.K. and Rottman, F.M. (1986). A potential variant of bovine growth hormone resulting from non-splicing of an intron. Federation Proceedings **45**, 1703.

14 Hart, I.C. (1983). Endocrine control of nutrient partition in lactating ruminants. Proceedings of the Nutritional Society **42**, 181-194.

15 Hart, I.C. (1988). Altering the efficiency of milk production of dairy cows with somatotropin. In Nutrition and Lactation in the Dairy Cow (Garnsworthy, P.C., ed.), pp 232-247. Butterworths, London, Boston.

16 Hart, I.C., Chadwick, P.M.E., James, S. and Simmonds, A.D. (1985). Effect of intravenous bovine growth hormone or human pancreatic growth hormone-releasing factor on milk production and plasma hormones and metabolites in sheep. Journal of Endocrinology **105**, 189-196.

17 Hart, I.C. and Johnsson, I.D. (1986). Growth hormone and growth in meat producing animals. In Control and Manipulation of Animal Growth (Buttery, P.J., Lindsay, D.B. and Haynes, N.B., eds), pp 135-159. Butterworths, London.

18 Johnsson, I.D. and Hart, I.C. (1986). Manipulation of milk yield with growth hormone. In Recent Advances in Animal Nutrition (Haresign, W. and Cole, D.J.A., eds), pp 105-123. Butterworths, London.

19 Krivi, G.G., Salsgiver, W.J., Staten, N.R., Hauser, S.D., Rowold, E., Kasser, T.R., White, T.C., Eppard, P.J., Lanza, G.M. and Wood, D.C. (1988). Identification of residues of somatotropin involved in receptor binding and biological activity. Proceedings of the Endocrine Society 70th Annual Meeting, June 8-11 1988, New Orleans, U.S.A., 257A.

20 Lanza, G.M., Krivi, G.G., Bentle, L.A., Eppard, P.J., Kung, L., Hintz, R.L., Ryan, R.L. and Miller, M.A. (1988). Comparison of the galacotopoietic activity of several recombinant bovine somatotropin variants and pituitary derived bovine somatotropin. Proceedings of the Endocrine Society 70th Annual Meeting, June 8-11 1988, New Orleans, U.S.A., 242A.

21 Lee, M.O. and Schaffer, N.K. (1934). Anterior pituitary growth hormone and the composition of growth. Journal of Nutrition **7**, 337.

22 Linzell, J.L. (1966). Measurement of venous flow by continuous thermodilution and its application to measurement of mammary blood flow in the goat. Circulation Research, **18**, 745-754.

23 Linzell, J.L. (1974). Mammary blood flow and methods of identifying and measuring precursors of milk. In Lactation (Larson, B.L. and Smith, V.R., eds), vol. 1, pp 143-225. Academic Press, New York.

24 Machlin, L.J. (1973). Effect of growth hormone on milk production and feed utilisation in dairy cows. Journal of Dairy Science **56**, 575-580.

25 McDowell, G.H., Gooden, J.M., Leenanuruksa, M.J. and English, A.W. (1987). Effects of exogenous growth hormone on milk production and nutrient uptake by muscle and mammary tissues of dairy cows in mid-lactation. Australian Journal of Biological Science **40**, 295-306.

26 Peel, C.J. and Bauman, D.E. (1987). Somatotropin and lactation. Journal of Dairy Science **70**, 474-486.

27 Peel, C.J., Fronk, T.J., Bauman, D.E. and Gorewit, R.C. (1983). Effect of exogenous growth hormone in early and late lactation on lactational performance of dairy cows. Journal of Dairy Science **66**, 776-782.

28 Staten, N.R., Hauser, S.D., Rowold, E., Kasser, T.R., White, T.C., Eppard, P.J., Lanza, G.M. and Wood, D.C. (1988). Identification of residues of somatotropin

involved in receptor binding and biological activity. Proceedings of the Endocrine Society 70th Annual Meeting, June 8-11 1988, New Orleans, U.S.A.

29 Young, F.G. (1947). Experimental stimulation (galactopoiesis) of lactation. British Medical Bulletin, **5**, 155-161.

GROWTH PROMOTING PROPERTIES OF RECOMBINANT GROWTH HORMONE

J.M. Pell

AFRC Institute for Grassland and Animal Production, Hurley, Maidenhead, Berkshire SL6 5LR, U.K.

INTRODUCTION

Greater economic efficiency for meat production may be gained by increasing lean tissue growth with minimal concomitant fat accretion. Therefore, a major aim of current research in animal production is to attempt to increase growth rates in farm animals so that their maximum "genetic potential" for muscle growth may be achieved. It is well established that growth hormone (GH) is not only essential for normal growth in young animals but that it is also important for the partitioning of nutrients between muscle and adipose tissue (2). However, administration of exogenous GH to farm animals is a very different physiological situation than that in which GH action is usually investigated; they are already in a normal hormonal environment rather than the hypopituitary state usually associated with GH treatment. Our basic objective is to induce additional growth in a normal animal rather than to investigate the correction of hormonal deficiency.

THE PHYSIOLOGICAL ROLE OF ENDOGENOUS GH

Even though GH is necessary for normal growth it has been difficult to demonstrate any correlation between endogenous plasma GH concentrations and growth rate in intact animals (for review, see Ref. 4) and this is probably due to a combination of several factors. GH concentrations are episodic and may fluctuate on a minute by minute basis (5) and therefore it is difficult to quantify GH status; indeed, in ruminants GH concentrations are inversely related to nutritional well-being and are greater in the fasted than in the fed state (1). GH acts in concert with and via several other anabolic hormones, notably insulin-like growth factor-I (IGF-I, 13, 28). Many aspects of metabolism are influenced by GH and it is almost impossible to consider any one, such as growth, in isolation. Lastly, growth itself is such a complex process that it is difficult to provide "ideal" conditions so that an animal can respond maximally to GH.

It is appropriate, therefore, to consider GH as having both short- and long-term actions in vivo. In the long-term, it has a vital homeorhetic function (2) to channel nutrients towards some target process, such as growth or lactation, whenever possible. Its short-term function is "protein-sparing"; for example, during periods of food deprivation GH acts to preserve body protein pools at the expense of fat which is mobilized, presumably for the provision of energy. It is attractive to expect that, in well-fed animals, additional exogenous GH might tip this balance further and induce a preferential accumulation of body protein.

EXOGENOUS GH: GROWTH AND BODY COMPOSITION

Recombinant DNA technology has made it possible to produce large amounts of pure GH for administration to young animals. Under certain circumstances, exogenous GH has induced increases in liveweight gain of 22% in lambs fed ad libitum (14). Generally, though, responsiveness to GH in terms of liveweight gain has been variable in ruminants and sometimes increased rates of whole body growth do not occur (for review see Refs. 11, 15). This variability is probably due to the influence of other factors and particularly to the quantity and quality of feed offered to the animals (discussed in a

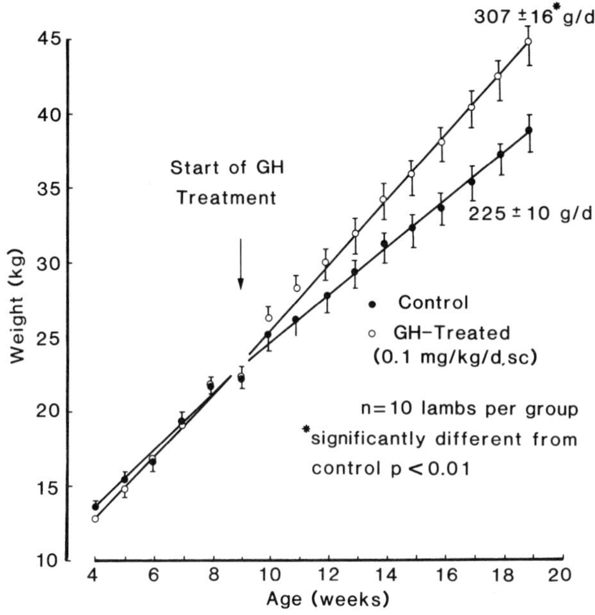

Fig. 1 Liveweights of control and GH-treated lambs

later section).

In an attempt to reduce animal variability twin lambs were used for our initial studies; one served as a control and the other was treated with recombinant bovine GH at a dose of 0.1 mg/kg/day (subcutaneous) from 9 weeks of age until slaughter at approximately 20 weeks. The lambs were offered a 16% crude protein diet at an intake of 40 g per kg liveweight. (Unless stated as otherwise, these conditions have been used for all of our studies on the action of exogenous GH.) During the pre-treatment period, no difference in growth rate was observed between the two groups of animals. After treatment began the GH-treated group grew, on average, 36% faster than their controls (Fig. 1).

The body composition of the lambs is illustrated in Fig. 2. Tissue weights were expressed as a fraction of final liveweight (and therefore normalized) and the difference between the fractional weights for the GH-treated and control animals is presented as a percentage change from control values. GH treatment induced a significant increase in the fractional weights of some visceral tissues (intestines, 15%; liver, 12%) whilst a concomitant decrease of approximately 30% was observed for visceral fat. "Protein-type" tissues (eg heart) tended to increase in fractional weight. No change in recovery of cold carcass was observed between GH-treated and control lambs but analysis of minced carcass showed a significant decrease in fat content (12.5 ± 4.2%, mean ± SEM, $P < 0.05$, $n = 7$ per group) with only a small increase in protein content (3.3 ± 2.1%). However, 9 skeletal muscles (including the gluteobiceps and vastus lateralis) were dissected from the hindquarters of these carcasses and their combined weights were significantly increased in GH-treated lambs when expressed as a fraction of cold carcass weight.

THE INTERACTION BETWEEN GH AND OTHER ANABOLIC HORMONES

The relationship and cooperation between GH, IGF-I and other anabolic hormones (such as insulin) is not yet fully understood. However, GH does act both directly and via IGF-I to mediate its anabolic actions in bone (13); whether this is also the case for other tissues, such as skeletal muscle, remains to be established.

Total plasma IGF-I concentrations during GH-treatment of the lambs, described in the previous section, are illustrated in Fig. 3, left panel. Even though there was little change during development in control lambs, there was an immediate and sustained increase in plasma IGF-I concentrations for GH-treated lambs. The physiological relevance of this dramatic increase must be considered. Most of the IGF-I in blood is associated with binding proteins (3) and probably inactive; whether long-term GH treatment alters the ratios of free and protein-bound IGF-I remains to be established. Plasma IGF-I may represent an endocrine response to GH: in this context the production of growth factor from a specific tissue, such as the liver, and its transport in blood to sites of action. Alternatively, blood may be a vehicle for the disposal of IGF-I which has been produced locally by many tissues.

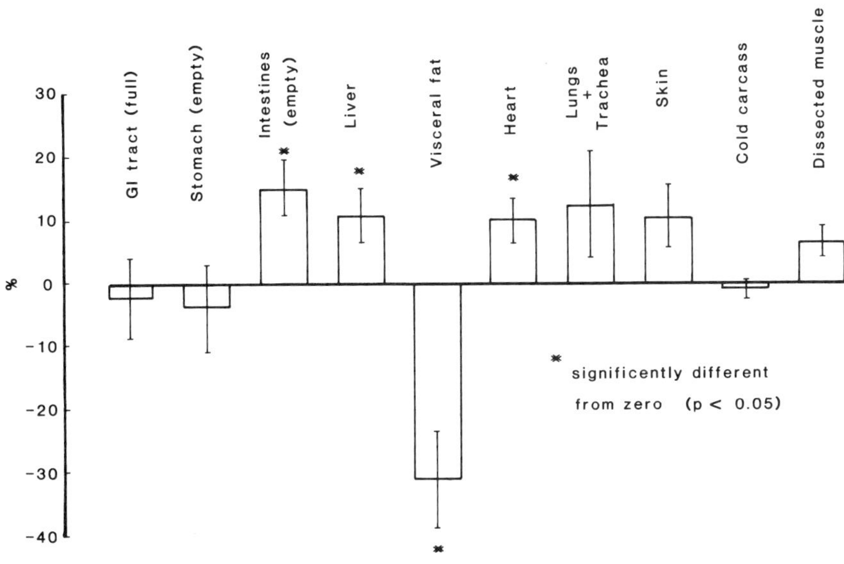

Fig. 2 Change in body composition of GH-treated lambs

Liver is one of the most abundant sources of mRNA for IGF-I (21). Dot blot hybridization studies (using a cDNA probe for human mRNA) on samples of liver were performed on five of the sets of control and GH-treated twin lambs. Relative abundance of mRNA (expressed as a percentage of the most abundant tissue) was always decreased in the GH-treated twin (control, 74.3 ± 6.9%; GH-treated, 63.2 ± 3.4%). This implies that the liver is not the major source of the additional IGF-I found in the plasma of

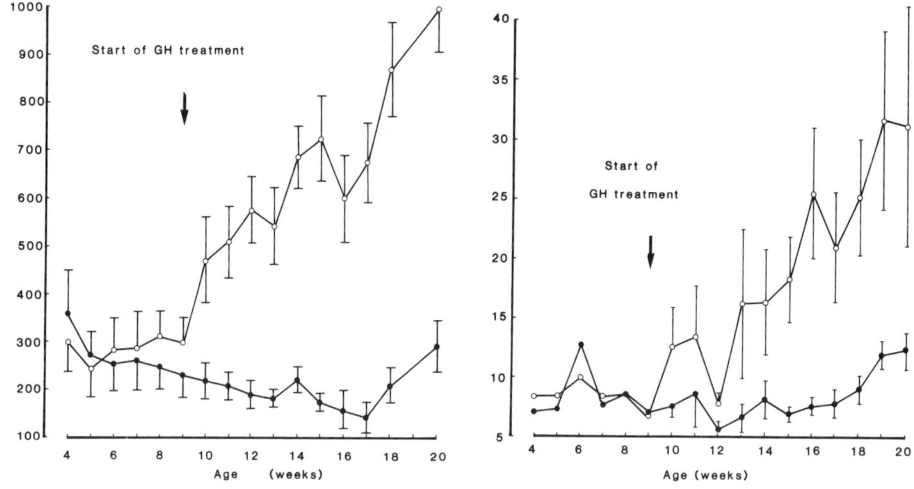

Fig. 3 Total plasma IGF-I (ng/ml, left panel) and insulin (μU/ml, right panel) concentrations in jugular blood sampled from control (solid circles) and GH-treated (open circles) lambs.

GH-treated lambs unless the activity of the mRNA is increased or that the methodology assays inappropriate mRNA.

Plasma insulin concentrations increased in GH-treated lambs (Fig. 3, right panel) even though the magnitude of the increase was very variable. This elevated insulin concentration is, in part, due to the diabetogenic action of GH (10). It remains to be established whether insulin has a function in optimizing the anabolic response to GH since GH does not appear to be anabolic in the absence of insulin (30).

THE EFFECTS OF GH ON METABOLISM BY SPECIFIC TISSUES

The increases in net muscle and/or whole body growth rates which are induced by administration of exogenous GH could be due to one or a combination of several metabolic factors:

1 an increase or a change in the pattern of nutrient absorption from the diet;
2 an alteration in their subsequent hepatic metabolism so that more or different substrates are passed on to peripheral tissues;
3 a change in metabolism by target tissues (skeletal muscle, bone, adipose tissue).

Gastrointestinal tract

There is evidence that the increased whole body growth that can be induced by immunization against somatostatin is due, at least in part, to a slowing of digesta flow through the gastrointestinal tract and therefore a possible increase in nutrient absorption (7). However, total mean retention times of both the solid and liquid phase components of digesta were unaffected by GH treatment (controls; solid phase, 44.3 ± 3.2

h; liquid phase, 35.3 ± 2.9 h; treatment; solid phase, 41.4 ± 1.0 h, 33.9 ± 1.5 h; means ± SEM, n = 8).

To investigate nutrient absorption more fully, catheters were inserted into an arterial supply and the portal vein of control and GH-treated twin lambs (16). No changes in net metabolite uptake into the portal vein could be detected. Therefore, it may be concluded that GH exerts little effect on nutrient absorption from the gastrointestinal tract.

Liver

GH-treatment does appear to influence the manipulation of nutrients by the liver. Hepatic metabolism was investigated in lambs using, once again, the multicatheterized sheep preparation of Katz and Bergman (16). Net uptake of non-esterified fatty acids (NEFA) and output of 3-hydroxybutryrate and glucose all tended to increase in GH-treated lambs (Fig. 4). Thus, any NEFA which are mobilized from adipose tissue in response to GH may be metabolized by the liver. In addition, the increased peripheral glucose concentrations which are often observed in GH-treated sheep may be due to increased rates of gluconeogenesis as well as to insulin-resistance in extra-hepatic tissues.

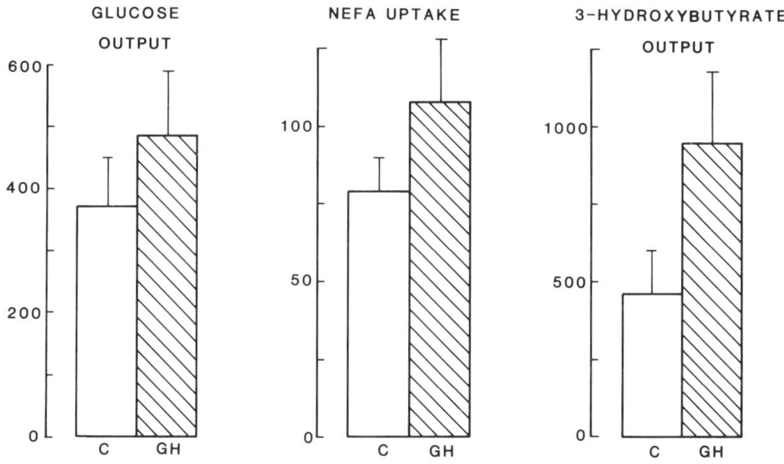

Fig. 4 Net hepatic metabolism (μmol/min of glucose, NEFA and 3- hydroxybutyrate in vivo (6) for control (open bars) and GH-treated (hatched bars) lambs; mean ± SEM, n = 6 per group.

The liver is the major site of GH-binding in animals and in sheep two receptors have been identified: high and low-affinity sites (8). The capacity and affinity of hepatic GH-binding sites were determined for six sets of control and GH-treated twin lambs using self-displacement of labelled bGH (9). No change in apparent affinity of either binding site could be detected in GH-treated lambs when compared with their controls, implying that receptor type was unaffected by GH treatment. However, receptor capacity was significantly increased ($P < 0.05$) by GH for both the high (control, 3.03 ± 0.82; GH-treated, 6.01 ± 1.46 pmol/g microsomal protein; means ± SEM) and low (control,

9.42 ± 3.50; GH-treated 13.9 ± 4.3 pmol/g) affinity receptor. When the GH-induced increase in liver size is also taken into account, total GH binding capacity was at least doubled. Since bGH was used for this study, the increase in binding capacity presumably represents "somatogenic" receptors (32). The functional significance of these additional GH receptors is, as yet, unclear.

Table 1 Lipid metabolism (μmol/2h/g lipid) in vitro for adipose tissue in control and GH-treated lambs. Small pieces (10 mg) of tissue were used to determine rates of lipogenesis and oxidation from [^{14}C] acetate (31, 33) and of lipolysis from glycerol release. Means ± SEM; n = 7 per group; significantly different from controls: *P < 0.05; **P < 0.01

	Lipogenesis		Lipolysis		Oxidation	
Fat depot	C	GH	C	GH	C	GH
Subcutaneous	45.2 ±8.9	18.5* ±3.7	5.79 ±0.91	6.87 ±0.75	2.69 ±0.50	1.06** ±0.35
Visceral	20.3 ±3.2	11.2* ±1.1	5.31 ±0.65	7.80* ±0.63	0.61 ±0.15	0.42 ±0.09

Adipose tissue

Long-term treatment of lambs with GH can have major net catabolic effects on both visceral and subcutaneous fat depots, significantly reducing their weight per unit of body weight (11). The mechanism of this net catabolic GH action was examined in more detail in lambs fed concentrate (16% crude protein) ad libitum. Control lambs were injected subcutaneously with saline buffer and treated lambs with 10 mg bGH every other day for 18 days. The effects on lipid metabolism in vitro are given in Table 1 for subcutaneous (sampled from the hindquarters) and visceral (omental) fat. The major action of GH was a dramatic inhibition of lipid synthesis. Even though subcutaneous fat is more labile and sensitive to nutritional state than is visceral fat (26), lipolysis was affected little by GH in this fat depot whereas it was significantly increased in omental fat. Oxidation of acetate was inhibited by GH treatment for both types of adipose tissue. Thus, GH appears to inhibit the synthesis of lipid rather than to promote its breakdown. It should be noted though, that this study was performed on lambs which were not nutritionally restricted in any way; the balance of lipid synthesis versus degradation in response to GH treatment could be different in restricted animals.

Muscle

In meat production skeletal muscle is, of course, the most important target tissue for GH action. In addition, the quality of the increased muscle mass which may be induced by GH must be considered. There is little point in partitioning nutrients into inappropriate protein pools, such as connective tissue, which would toughen the meat. Therefore the mechanism of GH action on skeletal muscle was investigated not just for total protein turnover but for collagen (the major protein of connective tissue) and non-collagen protein metabolism (22). Two muscles were studied: the gluteobiceps (mainly red-type, containing slow-twitch oxidative fibres) and the semitendinosus (mainly white-type, containing fast-twitch glycolytic fibres). Muscle protein is in a dynamic state; it is constantly being synthesized and degraded and therefore an increase in muscle growth could be due to a change in one or both of these processes. Non-collagen and collagen protein synthetic rates were therefore measured (17,18,22) and net muscle growth and non-collagen protein degradation were calculated.

Table 2 Muscle protein and RNA concentrations in control and GH-treated lambs; means ± SEM; n = 5 per group; significantly different from control animals; *P < 0.05.

	Muscle weight (g)		Non-collagen protein concentration (mg/g)		Collagen concentration (mg/g)		RNA concentration (μg/g)	
Muscle type	C	GH	C	GH	C	GH	C	GH
Gluteobiceps	300 ±7	342 ±20	116 ±2	114 ±2	5.72 ±0.65	6.21 ±1.01	682 ±7	725* ±9
Semitendinosus	109 ±4	122 ±7	120 ±2	117 ±1	3.39 ±0.22	3.89 ±0.33	627 ±14	685* ±15

Even though GH treatment induced a significant increase in net growth of muscle protein (Table 3), muscle weights were not significantly increased above control weights (Table 2) for the five sets of twin lambs studied. Total muscle non-collagen and collagen protein concentrations were also unaffected by GH treatment for both muscle types and the ratio of collagen to non-collagen protein remained constant. This implies that GH appeared to induce equivalent increases in net growth of both protein pools. Total RNA concentrations were significantly greater in GH-treated lambs for both the gluteobiceps and the semitendinosus. Since most of this RNA represents ribosomal RNA, GH induced an increase in the protein synthetic capacity of skeletal muscle.

The magnitude of the increased muscle growth rates in GH-treated lambs was similar for the gluteobiceps and semitendinosus (Table 3); this was mediated by different quantitative, though not qualitative, mechanisms. GH treatment induced a significant (30%) increase in both non-collagen and collagen protein synthesis rates in the gluteobiceps muscle, accompanied by a tendency for protein degradation to increase. Thus, the mechanism of the increased net muscle growth was due to an increase in protein turnover, with the increase in synthesis exceeding the increase in degradation. The change in protein synthetic rate was primarily due to an increase in the amount of ribosomal RNA, rather than to an increase in the efficiency of protein synthesis (RNA activity) from that RNA. Similar trends in protein turnover were also observed for the semitendinosus for non-collagen protein, although they did not reach statistical significance.

Thus, even though the gluteobiceps and semitendinosus of GH-treated lambs exhibited similar changes in net muscle growth when compared with control lambs, the gluteobiceps appeared to respond with a much greater change in protein turnover. The anabolic β-agonist, clenbuterol, also seems to have differential effects on muscle fibre type in rats; slow-twitch oxidative fibres are more sensitive to clenbuterol treatment, both with respect to fat content (19) and reversal of atrophy initially caused by denervation (20).

Table 3 Collagen and non-collagen protein metabolism in control and GH-treated lambs; means ± SEM; n = 5 per group; significantly different from controls: *P < 0.05.

	Non-collagen protein metabolism						Collagen protein synthesis (%/day)	
	Calculated muscle growth (%/day)		Protein synthesis (%/day)		Calculated protein degradation (%/day)			
Muscle type	C	GH	C	GH	C	GH	C	GH
Gluteobiceps	0.55 ±0.05	0.65* ±0.07	4.15 ±0.58	5.39* ±0.67	3.60 ±0.56	4.75 ±0.68	0.57 ±0.05	0.74* ±0.06
Semitendinosus	0.55 ±0.05	0.64* ±0.07	4.03 ±0.26	4.26 ±0.19	3.48 ±0.26	3.62 ±0.18	0.74 ±0.07	0.70 ±0.10

VARIABILITY OF RESPONSIVENESS TO EXOGENOUS GH

It has been difficult to predict, or in some cases even to demonstrate, an anabolic response to GH in terms of either liveweight gain or muscle growth (11, 15). The reasons for this remain obscure although it is widely recognized that the complexity of GH action and of the physiology of growth must be compounding reasons. No single factor has so far been identified as important; for example, plasma IGF-I concentrations may be increased several-fold following GH administration to lambs even though growth rates do not change (23). We therefore conducted an experiment to try and determine whether protein or energy intake were related to GH responsiveness (25).

Seventy-two lambs were randomly allocated to one of three dietary protein levels (12, 16 or 20% crude protein) which were fed at one of two energy levels (ad libitum or restricted to 30 g per kg liveweight). Within each dietary group lambs were injected daily with either buffered saline or bGH (0.1 mg/kg/day) from 9 weeks of age until slaughter at 19 weeks.

Analysis of variance of the average daily liveweight gains showed that GH induced a significant increase ($P < 0.01$) in restricted but not ad libitum fed lambs. It is also noteworthy that the response in the restricted lambs was independent of dietary protein level. At slaughter three skeletal muscles (gluteobiceps, vastus lateralis and semitendinosus) were dissected from the hindquarters of each lamb and visceral fat was also removed. Their weights are shown in Fig. 5, left and right panels respectively. Overall, GH treatment induced a significant increase ($P < 0.001$) in combined muscle weight for both restricted and ad libitum fed animals whereas visceral fat depots were significantly decreased ($P < 0.05$) only in ad libitum fed lambs. Thus, the anabolic response to GH in terms of muscle weight appears, at first sight, to be independent of dietary influence although its net catabolic action on visceral fat is dependent on energy intake. Of course, the doubling in visceral fat weight for control ad libitum fed lambs versus that for control restricted lambs must be a contributing factor. It is interesting that in ad libitum groups, the GH-treated lambs ate slightly, though significantly ($P < 0.05$), less than the controls (54.8 versus 52.3 g per kg liveweight). Plasma IGF-I concentrations for the last week of the study are shown in Table 4. Significant positive effects of GH treatment ($P < 0.001$), dietary energy intake ($P < 0.001$) and of dietary protein level ($P < 0.05$) were observed, which is in agreement with

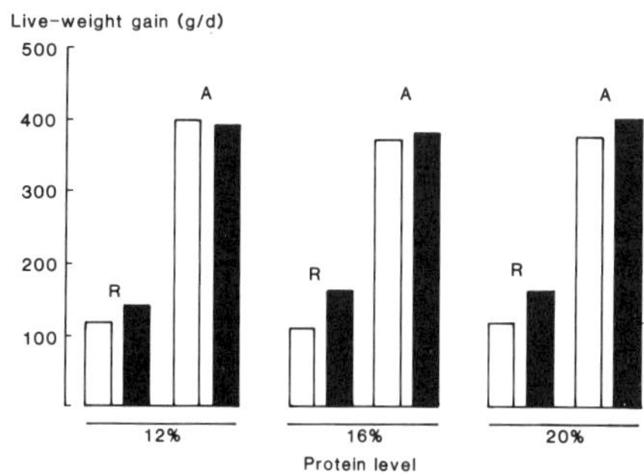

Fig. 5 Liveweight gains for control (open bars) and GH-treated (solid bars) lambs in different nutritional states.

Fig. 6 Total dissected muscle weight (three muscles, left panel) and total visceral fat weight (right panel) for control (open bars) and GH-treated (solid bars) lambs in different nutritional states.

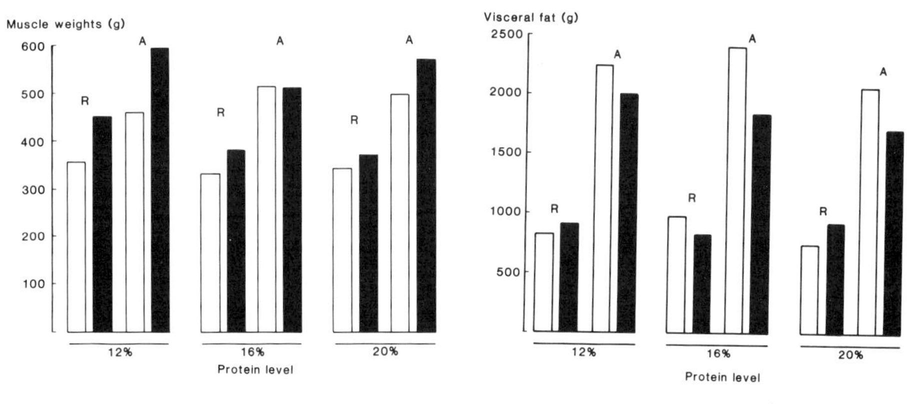

similar experiments in rats (27).The differential effects of GH on skeletal muscle and on visceral fat do, in part, explain the variability of responsiveness to GH treatment in terms of liveweight gain. Further interpretations may be made from a more detailed carcass analysis. Currently this experiment has not yielded any clear correlations between nutritional status and the response of muscle growth to GH, even though IGF-I concentrations were related to both protein and energy intake.

CONCLUSIONS: THE FUTURE

It is clear that administration of exogenous GH to lambs may induce widespread adaptations of tissue metabolism which are favourable for increasing the efficiency of meat production. However, the extent to which GH exerts its effects directly or via IGF-I has not been established unequivocally. Administration of exogenous hormone manufactured by recombinant DNA techniques is the beginning of many further advances in biotechnology which can be used to manipulate animal performance. One possibility is to change the activity, rather than the concentration, of GH which can be achieved by its

Table 4 Total IGF-I concentrations (ng/ml) in lambs in different nutritional states. Statistical analysis by analysis of variance (SE = 64); C, control; GH, GH-treated.

	Dietary protein level (%)					
	12		16		20	
Energy level	C	GH	C	GH	C	GH
Restricted	227	571	260	562	435	528
Ad libitum	509	776	666	806	711	877

immunological modification (12, 24). This type of approach could ultimately be incorporated into normal vaccination procedures similar to those used to confer disease resistance and could be more acceptable. It should be noted though that this approach would be long-term and therefore strategic manipulation of animal performance for short and specific periods of time would not be possible. In addition, until responsiveness to exogenous GH itself can be defined adequately, it will be difficult to assess the effectiveness of immunological approaches to increase the efficiency of animal production.

ACKNOWLEDGEMENTS

I would like to thank the following for their contributions to this article: P.C. Bates, D.E. Beever, H.L. Buttle, H.J. Coles, C. Elcock, E.M. Gill, R.L. Harding, D.J. Hathorn, J. Paradine, J.C. Saunders, A.D. Simmonds, A. Walsh.

REFERENCES

1 Bassett, J.M. (1974). Diurnal patterns of plasma insulin, growth hormone, corticosteroid and metabolite concentrations in fed and fasted sheep. Australian Journal of Biological Sciences **27**, 167-181.

2 Bauman, D.E., Eisemann, J.H. and Currie, W.B. (1982). Hormonal effects on partitioning of nutrients for tissue growth: role of growth hormone and prolactin. Federation Proceedings **41**, 2538-2544.

3 Butler, J.H. and Gluckman, P.D. (1986). Circulating insulin-like growth factor binding-proteins in fetal, neonatal and adult sheep. Journal of Endocrinology **109**, 333-338.

4 Davis, S.L., Hossner, K.L. and Ohlson, D.L. (1984). Endocrine regulation of growth in ruminants. In Manipulation of Growth in Farm Animals (Roche, J.F. and O'Callaghan, D. eds.) pp 151-178. Martinus Nijhoff, The Hague.

5 Davis, S.L., Ohlson, D.L., Klindt, J. and Anfinson, M.S. (1977). Episodic growth hormone secretory patterns in sheep: relationship to gonadal steroid hormones. American Journal of Physiology **233**, E519- E523.

6 Elcock, C., Buttle, H.L., Coles, H.J., Hathorn, D.J., Simmonds, A.D. and Pell, J.M. (1988). The effect of growth hormone on nutrient partitioning in lambs. Journal of Endocrinology **117** (Suppl), 59.

7 Fadlalla, A.M., Spencer, G.S.G. and Lister, D. (1985). The effect of passive immunization against somatostatin on marker retention time in lambs. Journal of Animal Science **61,** 234-239.

8 Gluckman, P.D., Butler, J.H. and Elliott, T.B. (1983). The ontogeny of somatotrophic binding sites in ovine hepatic membranes. Endocrinology **112,** 1607-1612.

9 Harding, R.L., Coles, H.J., Wallis, M. and Pell, J.M. (1987). Adaptations of hepatic growth hormone receptors following long term treatment of lambs with growth hormone. Journal of Endocrinology **115** (Suppl), 90.

10 Hart, I.C. (1983) Endocrine control of nutrient partition in lactating ruminants. Proceedings of the Nutrition Society **42,** 181-194.

11 Hart, I.C. and Johnsson, I.D. (1986). Growth hormone and growth in meat producing animals. In Control and Manipulation of Animal Growth (Buttery, P.J., Haynes, N.B. and Lindsay, D.B. eds.) pp 135-159. Butterworths, London.

12 Holder, A.T. and Aston, R. (1988). Antigen-antibody complexes that enhance growth. In Biotechnology in Growth Regulation (Heap, R.B., Prosser, C.G. and Lamming, G.E., eds.) p 167. Butterworths, London.

13 Isaksson, O., Jansson, J.-O. and Gange, I.A.M. (1982). Growth hormone stimulates longitudinal growth directly. Science **216,** 1237-1239.

14 Johnsson, I.D., Hart, I.C. and Butler-Hogg, B. (1985). The effects of exogenous growth hormone and bromocriptine on growth, body development, fleece weight and plasma concentrations of growth hormone, insulin and prolactin in female lambs. Animal Production **41,** 207-217.

15 Johnsson, I.D., Hathorn, D.J., Wilde, R.M., Treacher, T.T. and Butler-Hogg, B.W. (1987). The effects of dose and method of administration of biosynthetic bovine somatotrophin. Animal Production **44,** 405-414.

16 Katz, M.L. and Bergman, E.N. (1969). A method for simultaneous cannulation of the major splanchnic blood vessels of the sheep. American Journal of Veterinary Research **30,** 655-661.

17 Laurent, G.J. (1982). Rates of collagen synthesis in lung, skin and muscle obtained in vivo by a simplified method using [^{3}H]proline. Biochemical Journal **206,** 535-544.

18 Laurent, G.J., McAnulty, R.J. and Oliver, M.H. (1982). Anomalous tritium loss in the measurement of tissue hydroxy-[5-^{3}H]proline specific activity following chloramine-T oxidation. Analytical Biochemistry **123,** 223-228.

19 Maltin, C.A., Delday, M.I. and Reeds, P.J. (1986). Effect of a growth promoting drug, clenbuterol, on fibre frequency and area in hind limb muscles from young male rats. Bioscience Reports **6,** 293-299.

20 Maltin, C.A., Reeds, P.J., Delday, M.I., Hay, S.M., Smith, F.G. and Lobley, G.E. (1986). Inhibition and reversal of denervation-induced atrophy by the β-agonist growth promoter, clenbuterol. Bioscience Reports **6,** 811-818.

21 Murphy, L.J., Bell, G.I. and Friesen, H.G. (1987). Tissue distribution of insulin like growth factors 1 and 2 mRNA in the adult rat. Endocrinology **120,** 1279-1282.

22 Pell, J.M. and Bates, P.C. (1987). Collagen and non-collagen protein turnover in skeletal muscle of growth hormone-treated lambs. Journal of Endocrinology **115,** R1-R4.

23 Pell, J.M., Blake, L.A., Buttle, H.L., Johnsson, I.D., Simmonds, A.D. and Morrell, D.J. (1987). Insulin-like growth factors and growth hormone in sheep. Proceedings of the Nutrition Society **46,** 48A.

24 Pell, J.M., Elcock, C., Walsh, A., Trigg, T. and Aston, R. (1988). Potentiation of growth hormone activity using a polyclonal antibody of restricted specificity. In Biotechnology in Growth Regulation (Heap, R.B., Prosser, C.G. and Lamming, G.E., eds.) p 259. Butterworths, London.

25 Pell, J.M., Gill, M., Beever, D.E., Jones, A.R. and Cammell, S.B. (1989). Hormone and nutrient interaction in the control of growth: role of growth hormone (GH) and insulin-like growth factor-I (IGF-I). Proceedings of the Nutrition Society (in press).

26 Pothoven, M.A., Beitz, D.C. and Thornton, J.H. (1975). Lipogenesis and lipolysis in adipose tissue of ad libitum and restricted-fed beef cattle during growth. Journal of Animal Science **40,** 957-962.

27 Prewitt, T.E., D'Ercole, A.J., Switzer, B.R. and van Wyk, J.J. (1982). Relationship of serum immunoreactive somatomedin-C to dietary protein and energy in growing rats. Journal of Nutrition **112,** 144-150.

28 Salmon, W.D. and Daughaday, W.H. (1957). A hormonally controlled serum factor which stimulates sulphate incorporation by cartilage in vitro. Journal of Laboratory and Clinical Medicine **49,** 825-829.

29 Schlechter, N.L., Russell, S.M., Greenberg, S., Spencer, E.M. and Nicoll, C.S. (1986). A direct growth effect of growth hormone in rat hindlimbs shown by arterial infusion. American Journal of Physiology **250,** E231-E235.

30 Schweiller, E., Guler, H.P., Merryweather, J., Scandella, C., Maerki, W., Zapf, J. and Froesch, E.R. (1986). Growth restoration of insulin-deficient diabetic rats by recombinant human insulin-like growth factor-I. Nature **323,** 169-171.

31 Smith, R.W. and Walsh, A. (1984). Effect of lactation on the metabolism of sheep adipose tissue. Research in Veterinary Science **37,** 320-323.

32 Wallis, M. (1980). Receptors for growth hormone, prolactin, and the somatomedins. In Cellular Receptors for Hormones and Neurotransmitters (Schulster, D. and Levitski, A. eds.) pp 163-183. John Wiley and Sons, Chichester.

33 Yang, Y.T. and Baldwin, R.L. (1973). Preparation and metabolism in isolated cells from bovine adipose tissue. Journal of Dairy Science **56,** 350-365.

THE MECHANISMS BY WHICH PORCINE GROWTH HORMONE IMPROVES PIG GROWTH PERFORMANCE

T.D. Etherton

Department of Dairy and Animal Science, The Pennsylvania State University, University Park, PA 16802, U.S.A.

Chronic treatment of growing pigs with porcine growth hormone (pGH) markedly increases growth rate, improves feed efficiency (feed/gain), increases muscle growth and inhibits adipose tissue accretion. Growth hormone regulates a number of important metabolic and somatogenic events that include: elevation of plasma insulin-like growth factor I (IGF-I) concentration, stimulation of satellite cell proliferation, inhibition of glucose utilization by porcine adipocytes (due to a decrease in glucose transport, fatty acid synthesis and inhibition of several lipogenic enzymes) and antagonism of the stimulatory effects of insulin on glucose utilization in adipose tissue. These effects of GH on muscle and adipose tissue growth are coordinated in a manner that enables GH to alter the normal rate of nutrient partitioning between lipid and protein accretion to achieve the rather marked increase in protein deposition that occurs in parallel with the decrease in lipid deposition in pigs treated with exogenous pGH. In porcine adipose tissue, the metabolic effects of pGH are direct and are not mediated by IGF-I. The decrease in insulin sensitivity of porcine adipose tissue is not associated with any change in insulin binding or tyrosine kinase activity of the insulin receptor. The mechanisms by which pGH elicits its somatogenic effects in pigs are undoubtedly due in large part to the pGH-dependent elevation in serum IGF-I concentration. In association with the pGH-dependent rise in IGF-I there is a concurrent increase in a 150 kDa pGH-dependent IGF-I binding protein (IGF-BP) that acts as an inhibitor of IGF-I action. Further resolution of the mechanisms by which pGH affects pig growth and metabolism may lead to new strategies to enhance the biological potency of pGH or identify other ways of altering pig growth performance.

INTRODUCTION

An exciting era is evolving in animal agriculture. Numerous studies published during the past three years have shown that treating pigs with exogenous pGH dramatically increases pig growth performance (2-5, 10-12). The magnitude of response in these studies has varied somewhat primarily because of differences in experimental design (eg initial pig weight, length of study, breed, sex, dose of pGH used and difference in diet). Despite these differences, however, it has become apparent that pGH increases average daily gain (ADG) approximately 10 to 20%, improves feed efficiency (feed/gain; F/G) 15 to 35%, decreases adipose tissue mass and lipid accretion rates by as much as 50 to 80% and concurrently increases protein deposition by as much as 50%. Recently, it has been established that the effects of pituitary pGH on pig growth performance are mimicked by recombinant pGH (rpGH) (12). These responses have sparked interest in developing a rpGH-based growth promotant for the pork industry world-wide. Commercialization of this technology will enhance economic profitability of the producer and provide a leaner product for consumers. Since present U.S.A. dietary recommendations indicate that consumption of saturated fatty acids should be reduced to less than 10% of total calories (24), it is evident that perhaps the greatest benefit of pGH will be to provide leaner, more nutritious meat that will enable consumers to include meat in their diet and still achieve the dietary recommendations.

The recognition that treating pigs with pGH (either pituitary or recombinant) affects growth performance so noticeably has stimulated research directed toward increasing our understanding of how pGH works. As shown in Fig. 1, it is evident that pGH produces a wide variety of biological effects in numerous tissues. In this paper evidence will be discussed which indicates that pGH decreases glucose utilization and insulin sensitivity of porcine adipocytes and that these effects result in glucose-carbon being redirected to other tissues (likely muscle). As a result of the decline in lipid synthesis adipose tissue accretion rates decrease. This change in partitioning of glucose from adipose tissue to muscle may also be a means of providing additional energy to muscle to support the inceased rates of protein deposition that occur in pGH-treated pigs. In addition, the possibility of treating pigs with exogenous IGF-I as an alternative means to increase pig growth rate will be discussed.

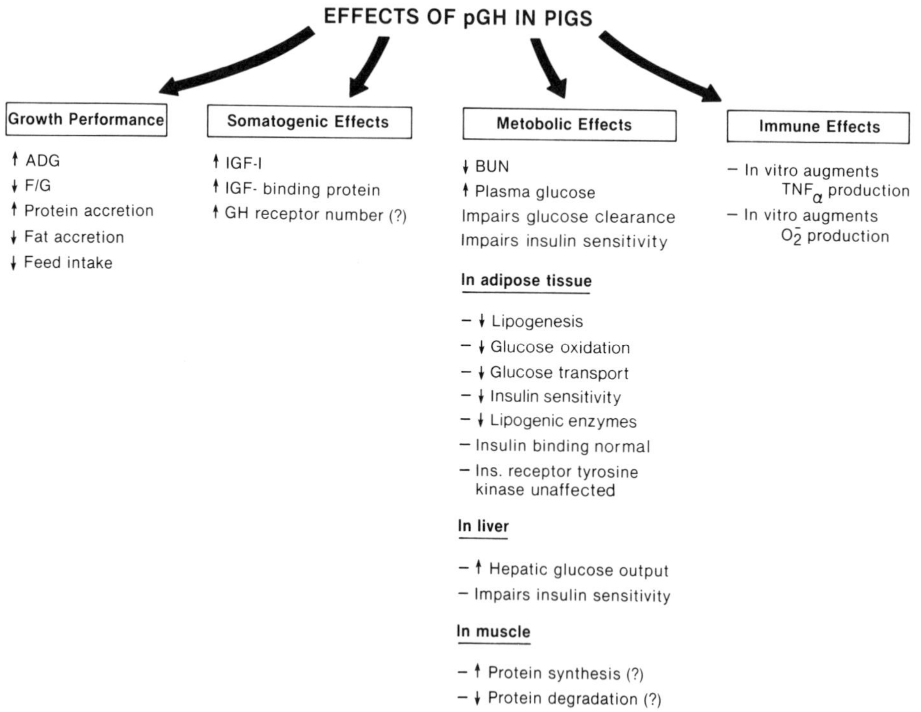

Fig. 1 Effects of treating pigs chronically with porcine growth hormone

ADG = average daily gain; F/G = feed/gain; BUN = blood urea nitrogen; TNF_α = tumor necrosis factor alpha

METABOLIC EFFECTS OF pGH

Nutrient partitioning is the coordinated uptake and release of nutrients by body tissues (9). We are not aware of any biochemical data that accurately describes the flow of nutrients between adipose tissue and skeletal muscle and relates this to growth rate of the respective tissues. Despite this, we can infer that large differences exist among meat animals because of the divergence seen in carcass composition, growth rate and feed efficiency. Furthermore, it is now apparent that pGH markedly affects nutrient partitioning in a manner which leads to changes in the normal pattern of nutrient deposition in adipose tissue and muscle and, hence, alters growth rates of these tissues (Fig. 2). In pigs treated with pGH adipose tissue accretion rates can be decreased by as much as 50 to 80%. This is associated with a marked redirection of nutrients away from lipid deposition toward protein deposition (Fig. 2). The quantity of glucose oxidized or used for lipid synthesis by adipose tissue (based on in vivo kinetic studies) in the pig has not been determined; however, it is likely that it is sizeable. This is based on the observed rates of lipid accretion in pigs (~ 180 to 280 g/day; 4), the estimated rate of glucose production (R_o; see 15) in the pig, the fact that most lipid deposited (~ 80%) in porcine adipose tissue is the result of de novo synthesis (16), the observation that glucose is the principal endogenous carbon source for fatty acid synthesis (19) and that adipose tissue is the preeminent tissue site of fatty acid synthesis in the pig (19). Based on this conceptual framework we have speculated that perhaps 20 to 40% of the glucose cleared per day is used for lipid synthesis in the pig (70 kg). Using these approximations it is evident that a significant quantity of glucose is redirected to other tissues when adipose accretion rates are reduced 50 to 80% by pGH.

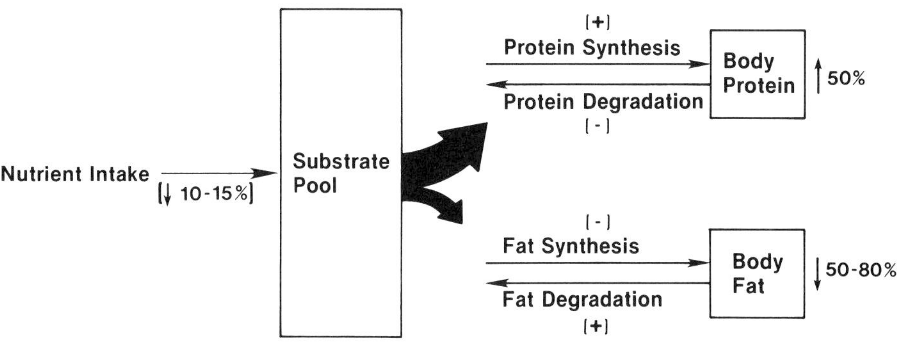

Fig. 2 The effects of treating pigs with porcine growth hormone on nutrient partitioning

---

In an attempt to establish the metabolic bases for the reduction in porcine adipose tissue growth we have examined the effects that pGH (both pituitary and recombinant) has on lipid metabolism (17, 28, 29, 31, 32). We have found that treating pigs for 7 days with pGH (70 μg/kg) markedly reduces fatty acid synthesis (50 to 70%, 31) and that this is associated with a marked decrease in the activity of several lipogenic enzymes (glucose-6-phosphate dehydrogenase, 6-phosphogluconate dehydrogenase, malic enzyme and fatty acid synthase; FAS) (17). The extent to which pGH can affect lipogenic enzyme activity is best illustrated by the reduction in FAS activity. After 7 days of pGH treatment FAS activity is not detectable in porcine adipose tissue (17).

In addition to the inhibitory effects that pGH has on lipogenesis in porcine adipose tissue, the ability of insulin to stimulate lipogenesis and glucose transport is also markedly decreased by pGH (17, 31). We have conducted studies looking at the effects of treating pigs with pGH (7 days with 70 μg/kg/d) on 3-0-methylglucose transport in isolated porcine adipocytes and found that both basal and insulin-stimulated glucose transport are decreased (17). Furthermore the percentage decrease (60 to 70%) in glucose transport rates agrees quite well with the observed decrease in lipogenic rate. These studies have established that pGH (both pituitary-derived and recombinant) is an insulin antagonist in porcine adipose tissue and has no insulin-like activity as suggested by studies with rat adipose tissue (14). It should be noted, however, that the insulin-like effects are only observed in tissue from GH-deficient rats or in cells incubated in the absence of GH (7). Thus, under normal circumstances (ie when GH is present) GH has no physiologically significant insulin-like effects in adipose tissue. This is further supported by the significant decreases observed for adipose tissue accretion in response to GH which certainly is not an insulin-like effect. Lastly, the effects that pGH has on glucose utilization and insulin sensitivity in porcine adipose tissue are intrinsic properties of the hormone since they are mimicked by rpGH.

The observations that pGH decreases adipose tissue growth and reduces lipogenic rates, lipogenic enzyme activities, glucose transport and insulin sensitivity of these processes is compelling evidence to support the idea that pGH directly affects these processes. To further resolve whether these effects are direct we have developed a system that enables us to culture adipose tissue explants for 48 h in a defined-medium while mantaining tissue viability (29, 32). These studies have shown that pGH potently (ED_{50} ~ 0.5 to 1 ng/ml) blunts the ability of insulin to maintain lipogenic capacity in a dose-dependent manner (12, 29, 32). Similar findings have been reported for ovine and bovine adipose tissue (8, 25). Recently, Glenn et al. (13) have shown that GH also suppresses glucose utilization and antagonizes insulin-stimulated glucose metabolism in 3T3-L1 adipocytes in a manner similar to that reported for porcine adipose tissue.

In our studies we have not observed any effects of pGH in acute incubations (2 h) with fresh adipose tissue. This indicates that pGH does not alter adipocyte metabolism by allosteric or covalent modification of intracellular enzymes or other regulatory proteins (eg the insulin receptor) but rather suggests that pGH affects these processes by reducing the number of glucose transporters and(or) lipogenic enzyme mass. There is no information available to indicate whether there is a pGH-dependent decrease in abundance of mRNA of the lipogenic enzymes and, if so, the mechanisms by which mRNA abundance is regulated.

The reduction in insulin sensitivity and responsiveness due to pGH may be the result of alterations in insulin binding. Studies we have done, however, with porcine adipocytes indicate that insulin receptor number and affinity are unaffected by treating pigs with pGH (17). This finding agrees with the report of Maloff et al. (18) for rat adipose tissue cultured with bovine or human GH. It is possible that insulin binding is normal and that the insulin receptor tyrosine kinase activity is impaired. To address this we measured tyrosine kinase activity in purified porcine adipocyte insulin receptors and found that pGH did not affect enzyme activity (17). The significance of this observation is that it suggests that the pGH-dependent blockade in insulin action occurs after this point. Based on the pleiotropic nature of insulin's actions (20), however, it is oversimplistic to expect that a single mechanism accounts for all of the actions of insulin; and accordingly, it might be expected that pGH affects more than one insulin signal transduction pathway. Evidence which supports this contention is the finding that pGH (chronic treatment) increases the antilipolytic response to insulin in pigs (35). As a result, it is possible to ascribe effects of pGH on insulin action that are divergent in porcine adipose tissue (ie a decrease in insulin stimulation of glucose uptake and lipogenesis versus the potentiation of the antilipolytic effect). The finding that pGH does not affect insulin binding to porcine adipocytes appears to be physiologically appropriate. If pGH blocked insulin action at the level of the receptor it would seem likely that stimulation of the antilipolytic effect would not occur and the diversity of biological effects lost.

The rates with which lipid is deposited is a function of the increment between synthesis and degradation (lipolysis). The majority of our work has focused on the effects that pGH has on glucose uptake and lipid synthesis because in our studies we have seen little effect of pGH on lipolysis in vitro or on free fatty acid (FFA) levels in vivo. This does not imply that FFA turnover is unaffected but simply that it has not been measured in pigs. Data are available that suggest an epinephrine challenge evokes a greater rise in circulating FFA in pigs treated with pGH (2). Despite this it is likely that when energy intake is high (ad libitum intake) the primary effect of pGH on adipose tissue accretion is on glucose uptake and lipid synthesis (2). It would be energetically favourable to redirect glucose to other tissues rather than expend ATP to transport glucose into the adipocyte, synthesize lipids and then hydrolyze the triglyceride to provide long-chain fatty acids for other tissues.

SOMATOGENIC EFFECTS OF pGH

IGF-I is a GH-dependent mitogen that has been shown to stimulate growth of rodents (21, 23) and the proliferation of satellite cells in skeletal muscle (6). Because of the important role that satellite cell proliferation plays in regulating postnatal muscle growth (see (1) for review) it is likely that the pGH-dependent increase in plasma IGF-I is important for the stimulation of muscle growth in pigs treated with pGH. We have considered the possibility that treating pigs with exogenous IGF-I might be an alternative means of increasing pig growth performance. Several observations, however, suggest that while IGF-I may increase growth rate in domestic animals (which has yet to be established) IGF-I treatment will not be comparable to pGH treatment because the metabolic effects that pGH has on adipose tissue are properties not mimicked by IGF-I (specifically, the effects of pGH on glucose uptake, metabolism and blunting insulin action) and it is these effects which are an important determinant in affecting adipose tissue accretion (hence, carcass composition) and feed efficiency.

In control pigs or pigs treated with pGH there is no detectable free IGF-I in blood (26). Approximately 70 to 80% of the immunoreactive IGF-I is bound to a 150 kDa IGF-BP, with the remainder associated with a 40 kDa IGF-BP (26). Incubation of porcine serum with $[^{125}I]$-IGF-I followed by fractionation over a FPLC sizing column has revealed that approximately 40% of the label is bound to the smaller IGF-BP with the majority of the remainder free (50 to 60%); little is bound to the 150 kDa complex (26). Thus, the high molecular weight protein is highly saturated with endogenous IGF-I and is the major IGF-I transport protein in pigs.

To gain insight into the physiology of the IGF-BPs we have purified an acid stable IGF-BP from porcine serum (27). This protein has a molecular weight of 45 kDa under reducing conditions and shares sequence homology with acid-stable IGF-BPs purified from rat and human serum (27). Polyclonal antiserum raised against this protein recognizes the 150 kDa IGF-BP in porcine serum indicating that the IGF-BP we have purified is a component of the larger protein and that this large IGF-BP is a heterogenous complex of subunit proteins. An RIA established for this IGF-BP in porcine serum has indicated that the 150 kDa IGF-BP is under pGH regulation (30). When pigs are treated with pGH acutely (10 to 1000 μg/kg BW) serum IGF-BP concentrations are not affected. However, chronic treatment (17 days) with pGH significantly increases IGF-BP. In the studies we have conducted it takes approximately 2 days for pGH to significantly increase IGF-BP concentration. Furthermore, the rise in IGF-BP is paralleled by an increase in IGF-I concentration. Hypophysectomy reduces IGF-BP concentrations by 60 to 70% (30). Daily treatment of pigs with IGF-I (either 4 or 8 mg/day) for 3 days does not affect IGF-BP concentration (30). Therefore, it appears that the 150 kDa IGF-BP concentration is regulated by pGH but not IGF-I. It remains to be established if chronic infusion of IGF-I affects IGF-BP concentration. It can be speculated that if chronic IGF-I infusion does not alter the concentration of 150 kDa IGF-BP that treatment of pigs with exogenous IGF-I will not be effective because IGF-I will be cleared rapidly (see below).

When exogenous IGF-I (recombinant human) is administered to pigs as a bolus injection it is cleared rapidly (t½ ~ 6 min) and causes significant hypoglycemia (33). The temporal profile of the elevation in serum IGF-I differs markedly from that observed after pGH administration where serum concentrations of pGH are elevated from 2 to ~ 12 h (length depends on dose of pGH injected, see Refs. 5 and 22). The apparent reason for the rapid clearance of exogenous IGF-I is that little is bound to the blood IGF-BPs because most of the IGF-I binding sites are occupied with endogenous IGF-I (26). It has been known for several years that free (unbound) IGF-I is an insulin mimic in rodents. In the pig, administration of IGF-I causes a marked hypoglycemia that is dose-dependent (33).

The observations that the GH-dependent IGF-BP concentration is not changed by exogenous IGF-I is evidence to support the argument that IGF-I treatment may not be as effective as treating pigs with pGH. In particular, the effects that pGH has on feed efficiency and carcass composition likely would not be reproduced. Other studies we have done provide further evidence for this hypothesis. Since adipose tissue plays such an important role in the adaptive response to exogenous pGH we asked whether this may be mediated by the elevation in IGF-I. In vitro, IGF-I is an insulin mimic in porcine adipose tissue and stimulates both lipogenesis and glucose oxidation (31, 34). This effect is blocked when physiological concentrations of the purified acid-stable IGF-BP are added to the buffer (34). This effect is highly specific because insulin action is not affected by addition of the IGF-BP. Our conclusion is that IGF-I in vivo does not mediate any effects of pGH with respect to adipose tissue because: (1) the effects of free IGF-I are insulin-like whereas the effects of pGH both in vivo and in vitro are to antagonize insulin action in adipose tissue and decrease adipose tissue growth rate; (2) the IGF-BPs block the insulin-like effects of free IGF-I; and (3) there is no detectable free IGF-I in porcine serum. It would appear that even if there were free IGF-I in blood or in the tissue that the insulin-like effects of IGF-I in porcine adipose tissue would be blunted because pGH treatment decreases adipose tissue sensitivity and responsiveness to IGF-I (31).

SUMMARY

A remarkable era has occurred in animal agriculture over the past eight years. The emergence of recombinant DNA technology has provided sufficient quantities of previously scarce proteins that have enabled scientists to test hypotheses not previously amenable to scrutiny. The emergence of molecular biology occurred over a time when it became clear that treating pigs with exogenous pGH markedly increased pig growth performance. During the past five years our understanding of how pGH works has increased significantly. Despite this, it is apparent that our understanding of the structural and functional aspects of the pGH receptor are rudimentary. In addition, very little is known about how the receptor generates signals nor the identity of those chemical signals that mediate the intracellular effects of pGH. Since pGH has many effects in a variety of tissues (see Fig. 1) it is reasonable to speculate that the signal transduction pathways by which pGH alters cell proliferation and metabolism may differ not only among target tissues but within target tissues. It is also clear that another major area of scientific scrutiny in growth biology is to elucidate the mechanisms by which IGF-I affects cell proliferation. In this context, it will be important to resolve the roles that IGF-BPs play in modulating IGF-I bioactivity.

REFERENCES

1 Allen, R.E. (1988). Muscle cell growth and development. In Designing Foods. Technical Options for the Marketplace. pp 142-162. National Academy Press, Washington, D.C.

2 Boyd, R.D. and Bauman, D.E. (1988). Mechanisms of action for somatotropin in growth. In Current Concepts of Animal Growth Regulation (Campion, D.R., et al., eds.). Plenum Publishing Corp., New York (in press).

3 Boyd, R.D., Bauman, D.E., Beermann, D.H., De Neergaard, A.F., Souza, L. and Butler, W.R. (1986). Titration of the porcine growth hormone dose which maximizes growth performance and lean deposition in swine. Journal of Animal Science **63** (Suppl. 1), 218.

4 Campbell, R.G., Steele, N.C., Caperna, T.J., McMurtry, J.P., Solomon, M.B. and Mitchell, A.D. (1988). Interrelationships between energy intake and exogenous porcine growth hormone administration on the performance, body composition and protein and energy metabolism of growing pigs weighing 25 to 55 kilograms body weight. Journal of Animal Science **66,** 1643-1655.

5 Chung, C.S., Etherton, T.D. and Wiggins, J.P. (1985). Stimulation of swine growth by porcine growth hormone. Journal of Animal Science **60,** 118-130.

6 Dodson, M.V., Allen, R.E. and Hossner, K.L. (1985). Ovine somatomedin, multiplication-stimulating activity, and insulin promote skeletal muscle satellite cell proliferation in vitro. Endocrinology **117,** 2357-2363.

7 Eden, S., Schwartz, J. and Kostyo, J.L. (1982). Effects of preincubation on the ability of rat adipocytes to bind and respond to growth hormone. Endocrinology **111,** 1505-1512.

8 Etherton, T.D., Evock, C.M. and Kensinger, R.S. (1987). Native and recombinant bovine growth hormone antagonize insulin action in cultured bovine adipose tissue. Endocrinology **121,** 699-703.

9 Etherton, T.D. and Walton, P.E. (1986). Hormonal and metabolic regulation of lipid metabolism in domestic livestock. Journal of Animal Science **63** (Suppl. 2), 76-88.

10 Etherton, T.D., Wiggins, J.P., Chung, C.S., Evock, C.M., Rebhun, J.F. and Walton, P.E. (1986). Stimulation of pig growth performance by porcine growth hormone and growth hormone-releasing factor. Journal of Animal Science **63,** 1389-1399.

11 Etherton, T.D., Wiggins, J.P., Evock, C.M., Chung, C.S., Rebhun, J.F., Walton, P.E. and Steele, N.C. (1987). Stimulation of pig growth performance by porcine growth hormone: Determination of the dose-response relationship. Journal of Animal Science **64,** 433-443.

12 Evock, C.M., Etherton, T.D., Chung, C.S. and Ivy, R.E. (1988). Pituitary porcine growth hormone (pGH) and a recombinant pGH analog stimulate pig growth performance in a similar manner. Journal of Animal Science **66,** 1928-1941.

13 Glenn, K.C., Rose, K.S. and Krivi, G.G. (1988). Somatotropin antagonism of insulin-stimulated glucose utilization in 3T3-L1 adipocytes. Journal of Cellular Biochemistry **37,** 371-383.

14 Goodman, H.M. and Coiro, V. (1981). Induction of sensitivity to the insulin-like action of growth hormone in normal rat adipose tissue. Endocrinology **108,** 113-119.

15 Gopinath, R. and Etherton, T.D. (1988). Effects of porcine growth hormone on glucose metabolism of pigs: II. Glucose tolerance, peripheral insulin sensitivity and glucose kinetics. Journal of Animal Science (in press).

16 Hood, R.L. and Allen, C.E. (1973). Lipogenic enzyme activity in adipose tissue during the growth of swine with different propensities to fatten. Journal of Nutrition **103,** 353-362.

17 Magri, K.A. (1988). Effects of growth hormone on glucose transport, lipogenic enzyme activities, insulin receptor binding and tyrosine kinase activity in pig adipocytes. M.S. Thesis, The Pennsylvania State University.

18 Maloff, B.L., Levine, J.H. and Lockwood, D.H. (1980). Direct effects of growth hormone on insulin action in rat adipose tissue maintained in vitro. Endocrinology **107,** 538-544.

19 O'Hea, E.K. and Leveille, G.A. (1969). Significance of adipose tissue and liver as sites of fatty acid synthesis in the pig and the efficiency of utilization of various substrates for lipogenesis. Journal of Nutrition **99,** 338-344.

20 Saltiel, A.R. and Cuatrecasas, P. (1988). In search of a second messenger for insulin. American Journal of Physiology **255,** C1-C11.

21 Schoenle, E., Zapf, J., Humbel, R. and Froesch, E. (1982). Insulin-like growth factor I stimulates growth in hypophysectomized rats. Nature **296,** 252-253.

22 Sillence, M.N. and Etherton, T.D. (1987). Determination of the temporal relationship between porcine growth hormone, serum IGF-I and cortisol concentrations in pigs. Journal of Animal Science **64,** 1019-1023.

23 Skottner, A., Clark, R., Robinson, I. and Frykland, L. (1987). Recombinant human insulin-like growth factor: Testing the somatomedin hypothesis in hypophysectomized rats. Journal of Endocrinology **112,** 123-132.

24 Target Levels and Current Dietary Patterns (1988). In Designing Foods. Animal Product Options in the Marketplace pp 45-62. National Academy Press, Washington D.C.

25 Vernon, R.G. (1982). Effects of growth hormone on fatty acid synthesis in sheep adipose tissue. International Journal of Biochemistry **14,** 255-258.

26 Walton, P.E. (1988). Purification and characterization of porcine growth hormone-dependent insulin-like growth factor binding protein. PhD Thesis, The Pennsylvania State University.

27 Walton, P.E., Baxter, R.C., Burleigh, B.D. and Etherton, T.D. (1988). Purification of the serum acid-stable insulin-like growth factor binding protein from the pig (Sus scrofa). Comparative Biochemistry and Physiology (in press).

28 Walton, P.E. and Etherton, T.D. (1986). Stimulation of lipogenesis by insulin in swine adipose tissue: Antagonism by porcine growth hormone. Journal of Animal Science **62,** 1584-1595.

29 Walton, P.E. and Etherton, T.D. (1987). The culture of adipose tissue explants in serum-free medium. Journal of Animal Science **65** (Suppl. 2), 25-30.

30 Walton, P.E. and Etherton, T.D. (1988). Effects of porcine growth hormone and insulin-like growth factor I (IGF-I) on immunoreactive IGF binding protein concentration in pigs. Journal of Endocrinology (in press).

31 Walton, P.E., Etherton, T.D. and Chung, C.M. (1987). Exogenous pituitary and recombinant growth hormones induce insulin and insulin-like growth factor I resistance in pig adipose tissue. Domestic Animal Endocrinology **4,** 183-189.

32 Walton, P.E., Etherton, T.D. and Evock, C.M. (1986). Antagonism of insulin action in cultured pig adipose tissue by pituitary and recombinant porcine growth hormone: Potentiation by hydrocortisone. Endocrinology **118,** 2577-2581.

33 Walton, P.E., Gopinath, R., Burleigh, B.D. and Etherton, T.D. (1988). Infusion of recombinant human IGF-I into pigs: Determination of circulating half-lives and chromatographic profiles. Hormone Research (in press).

34 Walton, P.E., Gopinath, R. and Etherton, T.D. (1988). Porcine insulin-like growth factor (IGF) binding protein blocks IGF-I action on porcine adipose tissue. Proceedings of the Society for Experimental Biology and Medicine (in press).

35 Wray-Cahen, C.D., Boyd, R.D., Bauman, D.E., Ross, D.R. and Fagin, K.D. (1987). Metabolic effects of porcine somatotropin (pST) in growing swine. Journal of Animal Science **65** (Suppl. 1), 261.

EVALUATION OF SOMETRIBOVE (METHIONYL BOVINE SOMATOTROPIN) IN TOXICOLOGY AND CLINICAL TRIALS IN EUROPE AND THE UNITED STATES

C.J. Peel, P.J. Eppard and D.L. Hard

Monsanto Europe S.A., Avenue de Tervuren 270, 1150 Brussels, Belgium

INTRODUCTION

Somatotropin purified from bovine pituitary glands consistently increases milk yields when administered to lactating dairy cows. However, there was no economical way to process sufficient pituitary glands for application in commercial dairying. Industry therefore attempted to isolate an active fragment of somatotropin which could be chemically synthesized but the programme was unsuccessful. Finally in this decade the entire somatotropin molecule of 191 amino acids can be readily produced in kilogram and even tonne quantities utilizing both recombinant DNA techniques and protein purification and formulation technology.

Prior to commercial marketing of somatotropin, it must be demonstrated that it is both effective and safe for the dairy cow, and that it also presents no risk to the consumer. Government authorities such as the Veterinary Products Committee in the United Kingdom and the Center for Veterinary Medicine (Food and Drug Administration) in the United States issue specific guidelines and evaluate the results of numerous tests conducted with somatotropins. These include evaluation of both the biology and the manufacturing of the product. In this paper, we present data from the biological testing of our specific molecule of somatotropin, sometribove, which has a single methionyl substitution on the amino end of the protein chain. Biological testing includes a chronic toxicology study in which up to 6 times the anticipated commercial dose was administered for two consecutive lactations, an acute toxicology study in which a massive dose 60 times the commercial level was administered during one week, and results of clinical trials in several countries where the commercial dose was evaluated under various dairy management conditions. Some data have been given at annual conferences during 1988 and the resulting abstracts are referenced throughout the paper.

TOXICOLOGY EXPERIMENTS

High doses of sometribove were administered to dairy cows to assess both the safety and margin for error in dosing of sometribove. In the most extreme experiment, cows were administered two doses of 15 gram of sometribove at a one week interval. This represented a lifetime (4 lactations) amount of sometribove at the commercial dose of 500 mg in one week. Perhaps not surprisingly, milk yield was increased over a 14 day period, the controls yielding 15.8 kg/day and the sometribove group 22.1 kg/day (12). Somatotropin concentrations were elevated from a basal level of 1 ng/ml to greater than 250 ng/ml. General health of the cows throughout this acute 14-day study was good, and examination of tissues and organs following slaughter on day 15 revealed no abnormalities.

A long term chronic toxicology study was also conducted in which cows received 600 mg, 1800 mg or 3000 mg sometribove every 2 weeks for two consecutive lactations. The 3000 mg dose is 6 times the anticipated commercial dose of 500 mg and provides for a rigorous assessment of cow health. Eighty-two Holstein cows commenced this study and a subset of 38 cows continued for the second consecutive lactation. Production increases were high as would be expected with high levels of somatotropin administration. In the first year, 3.5% FCM was increased by 8.3, 10.4 and 9.4 kg/day for doses of 600, 1800 and 3000 mg of sometribove every 2 weeks with injections commencing on week 9 and continuing to the end of lactation. Similar increases were observed during the treatment period in the cows which continued treatment for a second consecutive lactation. The improvement in apparent feed efficiencies (kg FCM/Mcal intake) ranged from 7 to 25 per cent for the sometribove treatment groups during the two years (4).

As this was foremost a safety study, comprehensive examinations of cow health including regular physical examinations, monitoring of clinical chemistry, and on selected cows a thorough gross and microscopic evaluation of tissues and organs of cows slaughtered at the end of the lactations were completed. There were no significant differences in clinical mastitis, somatic cell counts, or in minor health maladies such as digestive upsets or lameness, although there did appear to be a trend towards more problems at the highest dose of 3000 mg sometribove. In each year of study, the conception rate for the control group was 100% and the conception rates ranged from 67 to 89% for the sometribove groups with no pattern according to dose (3). There were no differences in ease of calving, incidence of parturient paresis, retained placenta or cystic ovaries among the four groups.

During the first weeks of sometribove administration, milk yields are increased immediately and the increase in voluntary feed intake lags behind the requirements for milk synthesis. Cows respond by increasing adipose mobilization reflected in increased plasma non-esterified fatty acids (NEFA's) and sparing protein catabolism, which results in lower concentrations of blood urea N and creatinine in the sometribove groups. There was no evidence of subclinical ketosis and both short and long term beta-hydroxybutyrate concentrations were not different from controls. Changes in NEFA's and blood urea N were transient and returned to normal concentrations when cows had adjusted feed intake upwards to more precisely match nutrient demands of the mammary gland.

Blood concentrations of somatotropin, glucose and insulin were elevated above the controls in a dose dependent fashion in the sometribove groups (9). The elevation of plasma glucose could be due to the antagonism of somatotropin to insulin action at the cellular level and/or increased hepatic gluconeogenesis. Blood insulin concentrations are then increased in response to the higher glucose levels. These physiological adaptations facilitate an increased flux rate of glucose to the mammary gland in support of the substantial increase in lactose synthesis and milk production with sometribove treatment.

Sometribove treatment generally did not affect the blood concentration of hormones primarily produced in the pituitary gland such as adrenocorticotropin (ACTH), prolactin, thyroid stimulating hormone (TSH), luteinizing hormone (LH), and follicle stimulating hormone (FSH), in the adrenal glands such as cortisol, or in the ovary/uterus such as progesterone and oestradiol.

Of the potential metabolic hormones, there was an elevation of thyroxine (T4) but no change in triiodothyronine (T3). However, there was no evidence of clinical hyperthyriodism and given that T3 and TSH were not changed the increase of T4 could reflect changes in receptor affinity or an increase in thyroid binding globulin which would slow thyroxine clearance and increase the blood thyroxine concentration.

Occasional differences were measured in serum concentrations of calcium, phosphorous and magnesium which were directionally opposed to the shifts in milk concentrations. This represents further physiological adaptations to sometribove and higher milk production. Leukocyte counts were generally unaffected but there was a mild decrease in the rate of erythrocyte accumulation through lactation in the sometribove groups. Given the abundant evidence of protein sparing this probably reflects another physiological economization of nitrogen and energy.

In conclusion, cows tolerated very well massive doses of sometribove (30 gram) and sustained high milk yields for two consecutive lactations when administered up to 3

Table 1 Clinical trial locations, principle investigators, cow numbers and management

Country or State (ref. no)	Institute	Principle Investigator (Monitor)	Cow Number and Management
Arizona (7)	University of Arizona	J.T. Huber (L. Kung)	80 Holsteins; open dirt pens. Individual feeding of hay and one of three total mixed rations
New York (1)	Cornell University	D.E. Bauman (V.K. Meserole)	80 Holsteins; tie stalls. Individual feeding of one of three total mixed rations
Missouri (11)	Monsanto Dardenne Technical Centre	W.A. Samuels (G.F. Hartnell)	126 Holsteins; tie stalls. Individual feeding of one of four total mixed rations
Utah (8)	Utah State University	R.C. Lamb (R.G. Hoffman)	72 Holstein; free-stall barn. Individual feeding of one of three total mixed rations
France	Sanders - Centre d'Expérimentation, Sourches	F. Vedeau (L.R. Schockmel)	58 Holstein-Friesians; open straw pens. Individual feeding. Maize silage and chopped hay, concentrates according to yield
Germany	Institut für Milcherzeugung der Bundesanstalt für Milchforschung, Kiel	H.O. Gravert and K. Pabst (C. Wollny)	60 Holstein-Friesians; tie stalls. Individual feeding of hay and grass silage, concentrates according to yield
Netherlands (10)	Instituut voor Veevoedingsonderzoek (IVVO) Lelystad	Y.S. Rijpkema (C.J. Peel)	64 Holstein-Friesians; free stall, straw bedding. Individual feeding of one of two total mixed rations of grass silage, maize silage and concentrates
United Kingdom	Institute for Grassland and Animal Production, Reading	R.H. Phipps (N. Craven)	90 Friesians; free stall, sand bedding. Individual feeding of one of three total mixed rations of grass silage, maize silage and concentrates

grams every 2 weeks. Some minor health disorders were evident at the highest dose of 3 grams but there was no evidence of significant health effects at the lower dosage levels (0.6 and 1.8 grams) indicating that a commercial dose of 0.5 gram every 2 weeks should be safe for dairy cows.

CLINICAL TESTING

Production variables

Sometribove has been evaluated in eight clinical trials in 5 countries utilising a total of 630 Holstein or Holstein-Friesian cows (Table 1). Management in these trials was good as indicated by the high milk yields of the control groups (Table 2) and the responses to sometribove were relatively uniform ranging from 2.7 to 5.2 kg additional milk per day. Typically the response was highest in the first 3 months of administration and declined towards the end of lactation. Milk composition was essentially uneffected by sometribove excepting for some cyclical changes at the start of treatment resulting from short-term changes in metabolism and energy balance. Such changes have commonly been observed in short-term studies, particularly when cows shift to a negative energy balance at the commencement of somatotropin administration. Dry matter intake gradually increased in cows receiving sometribove to peak around week 10 of treatment. Overall gross feed efficiencies were increased by 2.8 to 9.3% with the largest improvements in the first weeks of treatment and the increase being progressively reduced with duration of treatment as feed intake adjusted to a higher level. At the end of lactation there were no differences in liveweights although in countries in which cows had routine body condition scores (U.S., U.K. and France) there was a trend for the sometribove group to complete the lactation in a slightly lower body condition score.

Health

All experimental cows received physical examinations at least 3 times during lactation. Overall cow health was good and metabolic diseases were notably infrequent or absent. Although there was a numerically higher number of cows with clinical mastitis in the sometribove group during treatment (Table 3) by chance there was also a similarly higher incidence before treatment had commenced. In fact, the ratio of incidence in sometribove group to incidence in control group was 1.37 in the pretreatment period and 1.28 in the treatment period. Average somatic cells tended to increase in both the control and sometribove groups as lactation progressed. A significant increase in somatic cells was observed at 2 sites and in other trials there was no difference in somatic cells. Currently it is not known why there may be a site effect on changes in somatic cells. However in all countries the somatic cell counts of the groups were low and well below the levels expected in premium quality milk.

Reproduction

The sometribove groups had more days to conception, and hence longer intercalving intervals than the control groups. Conception rates also tended to be lower. Generally the sites with the largest difference between groups in days to conception were also the sites in which the milk production increase and hence energy balance changes were greatest (Table 4). The one exception was the Dardenne farm in Missouri but there the conception rate itself was significantly lower in the sometribove group. Given that substantial shifts in production and energy balance which characterised the commencement of sometribove treatment coincided with peak rebreeding activity, the observed effects on reproductive performance are not unexpected. More days to first insemination, lower conception rates to first service, more services per conception and increased calving interval have all been related to higher milk yields and associated changes in energy balance. Analyses of our data comparing groups of cows of similar milk yield in the

Table 2 The effect of formulated sometribove on fat corrected milk production (FCM), dry matter intakes (DMI), net energy intakes (NEI), body weights and both apparent and gross feed efficiencies during the period of treatment, weeks 9 - 41 of lactation

Site	Dose (mg)	FCM (kg/cow/day)	Diff FCM (kg/cow/day)	DMI (kg/day)	Body Weight (kg)	Feed Efficiency Apparent (FCM/DMI)	Feed Efficiency Gross[2] (FCM/NEI)
Arizona	0	26.6[a]		22.4	641	1.19	0.72
	500	28.8[b]	2.2	23.4	644	1.23	0.74
New York	0	27.3[a]		18.8[a]	574[a]	1.50[a]	0.94[a]
	500	30.4[b]	3.1	19.6[a]	563[b]	1.60[b]	0.99[b]
Missouri	0	23.8[a]		19.7[a]	612	1.20[a]	0.73[a]
	500	29.0[b]	5.2	21.8[b]	609	1.33[b]	0.79[b]
Utah	0	25.3[a]		20.2[a]	605	1.25[a]	0.76[a]
	500	29.0[b]	3.7	21.7[b]	611	1.33[b]	0.80[b]
France	0	21.9[a]		18.0[a]	608	1.22[a]	0.75[a]
	500	25.8[b]	3.9	19.0[b]	608	1.34[b]	0.82[b]
Germany	0	21.7[a]		16.9[a]	604	1.27[a]	0.81[a]
	500	25.3[b]	3.6	18.0[b]	602	1.38[b]	0.85[b]
Netherlands	0	24.9[a]		19.9[a]	598	1.24[a]	0.79[a]
	500	29.5[b]	4.6	21.7[b]	597	1.36[b]	0.85[b]
United Kingdom	0	19.3[a]		15.1[a]	566	1.28[a]	0.112
	500	23.0[b]	3.7	16.8[b]	574	1.36[b]	0.118

[1] FCM is 3.5% FCM (U.S. trials) and 4.0% FCM (European trials). Results are reported as least square means adjusted for pre-treatment responses.

[2] Values are on a net energy basis at all sites except the United Kingdom where values are on a metabolizable energy basis.

[a,b] Means with different superscripts within a site are significantly different ($P < 0.05$).

---

control and sometribove groups have indicated that in the 4 trials conducted in the United States differences in intercalving interval were entirely due to milk yield per se and not sometribove, but there was a difference in conception rates between groups (6). However in the United States trials breeding was limited to the first half of the treatment period. In the 4 European clinical trials in which the breeding period was not restricted no differences were observed in cows confirmed pregnant (control group 86.6% and sometribove group (85.3%).

Table 3 Number of cows with clinical mastitis and mean somatic cell counts in the pre-treatment and treatment periods

Site	Number of cows with Clinical Mastitis				Somatic Cells ($x 10^3$/ml)			
	Pretreatment[1]		Treatment		Pretreatment[1]		Treatment	
	0	500	0	500	0	500	0	500
Arizona	2	4	10	14	167	184	193	250
New York	0	1	4	14	55	96	97[a]	221[b]
Missouri	18	20	17	19	179	144	202	250
Utah	2	3	11	7	150	65	284	191
France	4	10	5	8	153	172	153	178
Germany	5	6	3	2	140	124	192	264
Netherlands	1	0	5	7	107	148	137	186
United Kingdom	3	4	3	3	60	60	133[a]	196[b]

[1] Animals are grouped according to subsequent treatment (mg bST)

[a,b] Means with different superscripts within a site are significantly different ($P < 0.05$)

FARM TRIALS

Trials have been conducted with commercial herds in the United Kingdom and milk production responses have ranged from 2.5 to 5.0 kg/day which is similar to the range observed in clinical trials. Feed management has generally been concentrates according to milk yield being fed in the milk parlour and ad libitum access to forages. No effects on either health or reproduction have been observed in these farm trials.

In a simple flat rate feeding system in which all cows received 7.9 kg/day concentrates during the housed winter feeding period, milk production was increased by 4.8 kg/day. Silage intake increased only slightly so cows were utilizing body reserves for much of this additional milk. When the cows were turned out to graze spring pasture, and concentrate feeding was stopped, the sometribove group both maintained an increase in milk production of 4.5 kg/day and replenished body reserves such that body weights of the groups were similar by the end of lactation (5).

Table 4 The effect of formulated sometribove on 3.5% fat-corrected milk production (FCM), energy balance (EB) and reproductive performance during the first 84 days of treatment

Site	Dose (mg)	N	3.5% FCM[1] (kg/cow/day)	Diff.[2] (FCM kg/cow/day)	EB[1,3] (Mcal/day	Change in EB[2] (MCal/day)	Serv. per Concep.	Conc. Rate (%)	Days Open
Arizona	0	41	31.6[a]		7.3[a]		1.8	88	79[a]
	500	40	35.8[b]	4.2	4.6[b]	-2.7	1.9	78	98[b]
New York	0	42	31.4[a]		0.9[a]		2.0	88	87
	500	42	34.5[b]	3.1	-0.2[b]	-1.1	2.3	83	98
Missouri	0	63	30.4[a]		8.1[a]		2.3	95[a]	82
	500	63	36.2[b]	5.8	4.8[b]	-3.3	2.0	79[b]	77
Utah	0	36	29.8[a]		7.2[a]		2.7	83	126
	500	37	33.1[b]	3.3	5.2[b]	-2.0	2.3	76	119
France	0	28	28.8[a]		2.4[a]		1.8	82	87
	500	30	33.3[b]	4.5	0.7[b]	-1.7	1.9	80	94
Germany	0	30	27.3[a]		1.7[a]		2.1	83	101
	500	30	32.1[b]	4.8	0.2[b]	-1.5	2.4	80	105
Netherlands	0	32	31.9[a]		4.1[a]		1.6	84	104
	500	32	37.5[b]	5.6	1.7[b]	-2.4	2.4	84	127
United	0	45	25.1[a]		4.3[a]		2.0	96	91[a]
Kingdom	500	45	30.2[b]	5.1	0.7[b]	-3.6	2.5	91	117[b]

[1] Results are reported as least-square means adjusted for pretreatment responses

[2] Numerical difference between placebo and treatment means

[3] Values are on a net energy basis at all sites except United Kingdom where values are on a metabolizable energy basis

[a,b] Means with different superscripts within a site are significantly different ($P < 0.05$)

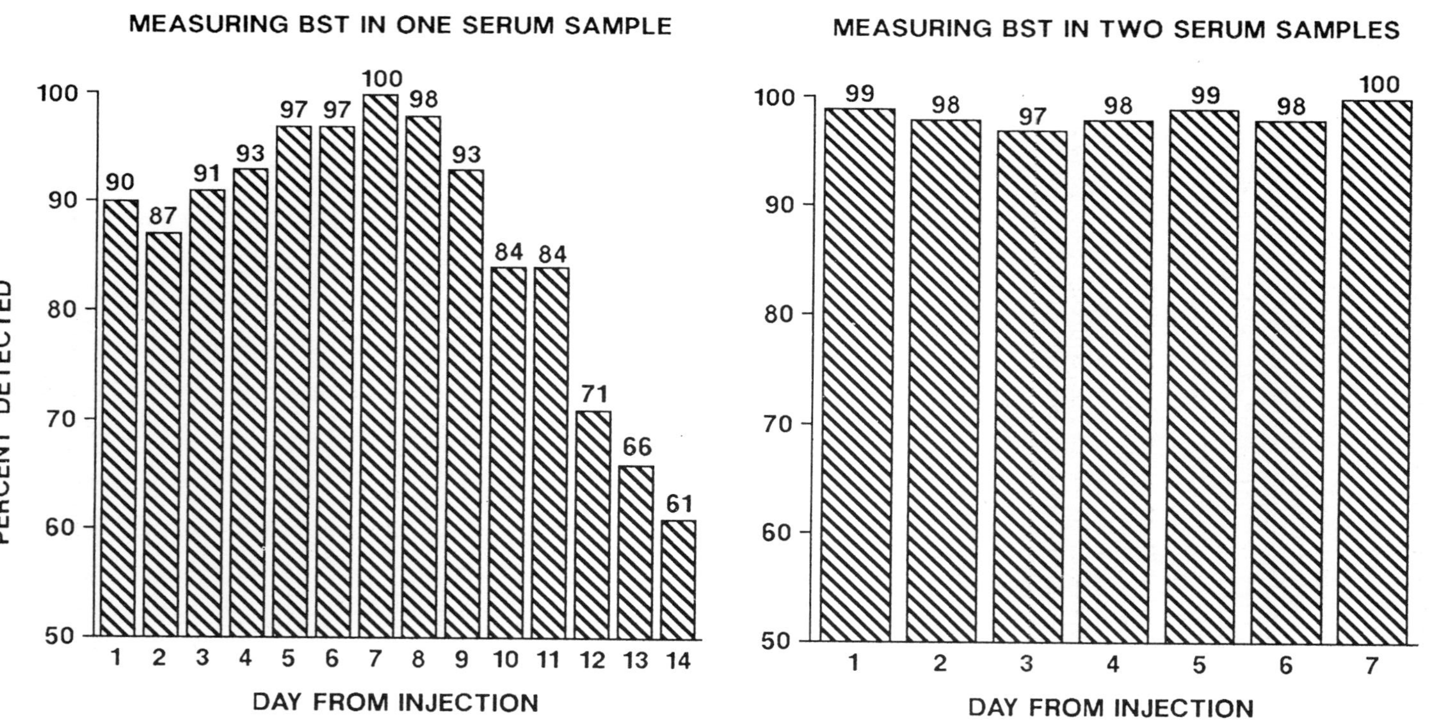

Fig. 1 Ability to detect Sometribove treated cows by the measurement of bST in serum. Animals were injected i.m. with 500 mg Sometribove. Courtesy of R.J. Collier

SOMETRIBOVE AND THE DETECTION OF TREATED COWS

There has been concern expressed by some in national herd recording or sire evaluation schemes as to the effects of bST on the integrity of milk records utilized for the identification of superior cows and bulls. Based on published clinical trial data, computer simulations have been conducted on the possible scenarios of bST application (2). These simulation models have indicated that if bST treatment is recorded, then effects on sire rankings and the accuracy of sire proofs is minimal. Obviously the integrity and responsibility of dairy farmers are essential pre-requisites in any national recording scheme and in the vast majority of cases farmers would voluntarily record cows receiving bST. However, should sometribove be used to manipulate records we believe a single blood test or better still two blood tests one week apart would enable detection of individual sometribove treated cows (Fig. 1). Of course the reliability of detection would quickly reach 100% accuracy if a group of cows had been treated. Somatotropin concentrations in milk are not changed during sometribove administration so sampling and testing of the milk cannot be effectively used.

CONCLUSION

The somatotropins are the first major products produced by recombinant DNA techniques for the animal industries. Sometribove which is just one of the somatotropins has been rigorously evaluated in large numbers of cows and in a range of management conditions in different countries. Evaluation of high doses in a chronic toxicology study for two consecutive lactations has been completed. Clinical evaluation at the commercial dose of 500 mg is continuing for multiple lactations at several sites. Sometribove has been shown to be effective in capital intensive systems utilising complete feeds and much simpler feeding systems such as flat rate feeding of concentrates and grazing pasture. As productivity enhancers the somatotropins are controversial given the food surpluses in much of the Western world. However progress in agriculture has always been based on improved productivity which in turn has provided the consumer with high quality and inexpensive food. The somatotropins if proven to be safe to both the consumer and the animal should assist in continuing this process.

REFERENCES

1 Bauman, D.E., Hard, D.L., Crooker, B.A., Erb, H.N. and Sandles, L.D. (1988). Lactational performance of dairy cows treated with a prolonged-release formulation of methionyl bovine somatotropin (sometribove). Journal of Dairy Science **71** (Suppl. 1), 205 (abstract P260).

2 Burnside, E.B. (1987). Impact of somatotropin and other biochemical products on sire summaries and cow indexes. Journal of Dairy Science **70**, 2444.

3 Cole, W.J., Eppard, P.J., Lanza, G.M., Hintz, R.L., Madsen, K.S., Franson, S.E., White, T.C., Ribelin, W.E., Hammond, B.G., Bussen, S.C., Leak, R.K. and Metzger, L.E. (1988). Response of lactating dairy cows to multiple injections of sometribove, USAN (recombinant methionyl bovine somatotropin) in a prolonged release system. Part II Health and reproduction. Journal of Dairy Science **71** (Suppl. 1), 184 (abstract P196).

4 Eppard, P.J., Lanza, G.M., Hudson, S., Cole, W.J., Hintz, R.L., White, T.C., Ribelin, W.E., Hammond, B.G., Bussen, S.C., Leak, R.K. and Metzger, L.E. (1988). Response of lactating dairy cows to multiple injections of sometribove, USAN (recombinant methionyl bovine somatotropin) in a prolonged release system. Part I Production Response. Journal of Dairy Science **71** (Suppl. 1), 184 (abstract P195).

5 Furniss, S.J., Stroud, A.J., Brown, A.C.G. and Smith, G. (1988). Milk production, feed intakes and weight change of autumn calving, flat rate fed dairy cows given two weekly injections of recombinantly derived bovine somatotropin (BST). Proceedings of the British Society of Animal Production Winter Meeting, Paper No 1.

6 Hard, D.L., Cole, W.J., Franson, S.E., Samuels, W.A., Bauman, D.E., Erb, H.N., Huber, J.T. and Lamb, R.C. (1988). Effect of long term sometribove, USAN (recombinant methionyl bovine somatotropin), treatment in a prolonged release system on milk yield, animal health and reproductive performance-pooled across four sites. Journal of Dairy Science **71** (Suppl. 1), 210 (abstract P273).

7 Huber, J.T., Willman, S., Marcus, K., Theurer, C.B., Hard, D.L. and Kung, L. (1988). Effect of sometribove (SB), USAN (recombinant methionyl bovine somatotropin) injected in lactating cows at 14-d intervals on milk yields, milk composition and health. Journal of Dairy Science **71** (Suppl. 1), 207 (abstract P264).

8 Lamb, R.C., Anderson, M.J., Henderson, S.L., Call, J.W., Callan, R.J., Hard, D.L. and Kung, L. (1988). Production response of holstein cows to sometribove USAN (recombinant methionyl bovine somatotropin) in a prolonged release system for one lactation. Journal of Dairy Science **71** (Suppl. 1), 208 (abstract P267).

9 Lanza, G.M., Eppard, P.J., Miller, M.A., Franson, S.E., Ganguli, S., Hintz, R.L., Hammond, B.G., Bussen, S.C., Leak, R.K. and Metzger, L.E. (1988). Response of lactating dairy cows to multiple injections of sometribove, USAN (recombinant methionyl bovine somatotropin) in a prolonged release system. Part III Changes in circulating analytes. Journal of Dairy Science **71** (Suppl. 1), 184 (abstract P197).

10 Rijpkema, Y.S., van Reeuwijck, L.A., Peel, C.J. and Mol, E. (1987). Responses of dairy cows to long-term treatment with somatotropin in a prolonged release formulation. Proceedings of the 38th Annual Meeting of EAAP 428 (abstract N 5.8).

11 Samuels, W.A., Hard, D.L., Hintz, R.L., Olsson, P.K., Cole, W.J. and Hartnell, G.F. (1988). Long term evaluation of sometribove, USAN (recombinant methionyl bovine somatotropin) treatment in a prolonged release system for lactating cows. Journal of Dairy Science **71** (Suppl. 1), 209 (abstract P271).

12 Vicini, J.L., De Leon, J.M., Cole, W.J., Eppard, P.J., Lanza, G.M., Hudson, S. and Miller, M.A. (1988). Effect of acute administration of extremely large doses of sometribove, USAN (recombinant methionyl bovine somatotropin), in a prolonged release formulation on milk production and health of dairy cows. Journal of Dairy Science **71** (Suppl. 1), 168 (abstract P147).

GROWTH FACTORS

GROWTH PROMOTION USING RECOMBINANT INSULIN-LIKE GROWTH FACTOR-I

H.P. Guler, J. Zapf, K. Binz and E.R. Froesch

Metabolic Unit, University Hospital, 8091, Zürich, Switzerland

INTRODUCTION

Growth and development is controlled by heredity, nutrition and hormones. These determinants act together to give rise to adult body size. Growth hormone and insulin-like growth factors appear to be of particular importance; their absence leads to dwarfism. Growth hormone in excess results in high serum levels of insulin-like growth factor-I and subsequently causes gigantism or acromegaly. Biotechnology made it possible to study growth promoting properties of recombinant human insulin-like growth factor-I in different animal models.

HYPOPHYSECTOMIZED RATS

Growth promotion by insulin-like growth factor-I (IGF-I) has been repeatedly demonstrated during a short-term treatment of 6 days in hypophysectomized and diabetic rats (6, 8, 9). These animals stop growing and have low serum levels of growth hormone and IGF-I in common. Under these experimental conditions IGF-I stimulates growth in a dose-dependent manner (4, 9). Although these results strongly support the somatomedin hypothesis the question of whether growth hormone might be important at one step of regulation of longitudinal growth remained unanswered. The group of Isaksson (7) proposed a sequential action of growth hormone and IGF-I on chondrocytes in the growth plates. According to their hypothesis "longitudinal bone growth will be the result of both cell differentiation, directly stimulated by GH, and clonal expansion of cells in the proliferative layer of the growth plate due to the local production of growth factors". During a prolonged period of time, absence of GH in the presence of IGF-I should therefore lead to growth arrest due to a limited proliferative capacity of the cells. In order to test this argument the IGF-I infusion period was extended to 18 days (4).

Maximally effective doses of recombinant human insulin-like growth factor-I (rhIGF-I, 300 ug/day) and recombinant human growth hormone (rhGH, 200 mU/day) were infused subcutaneously by Alza mini-pumps in hypophysectomized (hypox) rats (body weight around 130 g at the beginning of the infusion). Saline-treated hypox rats served as controls. When the animals were killed total body weight gain, accumulated longitudinal bone length (as determined by intravital tetracyclin staining), organ weights and blood glucose were determined.

Table 1 Growth and metabolic indices in hypox rats infused for 18 days with rhIGF-I, rhGH or saline

	Body weight gain (gram)	Serum IGF-I (ng/ml)	Bone length (μm)	Blood glucose (mg%)	Food intake (g/day)	Water intake (g/day)
rhIGF-I	39.2±4.1	402±158	666± 36*	118±20	8.7	21.7
rhGH	40.1±9.9	48± 25	945±194	97± 9	7.8	23.7
saline	3.0±4.1	12± 3	73± 12	121±14	5.9	20.3

Values are given as means ± SD of four rats per group
* $P > 0.05$ vs. rhGH treated rats

The following observations were made: (1) body weight and accumulated longitudinal bone length of both the rhIGF-I and rhGH-infused animals showed a similar increase (Fig. 1, Table 1); (2) IGF-I increased the weights (expressed as % of total body weight) of the spleen and the kidneys more than growth hormone (Table 2); (3) in contrast to saline- and GH-treated animals, the weight of the fat pads decreased during IGF-I infusion (Table 2). Thus, IGF-I and GH had quantitatively different effects on certain organs.

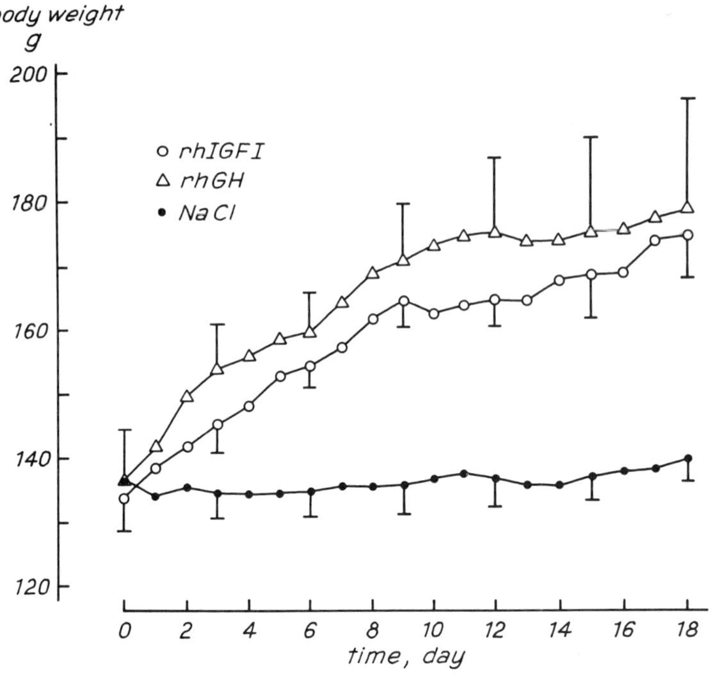

Fig. 1 Body weight of hypophysectomized rats infused with rhIGF-I, rhGH or saline for 18 days. N = 4 in each group. Mean ± SD. (From Guler et al., Ref. 4).

The increase in kidney weight may have a functional significance; continuous subcutaneous infusion of rhIGF-I in healthy adults increases creatinine clearance up to 30% above preinfusion levels (3). The effects are reversible when the infusion is stopped. Urinary excretion of albumin in these subjects remains normal.

Table 2 Organ weights (% of total body weight) of hypox rats infused for 18 days with rhIGF-I, rhGH or saline

	Spleen	Kidney	Fat pads
rhIGF-I	0.386 ± 0.034**	0.680 ± 0.054***	0.297 ± 0.064+
rhGH	0.231 ± 0.067	0.465 ± 0.017	0.384 ± 0.068
saline	0.178 ± 0.010	0.487 ± 0.033	0.512 ± 0.046

Values are given as means ± SD of four rats per group.
**P < 0.01, ***P < 0.001 vs. rhGH treated rats, +P < 0.01 vs. saline treated rats

YOUNG MINI-POODLES

Serum levels of IGF-I parallel body size in poodles of different breeds; mini-poodles have low IGF-I serum levels compared to king poodles (2). Growth hormone responses to clonidine were reported to be the same as in king poodles (2). We therefore tested whether mini-poodles infused with rhIGF-I grew faster and became taller than controls (5).

One litter of four mini-poodles was infused subcutaneously with rhIGF-I (6 mg/day/dog) from the age of 91 to 221 days (a total of 130 days). Four mini-poodles from the same breeder served as controls. IGF-I serum levels in the infused dogs rose three to four times above normal. Body weight increased to the same extent in both groups (Fig. 2). Accumulated longitudinal bone growth as determined by serial X-rays of the left foreleg (performed every other week) tended to be greater in the IGF-I infused animals, but the difference was not statistically significant. However, the adapted body mass index (calculated as body weight divided by the square of the radial length) decreased in the rhIGF-I infused animals and increased in the controls ($P < 0.05$). Growth hormone responses to clonidine or growth hormone releasing hormone were greatly suppressed during the IGF-I infusion. Hence, long-term infusion of IGF-I exerts a negative feedback regulation on endogenous growth hormone secretion as proposed earlier by others (1). Serum levels of IGF-II were depressed, an effect that was also observed in healthy adults during IGF-I infusion (3). Serum creatinine levels in the IGF-I infused animals were significantly lower than in the controls. This suggests a prolonged increase of creatinine clearance (3).

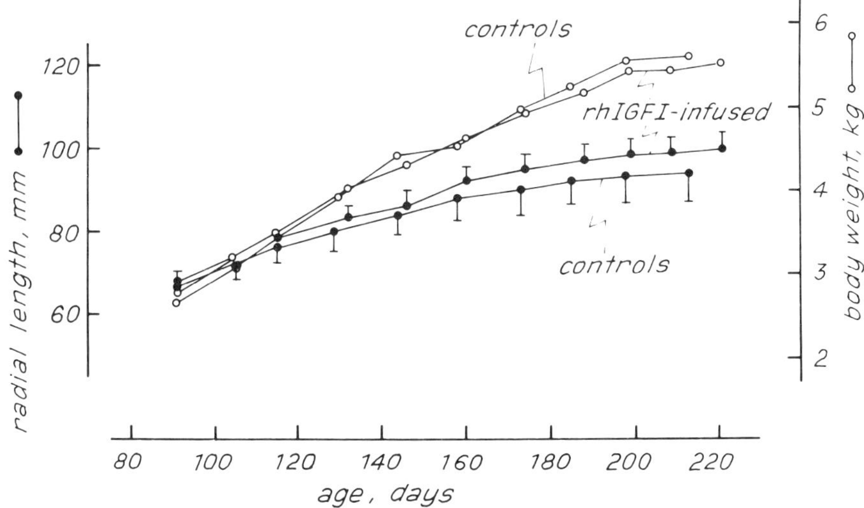

Fig. 2 Radial length and body weight of four mini-poodles infused with rhIGF-I and of four control infusions. Means ± SD.

SUMMARY AND CONCLUSIONS

Systemic infusion of IGF-I increases body weight and longitudinal bone growth in hypophysectomized rats but not in the endocrine competent mini-poodle. Negative feedback regulation of IGF-I on GH secretion and low IGF-II serum levels may contribute to their unaffected growth performance. IGF-I infused mini-poodles did not show signs of acromegaly. The effects of IGF-I and GH on several organs are quantitatively different. IGF-I particularly stimulates the weights of the spleen and kidney. It decreases the weight of epididymal fat tissue in hypox rats and may influence body composition of mini-poodles. IGF-I has marked renal effects.

REFERENCES

1 Abe, H., Molitch, M.E., van Wyk, J.J. and Underwood, L.E. (1983). Human growth hormone and somatomedin C suppress the spontaneous release of growth hormone in unanaesthetised rats. Endocrinology **113,** 1319-1324.

2 Eigenmann, J.E., Patterson, D.F. and Froesch, E.R. (1984). Body size parallels insulin-like growth factor I levels but not growth hormone secretory capacity. Acta Endocrinologica **106,** 448-453.

3 Guler, H.P., Schmid, Ch., Zapf, J. and Froesch, E.R. (1988). Effects of recombinant human insulin-like growth factor I on insulin secretion and renal function in normal human subjects. (Submitted).

4 Guler, H.P., Zapf, J., Scheiwiller, E. and Froesch, E.R. (1988). Recombinant human insulin-like growth factor I stimulates growth and has distinct effects on organ size in hypophysectomized rats. Proceedings of the National Academy of Sciences **85,** 4889-4893.

5 Guler, H.P., Binz, K., Eigenmann, E., Jäggi, S., Zimmerman, D., Zapf, J. and Froesch, E.R. (1988). Long-term infusion of recombinant human insulin-like growth factor I in young mini-poodles. (Submitted).

6 Hizuka, N., Takano, K., Asakawa, K., Miyakawa, M., Tonaka, I., Honkawa, R., Hasegawa, S., Mikasa, Y., Saito, S., Shibaski, T. and Schizume, K. (1987). In vivo effects of IGF-I in rats. Endocrinologia Japanica **34,** 115-121.

7 Isgaard, J., Nilsson, A., Lindahl, A., Jansson, J-O. and Isaksson, O.G.P. (1986). Effects of local administration of GH and IGF-I on longitudinal bone growth in rats. American Journal of Physiology **250,** E367-E372.

8 Scheiwiller, E., Guler, H.P., Merryweather, J., Scandella, C., Maerki, W., Zapf, J. and Froesch, E.R. (1986). Growth restoration of insulin-deficient diabetic rats by recombinant human insulin-like growth factor I. Nature **323,** 169-171.

9 Schoenle, E., Zapf, J., Humbel, R. and Froesch, E.R. (1982). Insulin-like growth factor I stimulates growth in hypophysectomized rats. Nature **296,** 252-253.

THE DIRECT EFFECTS OF GROWTH HORMONE ON CHONDROGENESIS AND OSTEOGENESIS

Z. Hochberg[1], G. Maor[2], D. Lewinson[2] and M. Silbermann[2]

[1] Department of Pharmacology and [2] Laboratory for Musculoskeletal Research, Rappaport Family Institute for Research in the Medical Sciences and Faculty of Medicine, Technion-Israel Institute of Technology, Haifa, Israel

Current concepts describe the effect of growth hormone (GH) on longitudinal bone growth as a direct effect on proliferation and maturation of the proximal cells in the growth plate column. Data are provided and indirect indicators cited to expand this mechanism. It is suggested that GH may contribute to growth plate growth and ossification through its effects on the entire growth plate columns and on the ossification front, and by indirect effects through the bone marrow and cortical bone.

INTRODUCTION

Growth hormone (GH) has been recognized for several decades as the main stimulator of longitudinal bone growth (4, 34). This effect of GH was formerly considered to be mediated by insulin-like growth factor I (IGF-I), of hepatic origin (7, 10, 28). The IGF-I hypothesis has recently been challenged by several reports referring to direct effects of GH at the site of the epiphyseal growth plate. Direct administration of GH to the epiphysis and infusion of GH into a femoral artery stimulated the growth of long bones (14, 29, 30). These in vivo studies have been strengthened by in vitro experiments which documented stimulatory effects of GH on chondrocytic proliferations (3, 19, 20).

The aim of the present report is to summarize the current concepts concerning the direct effects of GH on longitudinal growth of long bones and to add further experimental evidence to support these concepts. Further, it intends to open additional vistas on the effects of GH on the epiphyseal growth plate at the ossification front and via the bone marrow.

MECHANISTIC THEORIES OF GH ACTION

Green et al. (13), while proposing their dual-effector theory, also proposed a direct action of GH to promote differentiation of precursor cells and an indirect action that is mediated by a locally produced IGF-I, which promotes cell multiplication in young differentiated cells. This theory, which was originally developed through the use of preadipose 3T3 cells (26, 27, 37), was suggested to apply to the cartilage tissue as well. It remains to be seen whether the dual-effector theory applies also to other cell types. In our experiments with bone marrow colonies, we have shown that in the process of granulopoiesis GH is dependent on local production of IGF-I in both its differentiative and proliferative effects (24).

Isaksson and colleagues have recently broadened the above-mentioned theory. Using a colony assay of tibial epiphyseal cartilage cells (18, 20), they noted that during the process of cell differentiation, cells that had been directly stimulated by GH underwent two parallel evolutions: the gene encoding for IGF-I was expressed, resulting in an increased synthesis of IGF-I, and, concomitantly, the cells became responsive to IGF-I. Through either an autocrine or a paracrine mechanism, such cells responded to the newly synthesized IGF-I with multiplication, to produce the cartilage columns in the epiphyseal growth plate (15).

GH EFFECTS ON ORGAN CULTURE OF MANDIBULAR CONDYLES

The cartilage of neonatal mouse mandibular condyle, a secondary-type cartilage, serves as a growth centre for the young growing mandible (8). When grown in vitro, it preserves the characteristic gradient of a growth centre with proliferative cells, followed by differentiation and maturation, to become hypertrophic chondrocytes (22, 32, 33). When incubated in vitro with GH, both the proliferative and the differentiative processes are stimulated (23). The outmost layer of proliferating cells grow significantly as evidenced by both morphometric analysis and [^{3}H]thymidine autoradiography. An enhanced growth of the chondroblastic and hypertrophic layers indicates a GH effect on the differentiation of precartilage cells. The optimal dose for the stimulatory effect of GH in vitro was 50 ng/ml, and a dose of 100 ng/ml was less effective (23). A similar dose-dependent relationship was previously reported (19). By 6 days in culture, GH-treated condyles were 50% longer than untreated controls. In addition, there was also a significant increase in the size of the calcified cartilage, providing further evidence for the effect of GH on cartilage maturation.

GH EFFECTS IN TISSUE CULTURE OF CHONDROPROGENITOR CELLS

This tissue culture system enables studies on the growth and differentiation of chondroprogenitor cells (21). Grown on a collagenous substrate, the chondroprogenitor cells of newborn mouse mandibular condyles mature to become chondroblasts within 24 hours. These cells later differentiate to become hypertrophic chondrocytes and to synthesize cartilage matrix. In vitro treatment of these cells with 50 ng/ml of human GH (hGH) brought about a pronounced increase in the size of the cartilaginous nodule, with a substantial increase in proliferation of the immature cells and in growth of the differentiated cartilage. Within 6 days in culture, the hGH-treated tissue developed calcified matrix and other bony elements: bone trabeculae, osteocytes, osteoblasts, and surprisingly enough, multinucleated cells that resembled osteoclasts. Thus, GH was able to transform chondroprogenitor cells, through a phase of cartilage nodule, into a bone tissue.

The next question is, therefore, whether a similar sequence of events occurs at the ossification front.

GH EFFECTS ON THE OSSIFICATION FRONT

The ossification front starts with the last row of hypertrophic cartilage cells (6) and extends through a zone of chondroclastic activity, reaching the vascular marrow and the osteoblasts originating from mesenchymal cells in the marrow tissue. All these elements are potential responders to GH. Indirect evidence for the role of GH in the development of these cells is the positive autoradiography of [^{125}I]hGH, which shows an uptake by cells along the ossification front of cultured mandibular condyles. Indeed, GH has been shown to exert a stimulatory effect on the multiplication of osteoblasts from rat, mouse, and chicken calvaria (9, 35, 36). Whether similar effects are exerted at the metaphyseal ossification front remains to be seen. Further potential targets for GH are vascular endothelial cells, osteoclasts, and the adjacent bone marrow. The appearance of osteoclasts in the chondroprogenitor tissue culture system may be the first suggestion of an effect of GH on osteclastic differentiation.

In a study of the effect of hypothyroidism on the growth plate, we have seen the old phenomenon: in the hypothyroid animal the epiphyseal growth plate cartilage is completely sealed off from the underlying bone marrow by mineralized tissue (17, 31). However, the hypothyroid rat is also GH deficient. When hGH is given to the hypothyroid animal, it enables the vascular marrow to invade the ossification front.

GH EFFECTS ON THE BONE MARROW

At the distal end of the ossification front, the vascular marrow invades the spaces that are formed by the resorption of mineralized cartilage cells. Only indirect evidence hints at a possible contribution of the marrow to longitudinal bone growth. Bab et al. (1, 2) reported that removal of tibial marrow induced increased formation of bone and cartilage in the distant mandibular condyle; they suggested that it was the regenerating bone marrow which produced a potent growth-promoting effect directed at osteogenic cells. The bone marrow is a target tissue for GH and IGF-I. Reports are available on the stimulatory effect of hGH and IGF-I on colony formation in vitro by primitive and relatively mature human marrow erythroid progenitors (5, 11, 12). We have recently found that in vitro granulopoiesis is also enhanced by GH and IGF-I (24) and that the hGH effect on both erythropoiesis and granulopoiesis is mediated by a paracrine action (24, 25). We have shown that the presence of monocytes is essential for the hGH but not for the IGF-I effect. Thus, the monocytes produce a paracrine factor which contributes to the proliferation and maturation of bone marrow cells. While injection of hGH at the metaphyseal end of the growth plate did not induce longitudinal growth of the plate (16), an accurate analysis of the components of the ossification front was not done. It remains to be seen whether bone marrow monocytes contribute to events of the ossification front.

CONCLUSIONS

After several years of scepticism, GH has been widely accepted as a direct stimulator of longitudinal bone growth. In vitro experiments have documented a direct effect of GH on the early differentiation of reserve cells and on the multiplication of young chondroblasts in the early phases of endochondral ossification. The hypothetical mechanism of GH action suggested by Isaksson et al. (15) is well supported by experimental results. In the present report we have tried to add evidence for the suggested theory, but also to ask further questions about the role of GH in the events taking place at the cartilage-bone interphase during the process of endochondral ossification. The data are scarce and inconclusive, but we feel that GH may contribute to endochondral ossification through its effects on the ossification front and the bone marrow.

ACKNOWLEDGEMENTS

This work was supported by grant 185-056 from the Society of Biotechnology Research (GBF), F.R.G., and by a Nordist grant for the study of growth.

REFERENCES

1 Bab, I., Gazit, D., Massarawa, A. and Sela, J. (1985). Removal of tibial marrow induces increased formation of bone and cartilage in rat mandibular condyle. Calcified Tissue International **37**, 551-555.

2 Bab, I., Gazit, D., Muhlrad, A. and Shteyer, A. (1988). Regenerating bone marrow produces a potent growth-promoting activity to osteogenic cells. Endocrinology **123**, 345-352.

3 Barnard, R., Haynes, K.M., Werther, G.A. and Waters, M.J. (1988). Immunohistochemical localization of growth hormone receptors in rabbit tibia. Endocrinology **122**, 2562-2569.

4 Cheek, D.B. and Hill, D.E. (1974). Effect of growth hormone on cell and somatic growth. In Handbook of Physiology (Knobil, E. and Sawyer, W.H., eds.), Section 7, vol. 4, part 2, pp 159-185. American Physiological Society, Bethesda, MD.

5 Claustres, M., Chatelain, P. and Sultan, C. (1986). Comparative effects of human growth hormone (hGH) and insulin-like growth factor I/somatomedin C (IGF-I/m.l.c.-C) on erythropoietic cultures from normal children (Abstract). Pediatric Research **20,** 1191.

6 Cowell, H.R., Hunziker, E.B. and Rosenberg, L. (1987). The role of hypertrophic chondrocytes in endochondral ossification and in the development of secondary centers of ossification. Journal of Bone and Joint Surgery (American Volume) **69,** 159-161.

7 Daughaday, W.H., Hall, K., Salmon, W.D., van der Bronde, J.L. and Van Wyk, J.J. (1972). Somatomedin: proposed designation for sulphation factor. Nature **235,** 107.

8 Durkin, J.F. (1972). Secondary cartilage: a misnomer? American Journal of Orthodontics **62,** 15-41.

9 Ernst, M. and Froesch, E.R. (1988). Growth hormone dependent stimulation of osteoblast-like cells in serum-free cultures via local synthesis of insulin-like growth factor I. Biochemical and Biophysical Research Communications **151,** 142-147.

10 Froesch, E.R., Schmid, C., Schwander, J. and Zapf, J. (1985) Actions of insulin-like growth factors. Annual Review of Physiology **47,** 443-467.

11 Golde, D.W. (1979). In vitro effects of growth hormones. In Growth Hormones and other Biologically Active Peptides (Pecile, A. and Muller, E.E., eds.), pp 52-62. Excerpta Medica, Amsterdam.

12 Golde, D.W., Bersch, N. and Li, C.H. (1976). Growth hormone: species-specific stimulation of erythropoiesis in vitro. Science **196,** 1112-1113.

13 Green, H., Morikawa, M. and Nixon, T. (1985). A dual-effector theory of growth-hormone action. Differentiation **29,** 195-198.

14 Isaksson, O.G.P., Janson, J.O. and Gause, I.A.M. (1982). Growth hormone stimulates longitudinal bone growth directly. Science **216,** 1237-1239.

15 Isaksson, O.G.P., Lindahl, A., Nilsson, A. and Isgaard, J. (1987). Mechanism of the stimulatory effect of growth hormone on longitudinal bone growth. Endocrine Reviews **8,** 426-434.

16 Isgaard, J., Nilsson, A., Lindahl, A., Jansson, J.-O. and Isaksson, O.G.P. (1986). Effects of local administration of GH and IGF-I on longitudinal bone growth in rats. American Journal of Physiology **250,** E367-E372.

17 Lewinson, D., Harel, Z., Shenzer, P., Silbermann, M. and Hochberg, Z. (1988). Effect of thyroid hormone and growth hormone on recovery from hypothyroidism of epiphyseal growth plate cartilage and its adjacent bone. Endocrinology (in press).

18 Lindahl, A., Isgaard, J., Nilsson, A. and Isaksson, O.G.P. (1986). Growth hormone potentiates colony formation of epiphyseal chondrocytes in suspension culture. Endocrinology **118,** 1843-1848.

19 Lindahl, A., Isgaard, J., Carlsson, L. and Isaksson, O.G.P. (1987). Differential effects of growth hormone and insulin-like growth factor I on colony formation of epiphyseal chondrocytes in suspension culture in rats of different ages. Endocrinology **121,** 1061-1069.

20 Lindahl, A., Nilsson, A. and Isaksson, O.G.P. (1987). Effects of growth hormone and insulin-like growth factor-I on colony formation of rabbit epiphyseal chondrocytes at different stages of maturation. Journal of Endocrinology **115,** 263-271.

21 Maor, G., von der Mark, K., Reddi, H., Heinegard, D., Franzen, A. and Silbermann, M. (1987). Acceleration of cartilage and bone differentiation on collagenous substrata. Collagen and Related Research **7,** 351-370.

22 Maor, G. and Silbermann, M. (1981). In vitro effects of glucocorticoid hormones on the synthesis of DNA in cartilage of neonatal mice. FEBS Letters **129,** 256-260.

23 Maor, G., Silbermann, M. and Hochberg, Z. (1988). Growth hormone directly enhances chondrogenesis and osteogenesis in vitro (Abstract). Hormone Research (Basel) (in press).

24 Merchav, S., Tatarsky, I. and Hochberg, Z. (1988). Enhancement of human granulopoiesis in vitro by biosynthetic insulin-like growth factor I/somatomedin C and human growth hormone. Journal of Clinical Investigation **81,** 791-797.

25 Merchav, S., Tatarsky, I. and Hochberg, Z. (1988). Enhancement of erythropoiesis in vitro by human growth hormone is mediated by insulin-like growth factor I. British Journal of Haematology (in press).

26 Morikawa, M., Green, H. and Lewis, U.J. (1984). Activity of human growth hormone and related polypeptides on the adipose conversion of 3T3 cells. Molecular and Cellular Biology **4,** 228-231.

27 Nixon, T. and Green, H. (1983). Properties of growth hormone receptors in relation to the adipose conversion of 3T3 cells. Journal of Cellular Physiology **115,** 291-296.

28 Salmon, W.D. and Daughaday, W.H. (1957). A hormonally controlled serum factor which stimulates sulfate incorporation in vitro. Journal of Laboratory and Clinical Medicine **49,** 825-836.

29 Schlechter, N.L., Russell, S.M., Spencer, E.M. and Nicoll, C.S. (1986). Evidence suggesting that the direct promoting effect of growth hormone on cartilage in vivo is mediated by local production of somatomedin. Proceedings of the National Academy of Sciences, U.S.A. **83,** 7932-7934.

30 Schlechter, N.L., Russell, S.M., Greenberg, S., Spencer, E.M. and Nicoll, C.S. (1986). A direct growth effect of GH in the rat hindlimb shown by arterial infusion. American Journal of Physiology **250,** E231-E235.

31 Scow, R.O., Simpson, M.E., Asling, C.W., Li, C.H. and Evans, H.M. (1949). Response by the rat thryo-parathyroidectomized at birth to growth hormone and to thyroxin given separately or in combination. Anatomical Record **104,** 445-463.

32 Silbermann, M. and Frommer, J. (1972). The specific nature of endochondral ossificiation in the mandibular condyle of the mouse. Anatomical Record **172,** 659-667.

33 Silbermann, M. and Lewinson, D. (1978). An electron microscopic study of the premineralizing zone of the condylar cartilage of the mouse mandible. Journal of Anatomy **125,** 55-70.

34 Simpson, M.E., Asling, C.W. and Evans, H.M. (1950). Some endocrine influences on skeletal growth and differentiation. Yale Journal of Biology and Medicine **23,** 1-27.

35 Slootweg, M.C., van Buul-Offers, S.C., Herrmann-Erlee, M.P.M. and Duursma, S.A. (1988). Direct stimulatory effect of growth hormone on DNA synthesis of fetal chicken osteoblasts in culture. Acta Endocrinologica (Copenhagen) **118,** 294-300.

36 Slootweg, M.C., van Buul-Offers, S.C., Herrmann-Erlee, M.P.M., van der Meer, J.M. and Duursma, S.A. (1988). Growth hormone is mitogenic for fetal mouse osteoblasts but not for undifferentiated bone cells. Journal of Endocrinology **116,** R11-R13.

37 Zezulak, K.M. and Green, H. (1986). The generation of insulin-like growth factor-I-sensitive cells by growth hormone action. Science **233,** 551-553.

GROWTH PROMOTION BY GROWTH HORMONE AND INSULIN-LIKE GROWTH FACTOR-I IN THE RAT

I.C.A.F. Robinson[1] and R.G. Clark[2]

[1] Division of Neurophysiology and Neuropharmacology, National Institute for Medical Research, The Ridgeway, Mill Hill, London NW7 1AA, U.K.
[2] Growth Physiology, Ruakura Agricultural Research Centre, Ministry of Agriculture and Fisheries, Private Bag, Hamilton, New Zealand

We have developed methods for chronic intravenous cannulation of normal, hypophysectomized, and GH-deficient rats. In hypophysectomized rats, GH stimulates growth when given by continuous infusion, but is more effective when given in a pulsatile manner which mimics the normal GH secretory pattern in the male rat. Some of the effects of GH are also pattern dependent in normal rats. Pulses of GH induced in female rats increase growth, whether they are elicited by GRF pulses or by intermittent infusions of somatostatin. Intravenous IGF-I also stimulates growth in hypophysectomized rats but is less effective than GH, even when given in combination with a small dose of GH. A new dwarf rat which shows selective GH insufficiency, also shows pattern-dependent growth in response to GH. In this new animal model, IGF-I is less effective than GH in stimulating bone growth, whereas it has a relatively greater effect on the kidney and spleen. The role of IGF-I and its binding proteins in the endocrine or local actions of GH on body growth continues to be the subject of active investigation and debate.

INTRODUCTION

The endocrine control of body growth in mammals is complex. There is no shortage of hormonal candidates (growth hormone (GH), insulin-like growth factors (IGFs), hypothalamic regulatory peptides (GH-releasing factor, GRF and somatostatin (SS), thyroid, adrenal or gonadal hormones), involved in normal growth and development, and it is likely that modern molecular biology techniques will reveal further cellular products which may be released to coordinate tissue and organ growth. In the rat, pituitary GH plays a crucial role in the stimulation of growth. Removal of the pituitary gland interrupts post-weaning growth, and although some growth can be elicited in the hypophysectomized rat by feeding, insulin, or other hormone treatments, it is replacement therapy with GH which restores growth towards normal in the hypophysectomized rat. Whilst this central role of GH is well established, there is considerable debate about whether GH itself promotes growth by acting directly on peripheral tissues, or indirectly by stimulating the production of IGF-I in the liver, raising the levels of IGF-I in the circulation from whence it mediates the actions of GH by stimulating peripheral tissue growth. In this short paper we review some of our recent studies on the promotion of growth by manipulating directly by i.v. infusions, the blood concentrations of GH and IGF-I in experimental models of GH deficiency in the rat.

THE HYPOPHYSECTOMIZED RAT

The surgically hypophysectomized rat is the classical animal model for studying the effects of growth promoting agents. Although hypophysectomy necessarily removes all the pituitary hormones, GH by itself does stimulate growth, so that the effects of GH on body weight gain and bone growth can be examined and quantified. Whilst daily injections of very large doses of exogenous GH will produce normal, or even supra-normal growth rates (12), it was known that more frequent GH injections at a given dose were more effective in stimulating weight gain and bone growth in hypophysectomized rats (17, 24). The physiological significance of this idea stems from the work of Tannenbaum and Martin (1976) who showed that the secretion of GH in the conscious male rat is extremely episodic with an ultradian rhythm of regular large GH secretory bursts occurring at roughly 3-hourly intervals (23).

We developed a method for maintaining long-term intravenous cannulae in young hypophysectomized rats, which when coupled with a computer-controlled pumping system, gave us the possibility of controlling the plasma GH profile over several days (Fig. 1a). Using intravenous administration, we could avoid the problems of subcutaneous absorption kinetics or degradation of GH, and so reintroduce into hypophysectomized rats, a plasma GH profile more reminiscent of the normal physiological secretory pattern in male rats. With the animals housed individually in metabolic cages, food and water intake, body weight and longitudinal bone growth could all be readily determined (4). Intravenous infusions of GH produced dose-dependent increases in several measures of growth in hypophysectomized rats, and was more effective than the same doses given subcutaneously. In several experiments we were able to show that a pulsatile GH infusion which mimicked the normal male pattern (1 pulse every 3 hr) was more effective in stimulating growth than the same dose of GH given as a continuous i.v. infusion (Fig. 1b). Furthermore, the response was frequency dependent with the rank order of potency being 9 > 3 > 1 pulses/day for the same total daily dose of GH. These experiments confirmed the importance of the pattern of administration of GH in stimulating growth in the GH-deficient hypophysectomized rat, but were not necessarily relevant to the normal rat. We were therefore curious to know whether such pattern-dependent actions of GH could also be demonstrated in the normal rat.

NORMAL RATS

Most of the work done on the control of the GH secretory pattern in normal rats has been carried out in males. There are fewer studies of female rats, though it has been established that the secretory pattern is different from that in males, with a more continuous irregular plasma GH profile (5, 11). This difference in the secretory pattern of GH may be one of the contributory factors to their growth rate being slower than that of their male littermates. We wished to try to induce a pulsatile pattern of GH in female rats, and adopted a similar experimental approach with i.v. infusions in conscious chronically-cannulated rats. Since these normal females had intact pituitaries we attempted to manipulate the endogenous GH secretory profile by patterned infusions of GRF, rather than by giving extra exogenous GH. Repetitive i.v. pulses of GRF every 3 h elicited a series of 3-hourly GH secretory bursts, in effect converting the pattern of GH exposure from a 'female' to a 'male' type of GH release. This treatment induced an increase in growth rate in normal female animals, as well as in male rats whose endogenous GRF has been depleted by neonatal monosodium glutamate administration (7, Fig. 2).

The interpretation of these experiments is complicated by the problem that the exogenous administration of GRF not only affects the release of GH directly, but also stimulates the production of further endogenous GH (1). One must therefore take into account the possibility that the pattern dependence of the growth response to GRF reflects an increase in the total amount of GH available for release, as well as the imposition of a particular GH plasma profile. To address this problem, we performed a similar type of experiment but this time using an agent that inhibits GH secretion. An i.v. infusion of somatostatin (SS) given to normal rats produces a powerful blockade of GH secretion, but this inhibition is only maintained in the presence of SS, so that within minutes of stopping the SS infusion there is a large rebound release of GH (6). Groups of normal female rats were therefore given i.v. SS infusions, either continuously or intermittently, with the infusion interrupted for 10-90 minutes every 3 h (8). Such

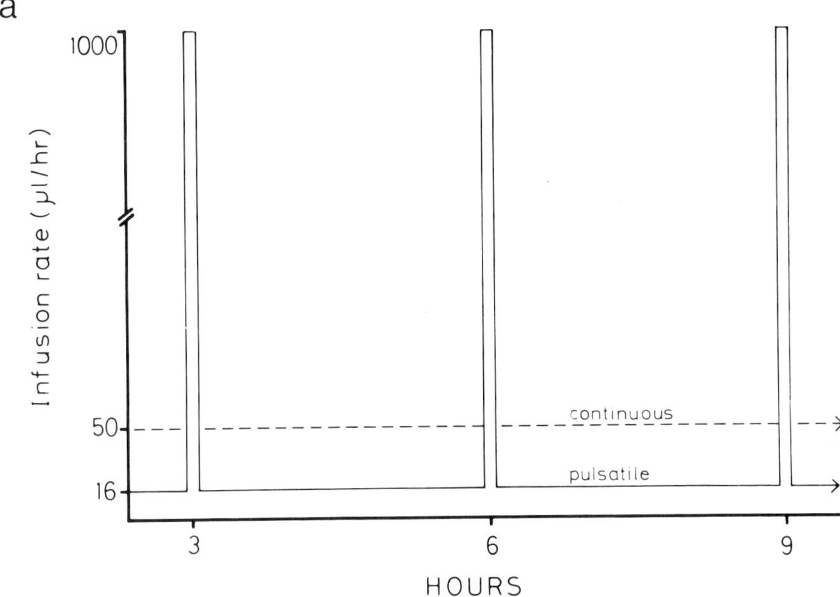

Fig. 1 (a) Computer-controlled multichannelled peristaltic pumps allow us to impose either a continuous or pulsatile pattern of administration of hormones directly into the jugular vein of chronically-cannulated conscious rats. (b) Body weight gain in hypophysectomized rats infused i.v. with human GH at 12 or 36 mU/day for 5 days either continuously (left panel) or in 3-hourly pulses (right panel). Data redrawn from Ref. 4.

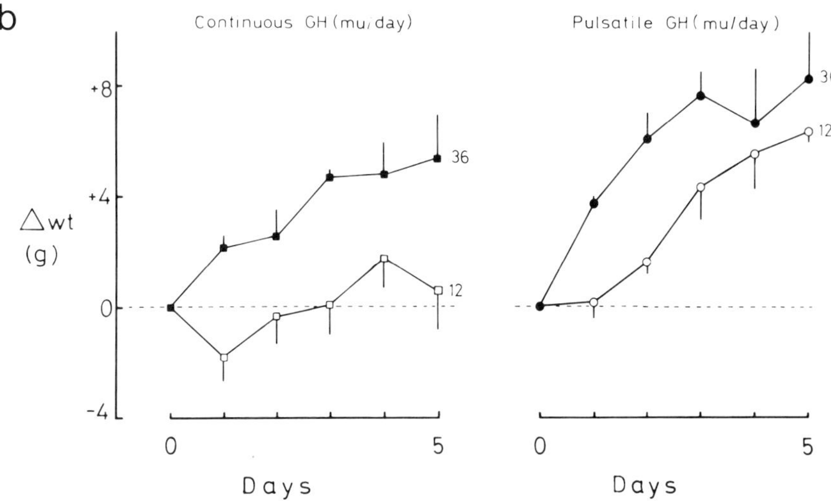

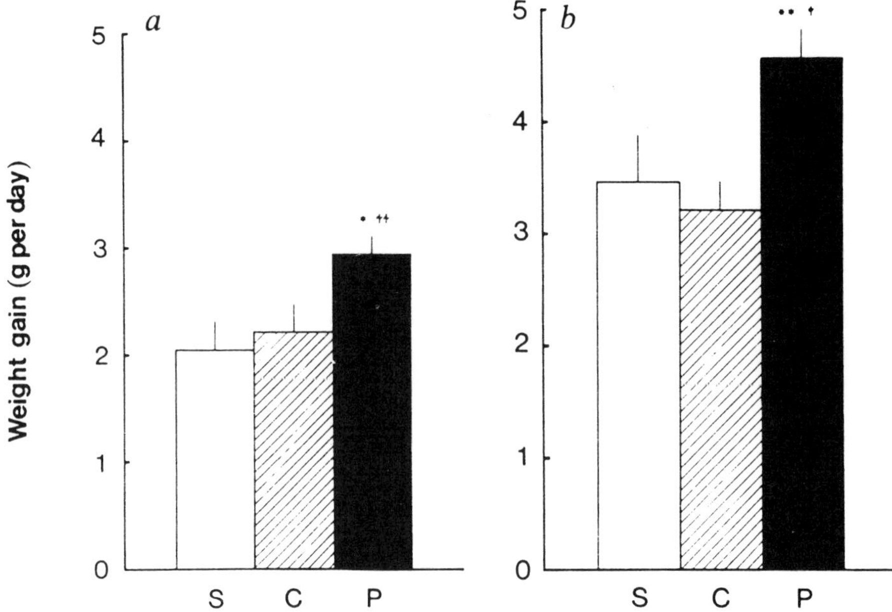

Fig. 2 Effects of continuous (C) or pulsatile (P) administration of GRF(1-29)amide on growth in normal female (a) or monosodium glutamate-treated male (b) rats. Continuous infusion of GRF (8 μg/day for 12 days) did not stimulate growth compared to saline (S) infusions, whereas the same daily dose given in 1 μg pulses every 3 hours did (7)

---

patterns of intermittent SS exposure elicit a series of rebound GH secretory bursts occurring at the same frequency as the interruptions (ie once every 3 h, Fig. 3). In this way we were again able to impose a 'male type' of GH plasma profile in normal female rats, and showed that this pattern stimulated growth compared with rats receiving the same total dose of SS without interruption; continuous SS had no effect on body weight and significantly suppressed bone growth (8). Thus in normal rats as well as in hypophysectomized animals, the growth response to GH depends on the temporal pattern of GH in the blood. It is therefore important to bear in mind the marked pattern dependence of the GH response in the rat, when comparing the potency of GH with other growth promoting agents (such as IGF-I) whose actions may differ in their dependence on the route or pattern of administration.

IN VIVO GROWTH-PROMOTING EFFECTS OF IGF-I

Having established a method for long term i.v. administration in hypophysectomized rats, we were interested to compare the growth-promoting potencies of IGF-I and GH using this method. Although the notion that IGF-I mediates the actions of GH was widely accepted, there had been very little direct evidence of growth promotion by IGF-I due to the difficulty of obtaining sufficient amounts of purified material for an in vivo bioassay. Earlier studies had used partially purified 'somatomedins' extracted from human plasma and administered to Snell dwarf mice (since their small body size should require much

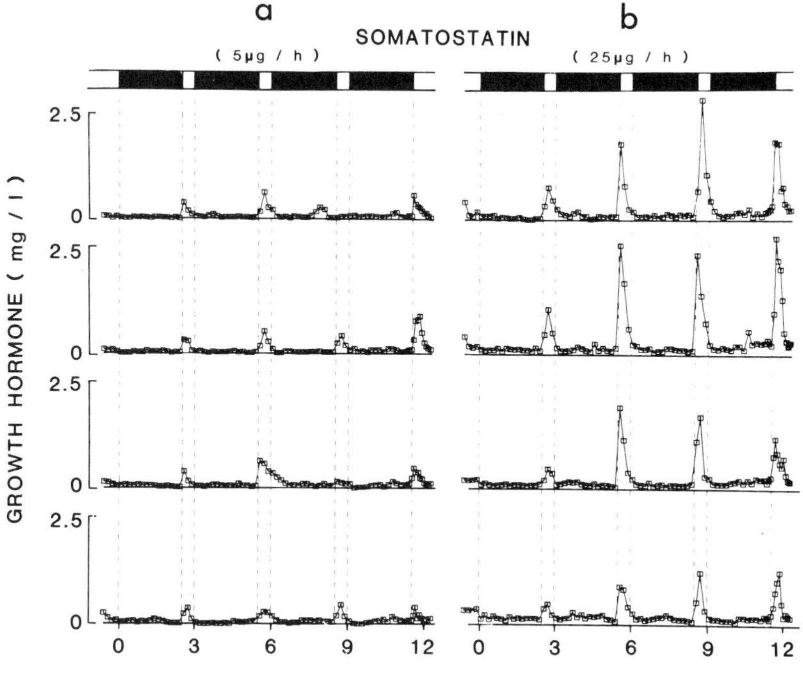

Fig. 3 Conscious female rats received i.v. infusions of SS (black bars) at 5 μg/h (a) or 25 μg/h (b), interrupted every 3 h for a 30 min period. Blood samples were withdrawn automatically and assayed for GH. Repetitive rebound releases of GH occurred in all the animals upon stopping the SS infusions (from Ref. 8).

---

less material than the rat (15, 25)). Although growth responses were obtained they were always smaller than those that could be induced by low doses of GH, though the latter preparations were much purer, of course. It was not until 1982 that Schoenle et al. (21) reported the results of subcutaneous infusions of purified human IGF-I in hypophysectomized rats. Although these experiments demonstrated a significant growth stimulation by IGF-I, the magnitude of the response was again disappointingly small, and whilst a number of possible explanations could be advanced (use of human instead of rat IGF-I, method or route of administration, too low a dose), IGF-I did not appear to be able to reproduce fully the growth promoting effects of GH in hypophysectomized rats. This is an important point, since it was necessary to show that circulating IGF-I levels could fully mediate the growth promoting actions of GH in order to substantiate the somatomedin hypothesis of GH action. In the last few years, the problems of scarcity of purified IGF-I have been overcome by the application of recombinant DNA technology, and we therefore decided to take advantage of the availability of biosynthetically produced IGF-I to compare the effects of GH and IGF-I by intravenous infusion.

Our first studies were carried out with recombinant methionyl IGF-I in the conscious hypophysectomized rat (22), and the results were very similar to those of Schoenle et al. (21). IGF-I produced a small but significant stimulation of body weight gain and bone growth (Fig. 4), but much less than those obtained with much lower doses of GH. Similar conclusions were arrived at in independent experiments with subcutaneous osmotic

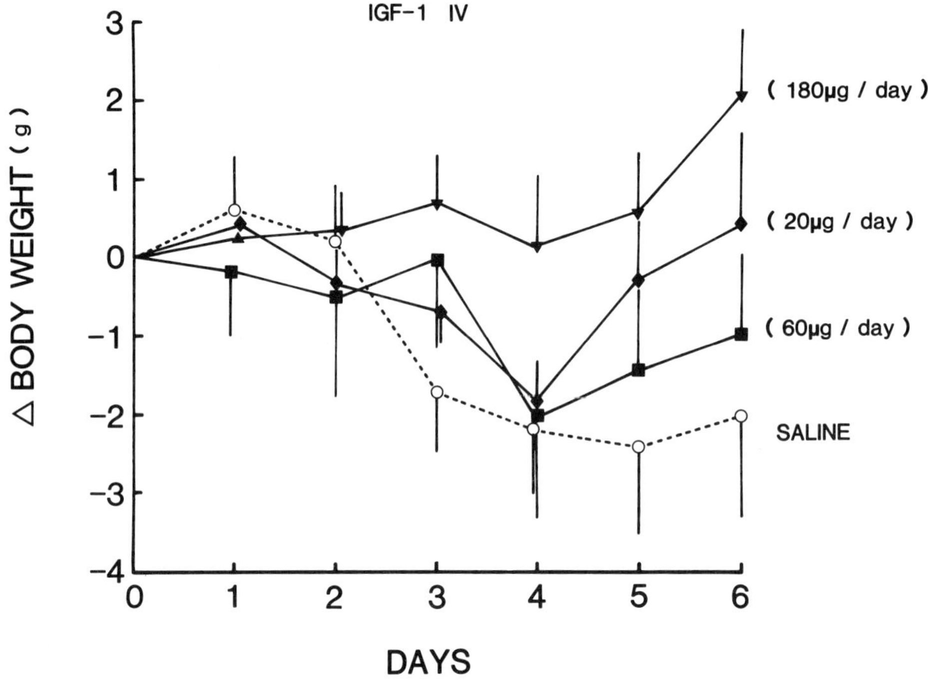

Fig. 4 Effects of recombinant methionyl human IGF-I given by i.v. infusion to hypophysectomized rats for 6 days at 0, 20, 60 or 180 μg/day. Only the highest infusion rate of IGF-I stimulated growth rate significantly. Data from Ref. 22.

---

minipump infusions or injections of GH or IGF-I; IGF-I was considerably less effective than GH in promoting growth in this animal model (22). Even when tested in combination with a small dose of GH which induced some growth, IGF-I was only poorly effective (Fig. 5). Furthermore, we obtained rapid growth in animals treated with GH whose circulating IGF-I levels were elevated, but below the levels of IGF-I in the animals infused with IGF-I which grew to a much lesser extent. Other recent studies of GH-induced growth also cast doubt on the direct relationship between growth and plasma IGF-I levels (12, 20). Again there are a number of possible explanations for these discrepancies, the major one being the unknown role of the carrier proteins for IGF-I, some of which are under GH control and therefore reduced in hypophysectomized animals. Nevertheless, in our view, the simplest explanation is that growth stimulation by GH in the hypophysectomized rat is not simply mediated via circulating IGF-I levels.

Using higher doses of IGF-I (300 μg/day) given for 18 days by continuous s.c. infusion, Guler et al. (1988) obtained increases in body weight and bone growth similar to those seen in response to GH given in a similar manner (13). The growth responses to both IGF-I and hGH tailed off in the latter half of the experiment, and in a shorter study (6 days) using different doses of GH and IGF-I, the dose-response curves were non-parallel and the maximal body weight and bone growth responses were lower for IGF-I than for GH. However, for the same total body weight gain, these authors noted marked increases in splenic and renal weight in IGF-I treated animals compared to those treated with GH

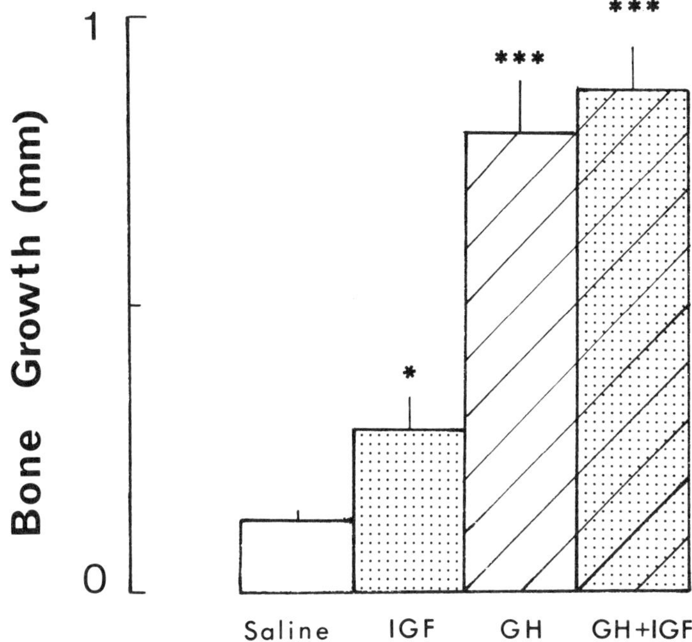

Fig. 5 Effects of priming hypophysectomized rats with bovine GH before treatment with IGF-I. One group of 12 rats were infused i.v. with bovine GH (10 mU/day) for 8 days, whilst another group received saline. After 4 days, IGF-I (120 μg/day) was added to the infusate for half the animals in each group. IGF-I stimulated bone growth ($P < 0.05$) compared to saline infused rats. GH infusion produced much greater bone growth ($P < 0.001$), which was not significantly increased by adding IGF-I to the infusate. Data from Ref. 22.

---

whereas the greatest longitudinal bone growth was observed in the GH-treated group. We had previously observed that bone growth was relatively poorly stimulated by IGF-I infusions compared with GH (22). This suggests that these peptides act independently in different tissues, perhaps on different cellular targets within these tissues, as has been suggested for longitudinal bone growth (16). It is possible that other differences between our experiments (22) and those of Guler et al. (13) explain the differing views of the potency of IGF-I as a growth promoting agent in this model. We used lower doses of methionyl IGF-I which may have been less potent than the recombinant human IGF-I used by Guler et al. On the other hand, these authors gave their peptides by subcutaneous infusion, whereas we have shown (4) that GH is much more effective when given by the intravenous route.

To a large extent, the problem of the relative weakness of IGF-I as a growth promoter via the endocrine route has been lessened by the realisation that IGF-I production occurs in many growing tissues (10), and the emphasis on the endocrine actions of IGF-I has therefore been somewhat overshadowed by the likelihood of a paracrine mode of action of IGF-I, perhaps acting in concert with GH. This should lead to a reconsideration of the role of the large amounts of IGF-I in the systemic circulation, most of which is not free, but non-covalently bound to carrier proteins which are themselves subject to endocrine and nutritional control (14). The interpretation of experiments involving the administration of IGF-I must therefore take into account the role of the circulating binding proteins for IGF-I. The presence of IGF-I binding proteins in the circulation undoubtedly alters the pharmacokinetics of IGF-I, and it is possible to argue that the binding proteins could either facilitate or inhibit the actions of IGF-I. They may act as a reservoir to maintain circulating IGF-I levels, but whether the binding proteins provide a steady supply of liver-derived IGF-I to peripheral tissues, or act to mop up free IGF-I and to prevent its widespread distribution in a free form is unknown. Some of these questions may be addressed when sufficient quantities of pure binding proteins become available for in vivo testing, but one recent report is also relevant to this problem. Cascieri et al. (1988) describe IGF-I analogues that retain biological potency but have greatly reduced affinity for the plasma binding proteins (2). IGF-I analogues that lacked affinity for binding proteins had an increased in vivo potency suggesting that the binding proteins are not necessary for effective presentation of the the peptide to its target tissues. The authors concluded that the binding proteins may normally inhibit the actions of circulating IGF-I, but before this conclusion is adopted, further studies are necessary to investigate whether such analogues of IGF-I can bind to other proteins in the vicinity of IGF-I receptors, or have otherwise altered kinetics of action at the receptors themselves.

STUDIES WITH GENETIC MUTANTS

An alternative approach to the study of the relative effectiveness of GH and IGF-I as growth promoters is to induce high levels of expression of these peptides in transgenic animals (eg. mice), which may lead to increased body size. This topic is covered by others in this volume, but one or two conclusions are relevant to our studies. In general the extra growth is greater for GH (whether induced by GRF transgenes or expressed by GH transgenes) than for IGF-I transgenes, but the analysis of the effects is complicated by the secondary effects of high gene expression, which is not subject to normal physiological or developmental regulation. Since in most cases, the transgenes are not under the control of the normal tissue specific promoters, they are expressed in many different tissues, and the high level of circulating IGF-I leads to a powerful suppression of endogenous GH production in the pituitary gland of the transgenic mouse. In the case of IGF-I transgenic mice this removal of endogenous GH may to some extent confound the effects of IGF-I on growth. In a recent study, it was shown that the induction of IGF-I expression in mice transgenic for GH preceded in time the gross increase in body weight that begins to occur at weaning (19). The authors concluded from this that the delay in growth in the transgenic mice until after weaning was due to an inability of the high circulating GH levels to stimulate IGF-I production. If this is so one might predict that mice transgenic for IGF-I would show increased growth during the first 3 weeks of life, especially since any secondary suppression of endogenous pituitary GH, would not confound the growth response during the 'GH-independent' post-natal period. However, it appears that mice transgenic for IGF-I show no difference in body weight gain until the 5th postnatal week (L. Mathews, personal communication, (18)). It would be interesting to know to what extent the relatively modest growth stimulation seen in mice expressing IGF-I could be enhanced by giving physiological amounts of GH to compensate for the suppression of endogenous GH synthesis which occurs in these mice (19).

From the foregoing it is clear that these approaches open up many new possibilities for creating extremely valuable new animal models for the endocrine manipulation of growth. For example, the creation of transgenic mice expressing the IGF-I binding proteins (especially if the transgenes were linked to promoters giving a restricted tissue specificity) may provide an insight into the role of these proteins in the organ-specific effects of IGF-I. On the other hand, detailed endocrine studies have not as yet been performed on these new mouse strains, so that statements about the cause of differences in growth based on tissue or body weights and single point blood samples at autopsy should be interpreted with some caution. The introduction of transgenes has generally led to models of increased growth; no less valuable are genetic variants exhibiting endocrine deficiency. There are a number of dwarf mouse strains which can be used for studies of the endocrinology of growth, and we have already referred to earlier studies which took advantage of the Snell dwarf mice which exhibit panhypopituitary dwarfism. However, these mice are too small to enable us to perform growth studies using our chronic cannulation techniques. We were therefore pleased to be able to characterise (with Dr. H.M. Charlton, Department of Human Anatomy, Oxford), a new mutant rat which shows dwarfism due to specific GH insufficiency.

DWARF RATS

The essential characteristics of these rats have been reported elsewhere (3). Pituitary GH levels are low but detectable, whereas other pituitary hormones are in the normal range. The dwarfism is not associated with major skeletal abnormalities, and can be corrected by exogenous GH treatment. The dwarf rats respond to GRF but the plasma GH response is reduced, reflecting the low pituitary GH content, and GRF treatment does not restore the pituitary GH content nor does it promote growth. The defect appears to lie in GH synthesis at the level of the somatotroph. Since the animals are otherwise healthy and fertile and the strain well established in several independent colonies,

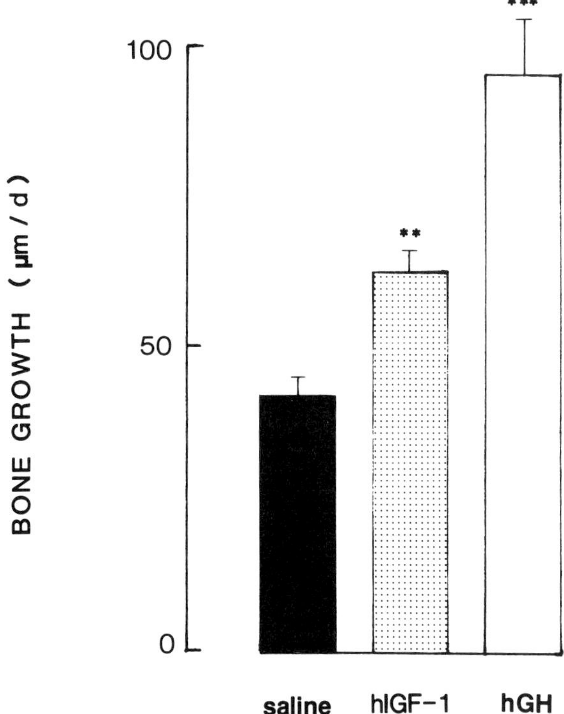

Fig. 6 Conscious male dwarf rats (3) were given continuous i.v. infusions of hGH (200 mU/day), recombinant human IGF-I (180 μg/day) or saline alone for 9 days. Both treatments increased body weight gain similarly, but hGH stimulated tibial bone growth to a much greater extent that did IGF-I. Data from Ref. 9.

this new dwarf rat appears to be an attractive alternative to the surgically hypophysectomized rat, as it provides a specific model of dwarfism due to GH insufficiency. In some of our first studies we therefore repeated the comparison of the growth-promoting effects of GH and IGF-I when given by i.v. infusion, but this time we used yeast-derived recombinant human IGF-I of high biological potency (> 14000 u/mg). The results of growth studies in this new dwarf rat confirmed and extended our earlier studies in hypophysectomized rats in several respects. Pulsatile infusions of GH were more effective than continuous infusions in stimulating both bone growth and body weight in the dwarf rats. In this case it is perhaps not surprising that continuous GH exposure was poorly effective in an animal that has a small but significant endogenous GH secretion, and which is already growing albeit at a slow rate. Interestingly, for the same body weight gain in two groups of dwarf rats given continuous i.v. infusions of hGH or IGF-I, we also observed a different pattern of tissue growth (9). Bone growth was relatively insensitive to IGF-I (Fig. 6), whereas this peptide caused a relative increase in renal and splenic growth. These effects were seen without any obvious changes in the pattern of IGF-I binding proteins in the dwarf rats, though as expected, GH treatment raised plasma IGF-I levels which are low in untreated dwarf rats. It is interesting that we were also able to observe tissue specific effects of IGF-I and GH, showing again that these peptides have different patterns of action when administered into the general circulation. The specific effects of IGF-I on kidney and splenic growth have also recently been observed in mice transgenic for IGF-I, though in this case it is possible that some of the tissue growth may be caused by local overexpression of the transgene, rather than via the increase in circulating levels of IGF-I.

CONCLUDING REMARKS

What conclusions can we draw from such experiments? Firstly, it is clear that in the rat, the growth promoting effects of GH depend on the pattern in which it is administered, and that a pulsatile plasma GH profile which resembles the normal physiological secretory pattern is more effective than a continuous exposure to GH. Although IGF-I has definite growth-promoting activity in the GH-deficient rat, it is far less effective than GH in our hands, and the two substances show qualitative differences in their patterns of action. One major action of GH is the stimulation of IGF-I production both in the liver and peripheral tissues, resulting in raised local tissue concentrations of IGF-I as well as increased circulating levels of this peptide. This is complicated by the presence of binding proteins which may act to modulate the effects of IGF-I, and certainly reduce the likelihood that short-term fluctuations in IGF-I production could be of physiological significance via the endocrine route. It is possible that the binding proteins act to target IGF-I to certain tissues, and it would be interesting to discover whether the newer IGF-I analogues that lack the ability to bind plasma proteins have a qualitative difference in the pattern of tissues that they stimulate. We should also not discount the possible role of circulating IGF-I in regulating GH production at a pituitary or hypothalamic level, perhaps on a timescale of hours or days rather than minute by minute. Since the administration of IGF-I into the bloodstream produces a very different pattern of tissue growth from that induced by GH (which also increases circulating IGF-I levels), we feel that IGF-I is primarily a local regulator of tissue growth acting in a paracrine fashion rather than mediating the effects of GH via an endocrine action. In our studies we have stressed the importance of the pattern of GH secretion or administration in stimulating growth. How this pattern is sensed by GH receptors and translated into the appropriate signals for IGF-I production, as well as other GH-dependent cellular processes which ultimately result in coordinated growth, remains for future study.

ACKNOWLEDGEMENTS

We would like to acknowledge the help of our many colleagues and co-workers, without whom many of our studies would not have been possible.

REFERENCES

1 Barinaga, M., Yamonoto, G., Rivier, C., Vale, W., Evans, R. and Rosenfeld, M.G. (1983). Transcriptional regulation of growth hormone gene expression by growth hormone releasing factor. Nature **306,** 84-86.

2 Cascieri, M.A., Saperstein, R., Hayes, N.S., Green, B.G., Chicchi, G.G., Applebaum, J. and Bayne, M.L. (1988). Serum half-life and biological activity of mutants of human insulin-like growth factor I which do not bind to serum binding proteins. Endocrinology **123,** 373-381.

3 Charlton, H.M., Clark, R.G., Robinson, I.C.A.F., Porter-Goff, A.E., Cox, B.S., Bugnon, C. and Bloch, B.S. (1988). Growth hormone deficient dwarfism in the rat: a new mutation. Journal of Endocrinology **119,** 51-58.

4 Clark, R.G., Jansson, J-O., Isaksson, O. and Robinson, I.C.A.F. (1985). Intravenous growth hormone: growth responses to patterned infusions in hypophysectomized rats. Journal of Endocrinology **104,** 53-61.

5 Clark, R.G., Carlsson, L.M.S. and Robinson, I.C.A.F. (1987). Growth hormone secretory profiles in conscious female rats. Journal of Endocrinology **114,** 399-407.

6 Clark, R.G., Chambers, G., Lewin, J. and Robinson, I.C.A.F. (1986). Automated repetitive microsampling of blood: growth hormone profiles in conscious male rats. Journal of Endocrinology **111,** 27-35.

7 Clark, R.G. and Robinson, I.C.A.F. (1985). Pulsatile administration of an amidated fragment of human growth hormone releasing factor produces growth in normal and GRF-deficient rats. Nature **314,** 281-283.

8 Clark, R.G. and Robinson, I.C.A.F. (1988). Paradoxical growth-promoting effects induced by patterned infusions of somatostatin in female rats. Endocrinology **122,** 2675-2682.

9 Clark, R.G., Robinson, I.C.A.F. and Skottner, A. (1988). Growth responses to human insulin-like growth-factor I (IGF-I) and human growth hormone (hGH) in a mutant rat. Journal of Physiology **398,** 80P.

10 D'Ercole, A.J., Stiles, A.D. and Underwood, L.E. (1984). Tissue concentrations of somatomedin C: further evidence for multiple sites of synthesis and paracrine or autocrine mechanisms of action. Proceedings of the National Academy of Sciences, U.S.A. **81,** 935-939.

11 Eden, S. (1979). Age and sex-related differences in episodic growth hormone secretion in the rat. Endocrinology **105,** 555-560.

12 Groesbeck, M.D., Parlow, A.F. and Daughaday, W.H. (1987). Stimulation of supranormal growth in prepubertal, adult plateaued, and hypophysectomized female rats by large doses of rat growth hormone: physiological effects and adverse consequences. Endocrinology **120,** 1963-1975.

13 Guler, H-P, Zapf, J., Scheiwiller, E. and Froesch, E.R. (1988). Recombinant human insulin-like growth factor I stimulates growth and has distinct effects on organ size in hypophysectomized rats. Proceedings of the National Academy of Sciences, U.S.A. **85,** 4889-4893.

14 Hall, K., Brismar, K. and Povoa, G. (1987). The role of somatomedin binding protein. In Growth hormone - basic and clinical aspects (Isaksson, O., et al., eds.), pp 415-424. Elsevier, Amsterdam.

15 Holder, A.T., Spencer, E.M. and Preece, M.A. (1981). Effects of bovine growth hormone and a partially pure preparation of somatomedin on various growth parameters in hypopituitary dwarf mice. Journal of Endocrinology **89,** 275-282.

16 Isaksson, O.G.P., Lindahl, A., Nilsson, A. and Isgaard, J. (1987). Cellular mechanisms for the stimulatory effect of growth hormone on longitudinal bone growth. In Growth hormone - basic and clinical aspects (Isaksson, O., et al., eds.), pp 307-319. Elsevier, Amsterdam.

17 Jansson, J-O., Albertsson-Wikland, K., Eden, S., Thorngren, K-G. and Isaksson, O.G.P. (1982). Effect of frequency of growth hormone administration on longitudinal bone growth and body weight in hypophysectomized rats. Acta physiologica Scandinavica **114,** 261-265.

18 Mathews, L.S., Hammer, R.E., Behringer, R.R., D'Ercole, A.J., Bell, G.I., Brinster, R.L. and Palmiter, R.D. (1988). Growth enhancement of transgenic mice expressing human insulin-like growth factor I. Endocrinology (in press).

19 Mathews, L.S., Hammer, R.E., Brinster, R.L. and Palmiter, R.D. (1988). Expression of insulin-like growth factor I in transgenic mice with elevated levels of growth hormone is correlated with growth. Endocrinology **123,** 433-437.

20 Orlowski, C.C. and Chernausek, S.D. (1988). Discordance of serum and tissue somatomedin levels in growth hormone-stimulated growth in the rat. Endocrinology **122,** 44-49.

21 Schoenle, E., Zapf, J., Humbel, R.E. and Froesch, E.R. (1982). Insulin-like growth factor I stimulates growth in hypophysectomized rats. Nature **296,** 252-253.

22 Skottner, A., Clark, R.G., Robinson, I.C.A.F. and Fryklund, L. (1987). Recombinant human insulin-like growth factor: testing the somatomedin hypothesis in hypophysectomized rats. Journal of Endocrinology **112,** 123-132.

23 Tannenbaum, G.S. and Martin, J.B. (1976). Evidence for an endogenous ultradian rhythm governing growth hormone secretion in the rat. Endocrinology **98,** 562-570.

24 Thorngren, K-G. and Hansson, L.I. (1977). Effects of administration frequency of growth hormone on longitudinal bone growth in the hypophysectomized rat. Acta Endocrinologica **84,** 497-511.

25 van Buul-Offers, S., and van den Brande, J.L. (1979). Effect of growth hormone and peptide fractions containing somatomedin activity on growth and cartilage metabolism of Snell dwarf mice. Acta Endocrinologica **92,** 242-257.

ACTION OF IGF-I ON MAMMARY FUNCTION

C.G. Prosser, I.R. Fleet and R.B. Heap

AFRC Institute of Animal Physiology and Genetics Research, Cambridge Research Station, Babraham, Cambridge CB2 4AT, U.K.

INTRODUCTION

The lactation cycle comprises an extensive period of growth and development in preparation for the secretion of milk (mammogenesis), the initiation (lactogenesis), and maintenance (galactopoiesis) of secretory activity and the cessation of milk secretion and regression of the mammary parenchyma (involution). Co-ordinating these events is a diverse set of hormonal stimuli, including steroid and polypeptide hormones and polypeptide growth factors. The purpose of this paper is to discuss recent evidence implicating a role for insulin-like growth factor I (IGF-I) in modulating mammary gland function. Since in many systems IGF-I and insulin will elicit the same biological responses it is relevant to begin by briefly reviewing the literature on insulin action in mammary function.

ROLE OF INSULIN IN MAMMARY FUNCTION

The results of in vitro experiments suggest insulin is an absolute requirement for mammary differentiation. As little as 0.2 nM is required for the induction of casein gene expression and casein production in mouse mammary explants (5, 35, 49). In the rat, although insulin has virtually no effect on the half-life of mRNA for the 25 kDa casein protein it is essential for its transcription (8). Insulin is also required for the maximal synthesis of casein in mammary explants from lactating cows (18) and α-lactalbumin secretion by mammary tissue from pregnant cows (19).

The requirement for insulin in mammary function in the intact lactating animal is less clear. During normal lactation in cows, milk yield and plasma insulin concentrations are inversely related (4, 25) although this may relate to different energy states of the animals. In goats, the close arterial infusion of insulin directly into the mammary gland causes a fall in milk yield (22, 27). This effect was presumably due to the depression of blood glucose, since when glucose was infused simultaneously milk yield was not altered (27). Mercer and Williamson (31) demonstrate clearly that insulin action in the mammary gland of the lactating rat is closely linked to the nutritional status of the animal and in particular the availability of carbohydrate. Thus these studies reflect the difficulty in discerning direct effects of insulin on the mammary gland in vivo without invoking secondary effects on whole body metabolism.

Studies of the induction of the glucose transport system in mouse mammary epithelial cells suggest that the mammary requirement for insulin is more important in the development/maturation phase than during established lactation. The glucose transport activity of mammary epithelial cells of 2 day lactating mice is 5-fold higher than that of cells from mid-pregnant mice (44). The basal rate of transport in the lactating cells is not acutely elevated by insulin and although the hormone does bring about an acute increase in cells from mid-pregnant mice this effect is small (40%) and probably not physiologically significant. Moreover when the basal rate of transport of cells from lactating animals is reduced 3-fold by fasting overnight, a 1 h exposure of the cells to insulin and 5 mM glucose only produces a 60% increase in activity (38). In contrast, the culture of mammary explants for three days with insulin, cortisol and prolactin elevates the basal rate of transport by the cells 400%, to the level of two day lactating mice (44). The data suggest insulin is more important for the development and maintenance of the glucose transport system of the mouse mammary epithelial cells than acute modulation of the system once lactation is established.

ROLE OF IGF-I IN MAMMARY FUNCTION IN VITRO

It was this proposed requirement for insulin in mammary gland development that led us to consider whether the related peptide, IGF-I, has a role in the regulation of mammary function. Insulin and IGF-I are structurally similar and elicit similar biological responses in a number of tissues (16, 36). The concentrations required to achieve these effects differ depending upon the tissue and the biological response concerned. For example, the specific biological activity of insulin is higher than IGF-I in the expression of the ovalbumin gene by the oviduct (15). However, IGF-I is much more potent than insulin in the induction of progestagen biosynthesis by granulosa cells (2).

With respect to the mammary gland an early report stated that a partially purified preparation of IGF-I from human serum failed to substitute for insulin in the accumulation of mouse casein mRNA (5). In a more recent study in which recombinantly-derived human IGF-I was used, the accumulation of β-casein mRNA in mouse mammary explants was induced 3-fold (43). The greater purity of the recombinantly-derived material presumably accounted for its greater biological activity in this system. However, the concentration required for half-maximal stimulation by IGF-I was 1.2 nM and only 0.2 nM for insulin.

Comparison of the concentration-activity profile for the induction of α-lactalbumin activity in the same system showed that the ED_{50} for IGF-I (5.9 nM) was approximately 20-fold higher than that of insulin (0.3 nM). In contrast, the induction of the glucose transport system in mouse mammary epithelial cells shared similar requirements for IGF-I and insulin (ED_{50} for insulin and IGF-I was 1.2 and 2.1 nM, respectively). In each case the maximum level induced by either insulin or IGF-I was the same. Neither epidermal growth factor (EGF) nor multiplication-stimulating activity (MSA), the rat equivalent of IGF-II, were capable of replacing insulin in the induction of these systems (5, 35, 44).

A cautionary note relates to the heterologous nature of these comparisons since hormones/growth factors from different species were tested on tissue from mice. As stated previously IGF-I was of human origin while insulin was extracted from porcine pancreas. It is possible that these compounds affect mouse mammary function differently compared with the homologous hormone/growth factor. Rat IGF activity for instance, is approximately twice as potent as human IGF-I in stimulating growth in hypophysectomized rats (46). Notwithstanding these limitations it is apparent that in the mammary gland IGF-I is an effective substitute for insulin in the induction of glucose transport activity and possibly casein synthesis, but not α-lactalbumin.

IGF-1 can also replace the supraphysiological doses of insulin that are required for the proliferation in vitro of mammary epithelial cells from mice (23), rats (14), sheep (51) and cows (47) whereas IGF-II was without significant effect. The doses of IGF-I required to produce the effects were very much lower than those of insulin.

IGF RECEPTORS

The production of a biological response by an agonist such as a hormone or growth factor requires the presence in target cells of specific receptors. Two distinct IGF receptors have been identified in a variety of tissues (36, 45). The type I IGF receptor preferentially binds IGF-I but can also bind insulin and IGF-II at high concentrations. The type II IGF receptor has a preference for IGF-II over IGF-I and does not recognize insulin. IGF-I can also bind to insulin receptors but the concentrations required are at least an order of magnitude above those of insulin. Comparison of the sequences for type I IGF and insulin receptor cDNAs reveal a close structural homology (50).

A more detailed account of the type I and type II receptors in the mammary gland is given by Collier and colleagues in this book so only a brief mention will be made here. Specific high affinity receptors for both IGF-I and IGF-II have been detected in ovine (13) and bovine mammary tissue (11). The type II receptors were relatively more abundant than type I receptors in tissue from both species.

While the role of the type I receptor in mediating the effects of IGF-I has been clarified that of the type II receptor in the mammary gland remains to be determined. Recent reports indicate the type II receptor is identical to the mannose-6-phosphate receptor, which targets lysosomal enzymes to the lysosome (33). The physiological significance of this dual specificity awaits further elucidation.

ROLE OF IGF-I IN MAMMARY FUNCTION IN VIVO

The consensus of the studies reviewed thus far indicate a role for IGF-I in mammary function in vitro. The presence of receptors specific for IGF-I in normally lactating mammary epithelium suggests that this factor also exerts a role in vivo. Further support for this may be derived from the ability of bovine growth hormone (bGH) to increase milk production in the ruminant. The subcutaneous injection of bGH to lactating cows causes a 10-40% increase in milk yield (3, 24, 37). The mechanism by which the galactopoietic effect of bGH is achieved is complex and involves the concerted action of a number of events in different tissues. A large part of the effect is due to the partitioning of essential nutrients toward the mammary gland (3, 24, 37), but alterations in the metabolism and secretory activity of the gland may also be involved.

Studies in vitro suggest bGH does not have a direct effect on mammary secretory function (17, 19, 48). In the intact animal, although the subcutaneous administration of bGH will increase milk yield its close arterial infusion into the mammary gland of the sheep and goat does not (29). The inability to detect significant specific binding of bGH by lactating mammary tissue implies the lack of a direct effect of bGH may be due to the absence of functional receptors for bGH in mammary tissue (1).

As has been suggested for other tissues GH action in the mammary gland may be mediated by IGF-I (35). In keeping with this concept the administration of bGH to lactating cows increases plasma concentrations of IGF-I 3-4 fold (10, 39). Blood flow to the mammary gland is also increased during bGH treatment (19, 39). Estimates of the amount of IGF-I that reaches the mammary gland of cows on the day before compared to the seventh day of bGH treatment are 24 and 116 nmol/min/udder half, respectively. A similar, though not as dramatic, increase in circulating IGF-I (42) and mammary blood flow (30) is also observed in the goat during bGH treatment.

The difficulty in interpreting such data is that greater than 95% of circulating IGF-I is bound to specific binding proteins. The major species has a molecular weight of 150 kDa while some IGF-I is associated with a 50 kDa protein (3, 39, 52). During GH treatment of lactating cows the 150 kDa binding protein species increases in parallel to those of IGF-I (39). The association of the IGF-I with these proteins restricts its movement out of the vascular space and attenuates its biological activity (32, 52).

IGF-I may not act exclusively as an endocrine factor. IGF-I is synthesized in a variety of cultured tissue slices and cells (36) and there are also reports of the GH-dependent production of mRNA for IGF-I in some tissues (28). Thus IGF-I may well function as an autocrine and/or paracrine factor and plasma levels may simply reflect its production by different tissues.

We have found that levels of immunoreactive IGF-I in mammary tissue of lactating goats treated with pituitary derived bGH for 4 days are 3-fold higher than those of untreated animals (Fig. 1). The tissue was taken 24-30 hours after the last injection of bGH when

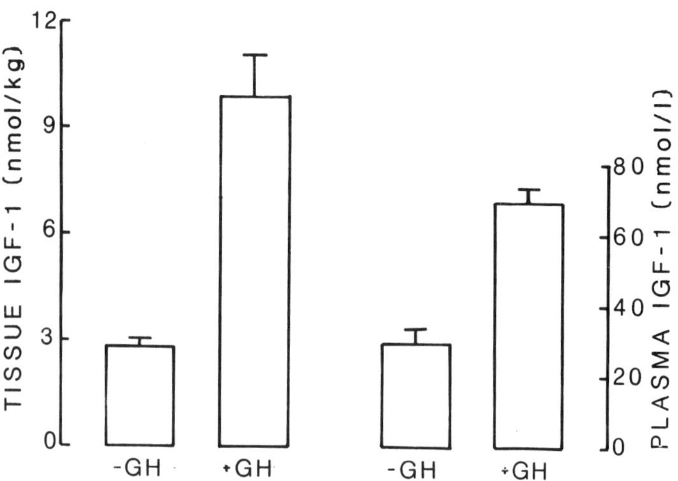

Fig. 1 Four lactating goats were given pituitary-derived bGH for four days (0.2 mg/kg/day; s.c.). Milk yield increased 15 ± 4% over this time. Mammary tissue was obtained by biopsy 7 days before treatment began and by autopsy 24 h after the last injection of bGH. IGF-I was measured by radioimmunoassay in acetic acid extracts of the tissue and that derived from blood trapped in the tissue was subtracted from the total (12).

---

circulating bGH had returned to basal values but IGF-I was still elevated 2-fold. Milk yield was increased an average of 15 ± 4% in treated animals.

Whether the tissue IGF-I was derived from the circulation or local synthesis is not known. Minor amounts of mRNA for IGF-I have been detected in the normally lactating mammary gland of the rat (34) and there is evidence for the synthesis of immunoreactive IGF-I by lactating bovine mammary explants in culture (7). However, there is as yet no evidence for increased mammary production during bGH treatment.

The galactopoietic effect of bGH is also accompanied by an increased secretion of IGF-I into milk. In the cow the changes in milk concentrations of IGF-I slightly preceded any changes in milk secretion rate (39). In goats, while four animals exhibited an increase in milk yield in response to bGH, one animal did not (42). Concentrations of IGF-I in the milk of this latter animal were not elevated in contrast to the other four. Plasma levels of IGF-I were 2-3 fold higher in all five animals. Presumably the greater secretion of IGF-I into milk of goats responding to bGH reflected the elevated tissue levels of IGF-I.

The significance of the increased tissue levels of IGF-I and augmented secretion of IGF-I into milk may be that they reflect the ability of IGF-I to regulate mammary gland function. In any case it would seem that the mammary epithelium is exposed to a greater amount of IGF-I during bGH treatment.

In an attempt to establish the origin of the IGF-I in milk, [^{125}I]IGF-I was infused directly into the gland of the goat via the pudic artery supplying that gland (40). Not only did [^{125}I] material appear in milk but approximately 23% of the total amount of label transferred was TCA precipitable, presumably reflecting intact [^{125}I]IGF-I. This intact material in milk co-eluted with authentic IGF-I on Sephacryl S-200. Comparison of the time course of appearance of [^{125}I]IGF-I into milk with that of EGF and oestrone sulphate (Table 1) suggests they probably share a similar pathway of secretion, that is via the golgi secretory vesicles within the mammary epithelial cells (20). The results suggest that at least some of the IGF-I secreted in milk may be derived from blood but further studies are required to establish what proportion of the total amount secreted this represents.

Table 1 Time of transfer of compounds from plasma into milk of goats

Compound	Transfer time (min)	Ref.
EGF	120	6
Oestrone sulphate	110	21
IGF-I	110	
Amino acids* (as milk protein)	170	21

*includes time required for synthesis of proteins within gland.

^{125}I-labelled EGF and IGF-I and ^{14}C-labelled oestrone sulphate and mixed amino acids were infused close-arterially for 1 h into the mammary gland of lactating goats. Milk was removed at frequent intervals with the aid of 200 mU of oxytocin and the time from the beginning of the infusion until maximum concentrations in milk were achieved was measured.

While the studies described above provide a summary of the interactions between bGH and IGF-I in the lactating ruminant, definitive proof that IGF-I modulates mammary function in vivo is lacking. More direct evidence for this effect may be obtained from studies of the surgically-prepared lactating goat. In these animals the blood vessels between the left and right glands were sectioned between double ligatures and a catheter was inserted into the pudic artery supplying one gland (41). Free, unbound hIGF-I (Ciba-Geigy, Basle) was infused directly into the gland which would be expected to increase local tissue concentrations. The opposite non-infused gland was exposed to infused IGF-I only after re-circulation. During this time most of the IGF-I would become complexed to high affinity binding proteins thereby restricting its movement out of the vascular space into the mammary tissue.

IGF-I was infused at a rate of 1.1 nmol/min for 6 h and this resulted in an increase in the rate of milk secretion by the infused gland of 30 $\pm$ 5% compared with that in the non-infused gland of 15 $\pm$ 4% ($P < 0.05$; Fig. 2). Additionally, concentrations of IGF-I in milk of the infused gland were increased 82 $\pm$ 22% compared with only 14 $\pm$ 9% ($P < 0.01$) in the milk of the non-infused gland. The small increase in the rate of milk secretion in both glands during saline infusion was not statistically significant and was due to the frequent milking regime and repeated oxytocin injections (26).

The increase in milk secretion during infusion of IGF-I was quite rapid, the first significant rise above that of the non-infused gland occurring 4 h after the beginning of the infusion. This implies that IGF-I stimulated the activity of pre-existing cells rather than cellular proliferation.

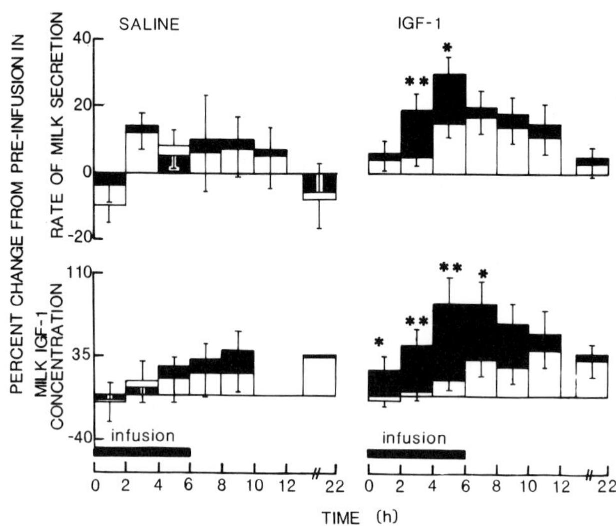

Fig. 2 Milk secretion rate (ml/2 h) and milk IGF-I concentration (nmol/l) of the infused (shaded) and non-infused (open) gland of three goats expressed as a percentage of the average values obtained over three days before the experiment. Both glands were milked out with the aid of 200 mU of oxytocin and results are the mean (± SEM) of one experiment on each of three goats for the saline infusion and two experiments on each goat for the IGF-I infusion. $^*P < 0.05$, $^{**}P < 0.01$; infused compared to non-infused glands. Adapted from Ref. 41.

The unilateral response suggests that IGF-I acted directly on the mammary epithelium to enhance milk synthesis.

Circulating levels of IGF-I increased from 32.4 ± 0.8 nmol/l before, to a maximum of 53.6 ± 1.9 nmol/l 4 h after infusion commenced. The levels remained high for at least 4 h after the infusion was stopped. Approximately 5% and 14% of the circulating IGF-I was present in the free form 2 and 4 h after the start of infusion, respectively. Thus the initial lack of response in the non-infused gland in terms of milk yield and IGF-I concentrations presumably reflects the reduced transfer and efficacy of re-circulated protein-bound IGF-1. This contrasts to the effect of the non-protein bound IGF-I in the infusate. Based on an infusion rate of 1.1 nmol/min and mammary blood flow of 378-543 ml/min, the infused gland would be exposed to an additional amount of free IGF-I at a concentration of 2-3 nmol/l.

It is interesting to note that blood flow through the infused gland was increased 44% during the infusion of IGF-I but not during saline infusion. As stated earlier mammary blood flow is increased during treatment of lactating cows (19,39) and goats (30) with bGH. Thus the possibility is raised that IGF-I is an important mediator of GH regulation of mammary blood flow. The mechanism of this effect may involve direct stimulation of the vascular system by IGF-I or an indirect stimulation via the mammary release of local vasodilatory compounds subsequent to the enhancement of mammary metabolic activity.

CONCLUDING REMARKS

The results of the studies described in this brief review support the idea of a role for IGF-I in the modulation of mammary gland development and function. While we have dealt with IGF-I as an individual mediator of specific events within the gland it is important to remember that their full expression requires the co-ordinated interaction with other hormonal stimuli. Some of the effects of IGF-I are similar to the actions of insulin, but these compounds have several distinct functions in the mammary gland as well. Notably, IGF-I appears to possess proliferative and differentiative properties whereas insulin appears to be mostly necessary for the maturation of the secretory cell.

Observation of the effects of IGF-I on mammary function in the intact animal should offer important insights into mechanisms of action of IGF-I in vivo and the influence of plasma binding proteins on its actions. Where previously the in vivo effects of IGF-I focused on the investigation of animals after endocrine imbalance or ablation (eg hypophysectomy), studies of mammary function outlined above utilize an experimental approach where no endocrine intervention is imposed.

ACKNOWLEDGEMENTS

We thank Mrs J. Hood for preparation of the manuscript and Mr A.J. Davis for technical assistance. The work was supported in part by the Joint Research Board of St. Bartholomew's Hospital, London and an AFRC Link Grant.

REFERENCES

1 Akers, R.M. (1985). Latogenic hormones: binding sites, mammary growth, secretory cell differentiation and milk biosynthesis in ruminants. Journal of Dairy Science **68**, 501-509.

2 Barano, J.L.S. and Hammond, J.M. (1984). Comparative effects of insulin and insulin-like growth factors on DNA synthesis and differentiation of porcine granulosa cells. Biochemical and Biophysical Research Communications **124**, 484-490.

3 Bauman, D.E. and Elliot, J.M. (1983). Control of nutrient partitioning in lactating ruminants. In Biochemistry of Lactation (Mepham, T.B., ed.), pp 437-468. Elsevier, Amsterdam.

4 Beck, N.F.G. and Tucker, H.A. (1978). Mammary arterial and venous concentrations of serum insulin in lactating dairy cows. Proceedings of the Society for Experimental Biology and Medicine **159**, 394-396.

5 Bolander, F.F. Jr., Nicholas, K.R., Van Wyk, J.J. and Topper, Y.J. (1981). Insulin is essential for accumulation of casein mRNA in mouse mammary epithelial cells. Proceedings of the National Academy of Science, U.S.A. **78**, 5682-5684.

6 Brown, K.D., Blakeley, D.M., Fleet, I.R., Hamon, M. and Heap, R.B. (1986). Kinetics of transfer of ^{125}I-labelled epidermal growth factor from blood into mammary secretions of goats. Journal of Endocrinology **109**, 325-332.

7 Campbell, P.G. and Baumrucker, C.R. (1988). Secretion of immuno-reactive insulin-like growth factor I and its binding protein from the bovine mammary

gland in vitro. Proceedings of the Endocrine Society 70th Annual Meeting, June 8-11, New Orleans, U.S.A. Abstract 510.

8 Chomczynski, P., Qasba, P. and Topper, Y.J. (1984). Essential role of insulin in transcription of the rat 25,000 molecular weight casein gene. Science **226,** 1326-1328.

9 Davis, S.R., Collier, R.J., McNamara, J.P., Head, H.M. and Sussman, W. (1988). Effects of thyroxine and growth hormone treatment of dairy cows on milk yield, cardiac output and mammary blood flow. Journal of Animal Science **66,** 70-79.

10 Davis, S.R., Gluckman, P.D., Hart, I.C. and Henderson, H.V. (1987). Effects of injecting growth hormone or thyroxine on milk production and blood plasma concentrations of insulin-like growth factors I and II in dairy cows. Journal of Endocrinology **114,** 17-24.

11 Dehoff, M.H., Elgin, R.G., Collier, R.J. and Clemmons, D.R. (1988). Both type I and II insulin-like growth factor receptor binding increase during lactogenesis in bovine mammary tissue. Endocrinology **122,** 2412-2417.

12 D'Ercole, A.J., Stiles, A.D. and Underwood, L.E. (1984). Tissue concentrations of somatomedin C: Further evidence for multiple sites of synthesis and paracrine or autocrine mechanisms of action. Proceedings of the National Academy of Sciences, U.S.A. **81,** 935-939.

13 Disenhaus, C., Belair, L. and Djiane, J. (1988). Characterization and physiological variations of IGF I and II receptors in the ewe mammary gland. Reproduction, Nutrition, Développement **28,** 241-252.

14 Ethier, S.P., Kudla, A. and Cundiff, K.C. (1987). Influence of hormone and growth factor interactions on the proliferative potential of normal rat mammary epithelial cells in vitro. Journal of Cellular Physiology **132,** 161-167.

15 Evans, M.I., Hager, L.J. and McKnight, C.S. (1981). A somatomedin-like peptide hormone is required during the oestrogen-mediated induction of ovalbumin gene transcription. Cell **24,** 187-191.

16 Froesch, E.R., Schmid, Chr., Schwander, J. and Zapf, J. (1985). Actions of insulin-like growth factors. Annual Review of Physiology **47,** 443-467.

17 Gertler, A., Cohen, N. and Maoz, A. (1983). Human growth hormone but not ovine or bovine growth hormone exhibit galactopoietic prolactin-like activity in organ culture from bovine lactating mammary gland. Molecular and Cellular Endocrinology **33,** 169-182.

18 Gertler, A., Weil, A. and Cohen, N. (1982). Hormonal control of casein synthesis in organ culture of the bovine mammary gland. Journal of Dairy Research **49,** 387-398.

19 Goodman, G.T., Akers, R.M., Friderici, K.H. and Tucker, H.A. (1983). Hormonal regulation of α-lactalbumin secretion from bovine mammary tissue cultured in vitro. Endocrinology **112,** 1324-1330.

20 Heap, R.B., Fleet, I.R., Hamon, M., Brown, K.D., Stanley, C.J. and Webb, A.E. (1986). Mechanisms of transfer of steroid hormones and growth factors in milk. Endocrinologica Experimentalis **20,** 101-118.

21 Heap, R.B., Hamon, M. and Fleet, I.R. (1984). Transport of oestrone sulphate by the mammary gland in the goat. Journal of Endocrinology **101,** 221-230.

22 Hove, K. (1978) Effects of hyperinsulinemia on lactose secretion and glucose uptake by the goat mammary gland. Acta Physiologica Scandinavica **104,** 422-430.

23 Imagawa, W., Spencer, E.M., Larson, L. and Nandi, S. (1986). Somatomedin-c substitutes for insulin for the growth of mammary epithelial cells from normal virgin mice in serum-free collagen gel cell culture. Endocrinology **119,** 2695-2699.

24 Johnsson, I.D. and Hart, I.C. (1986). Manipulation of milk yield with growth hormone. In Recent Advances in Animal Nutrition (Haresign, W. and Cole, D.J.A., eds), pp 105-123. Butterworths, London.

25 Koprowski, J.A. and Tucker, H.A. (1973). Bovine serum growth hormone, corticoids and insulin during lactation. Endocrinology **93,** 645-651.

26 Linzell, J.L. (1967). The effect of very frequent milking and of oxytocin on the yield and composition of milk in fed and fasted goats. Journal of Physiology **190,** 333-346.

27 Linzell, J.L. (1967). The effect of infusions of glucose, acetate and amino acids on hourly milk yield in fed, fasted and insulin-treated goats. Journal of Physiology **190,** 437-457.

28 Mathews, L.S., Norstedt, G. and Palmiter, R.D. (1986). Regulation of insulin-like growth factor I gene expression by growth hormone. Proceedings of the National Academy of Sciences, U.S.A. **83,** 9343-9347.

29 McDowell, G.H., Hart, I.C. and Kirby, A.C. (1987). Local intra-arterial infusion of growth hormone into the mammary glands of sheep and goats: effects on milk yield and composition, plasma hormones and metabolites. Australian Journal of Biological Science **40,** 181-189.

30 Mepham, T.B., Lawrence, S.E., Peters, A.R. and Hart, I.C. (1984). Effects of growth hormone on mammary function in lactating goats. Hormone and Metabolic Research **16,** 148-253.

31 Mercer, S.W. and Williamson, D.H. (1986). Time course of changes in plasma glucose and insulin concentrations and mammary-gland lipogenesis during re-feeding of starved conscious lactating rats. Biochemical Journal **239,** 489-492.

32 Meulig, C., Zapf, J. and Froesch, E.R. (1978). NSILA-carrier protein abolishes the action of non-suppressible insulin-like activity (NSILA-S) on perfused rat heart. Diabetologia **14,** 253-259.

33 Morgan, D.O., Edman, J.C., Standring, D.N., Fried, V.A., Smith, M.C., Roth, R.A. and Rutter, W.J. (1987). Insulin-like growth factor II receptor as a multifunctional binding protein. Nature **329,** 301-307.

34 Murphy, L.J., Bell, G.I. and Friesen, H.G. (1987). Tissue distribution of insulin-like growth factor I and II messenger ribonucleic acid in the adult rat. Endocrinology **120,** 1279-1282.

35 Nicholas, K.R., Sankaran, L. and Topper, Y.J. (1983). A unique and essential role for insulin in the phenotypic expression of rat mammary epithelial cells unrelated to its function in cell maintenance. Biochemical et Biophysica Acta **763,** 309-314.

36 Nissley, S.P. and Rechler, M. (1985). Insulin-like growth factors: Biosynthesis, receptors and carrier proteins. Hormonal Proteins and Peptides **12**, 128-203.

37 Peel, C.J. and Bauman, D.E. (1987). Somatotrophin and lactation. Journal of Dairy Science **70,** 474-486.

38 Prosser, C.G. (1988). Mechanism of the decrease in hexose transport by mouse mammary epithelial cells caused by fasting. Biochemical Journal **249,** 149-154.

39 Prosser, C.G., Fleet, I.R. and Corps, A.N. (1988). Increased secretion of insulin-like growth factor I into milk of cows treated with recombinantly derived bovine growth hormone. Journal of Dairy Research (in press).

40 Prosser, C.G., Davis, A.J., Fleet, I.R., Rees, L.H. and Heap, R.B. (1987). Mechanism of transfer of IGF-I into milk. Journal of Endocrinology **115** (Suppl.), 91.

41 Prosser, C.G., Fleet, I.R., Corps, A.N., Heap, R.B. and Froesch, E.R. (1988). Increased milk secretion and mammary blood flow during close-arterial infusion of insulin-like growth factor I (IGF-I) into the mammary gland of the goat. Journal of Endocrinology **117** (Suppl.), 248.

42 Prosser, C.G., Fleet, I.R., Hart, I.C. and Heap, R.B. (1987). Changes in concentrations of insulin-like growth factor I (IGF-I) in milk during bovine growth hormone treatment in the goat. Journal of Endocrinology **112** (Suppl.), 65.

43 Prosser, C.C., Sankaran, L., Hennighausen, L. and Topper, Y.J. (1987). Comparison of the roles of insulin and insulin-like growth factor I in casein gene expression and in the development of α-lactalbumin and glucose transport activities in the mouse mammary epithelial cell. Endocrinology **120,** 1411-1416.

44 Prosser, C.G. and Topper, Y.J. (1986) Changes in the rate of carrier-mediated glucose transport by mouse mammary epithelial cells during ontogeny: Hormone dependence delineated in vitro. Endocrinology **119,** 91-96.

45 Rechler, M.M. and Nissley, S.P. (1985). The nature and regulation of the receptors for insulin-like growth factors. Annual Review of Physiology **47,** 425-442.

46 Schoenle, E., Zapf, J., Hauri, Ch., Steiner, Th. and Froesch, E.R. (1985). Comparison of in vivo effects of insulin-like growth factors I and II and of growth hormone in hypophysectomized rats. Acta Endocrinologica Scandinavica **108,** 167-174.

47 Shamay, A., Cohen, N., Niwa, M. and Gertler, A. (1988). Effect of insulin-like growth factor I on deoxyribonucleic acid synthesis and galactopoiesis in bovine undifferentiated and lactating mammary tissue in vitro. Endocrinology **123,** 804-809.

48 Skarda, J., Urbanova, E., Becka, S., Houdebine, L.M., Delouis, C., Pichova, D., Picha, J. and Pilek, J. (1982). Effect of bovine growth hormone on development of goat mammary tissue in organ culture. Endocrinologia Experimentalis **16,** 19-31.

49 Topper, Y.J., Nicholas, K.R., Sankaran, L. and Kulski, J.K. (1984). Insulin biology from the perspective of studies on mammary gland development. In

Biochemical Actions of Hormones (Litwack, G., ed.), Vol 11, pp 163-192. Academic Press, New York.

50 Ullrich, A., Gray, A., Tam, A.W., Yang-Feng, T., Tsubokawa, M., Collins, C., Henzel, W., Le Bon, T., Kathuria, S., Chen, E., Jacobs, S., Francke, U., Ramachandran, J. and Fujita-Yamaguchi, Y. (1986). Insulin-like growth factor I receptor primary structure: Comparison with insulin receptor suggests structural determinants that define functional specificity. EMBO Journal **5**, 2503-2512.

51 Winder, S.J. and Forsyth, I.A. (1986). Insulin-like growth factor I (IGF-I) is a potent mitogen for ovine mammary epithelial cells. Journal of Endocrinology **108** (Suppl.), 141.

52 Zapf, J., Hauri, Ch., Waldvogel, M. and Froesch, E.R. (1986). Acute metabolic effects and half-lives of intravenously administered insulin-like growth factors I and II in normal and hypophysectomized rats. Journal of Clinical Investigation **77**, 1768-1775.

CHANGES IN INSULIN AND SOMATOMEDIN RECEPTORS AND UPTAKE OF INSULIN, IGF-I AND IGF-II DURING MAMMARY GROWTH, LACTOGENESIS AND LACTATION

R.J. Collier, S. Ganguli, P.T. Menke, F.C. Buonomo, M.F. McGrath, C.E. Kotts and G.G. Krivi

Monsanto Company, 700 Chesterfield Village Parkway, Chesterfield, Missouri 63198, U.S.A.

Receptors for insulin, IGF-I and IGF-II are present on growing, differentiating and lactating mammary epithelial cells. Scatchard analysis of these receptors indicate the presence of two receptor sites for insulin. In contrast, the type I and type II somatomedin receptors are composed of a single site in all physiological states examined. The type I somatomedin receptor cross-reacted with both insulin and IGF-II while the type II receptor only recognized IGF-II. Although the truncated form of IGF-I has a lower affinity for the type I receptor relative to native IGF-I it is more potent in stimulating mammary growth in vitro. This may be due to mediating effects of binding proteins. Mammary growth appears to be related to higher population of high affinity receptors for IGF-I in mammary tissue during pregnancy. Role of insulin, IGF-I and IGF-II during lactogenesis and galactopoiesis remains to be established.

BACKGROUND

The role of the insulin family in the regulation of mammary metabolism has always been paradoxical. Although it is believed that there is an insulin requirement for growth and differentiation of mammary tissue (7, 23, 32), the insulin requirement in lactating mammary tissue is much less clear (18, 22, 23). In ruminants there is a negative relationship between insulin concentrations during lactation and the level of milk production (18, 21). This is in spite of the fact that up to 60% of glucose turnover is taking place at the mammary gland (5). Although there is no apparent insulin requirement for glucose uptake by lactating mammary tissue there is a sizeable gradient of glucose concentration across the lactating mammary epithelial membrane (14) and it has been postulated that the rate of glucose transport into the cell may be the rate-limiting step of milk synthesis (10). Exogenous administration of insulin in lactating animals results in a decline in milk production as glucose is driven into peripheral tissue (18).

Possible alternate members of the insulin family that could be involved in the regulation of metabolism of mammary tissue are the somatomedins. Recently, it has been shown that there are both type I and type II receptors in bovine mammary tissue (12). Additionally, binding of IGF-I and IGF-II by mammary tissue of other species has been demonstrated (2, 13). Finally, both IGF-I and IGF-II are present in mammary secretions (9, 16, 23) implicating them in the regulation of mammary metabolism.

INSULIN

Insulin receptors are present in mammary tissue (6, 20) and composed of high and low affinity sites (20). In mouse mammary tissue, the high affinity sites appear to peak at mid-pregnancy (20), while the low affinity sites predominate during lactation (20). This was examined in our laboratory using mammary microsomal membranes prepared from mature female rats at various stages of gestation and lactation, Table 1.

Significant differences were observed in the binding capacity for the insulin receptors at various times of gestation and lactation. These were most noticeable when comparing virgin animals to days 7 and 14 of pregnancy and 7 and 14 of lactation. These results are shown in Table 1 and demonstrate the presence of high and low affinity sites during pregnancy and lactation in the rat. Low affinity sites were the dominant receptor population at both 7 days of pregnancy and 7 days of lactation. In addition, in this data set there appeared to be an increase in the high affinity receptor population during lactation. Also, at 7 days of pregnancy there was a greater population of low affinity sites than were detected at 7 days of lactation. Species differences in population of low and high affinity insulin receptor sites during various physiological states are likely.

Table 1 Effect of physiological state on numbers and affinities of binding sites for insulin, IGF-I and IGF-II

	$K_1 (M^{-1})$	$K_2 (M^{-1})$	R_1 fmoles/mg protein	R_2 fmoles/mg protein	R_1/R_2
		Insulin			
Virgin	5.63×10^9	2.58×10^8	14.4	8.0	1.8
7 day pregnant	3.3×10^9	2.1×10^7	10.9	518.0	0.04
7 day post gestation	3.5×10^9	1.4×10^8	21.8	98.0	0.22
		IGF-I			
Virgin	3.4×10^9	-	25.0	-	-
7 day pregnant	8.9×10^8	-	93.0	-	-
7 day post gestation	2.4×10^9	-	18.9	-	-
		IGF-II			
Virgin	3.8×10^{11}	-	0.35	-	-
7 day pregnant	3.7×10^8	-	556.0	-	-
7 day lactating	2.8×10^8	-	540.0	-	-

Mammary microsomal membranes were prepared from rats (n = 7) at different stages of gestation and lactation by differential centrifugation of mammary tissue homogenate in 0.25 M sucrose/0.025 M Tris buffer. Radiolabelled insulin, IGF-I and IGF-II were bound to mammary membranes by incubating radiolabelled ligand and 250 μg of membrane in the presence of various levels of cold ligand (0-100 ng/tube) in a total volume of 450 μl of 25 mM Tris buffer with 0.1% BSA. Incubation was carried out for 18 hours at 4°C. The membranes were separated by centrifugation and counted in a gamma-counter. Binding data were analyzed by Scatchard analysis.

In Table 2, there is a comparison of insulin receptor sites in mammary tissue of the rat and dairy cow. Two receptor populations were also detected in mammary tissue of the dairy cow and these receptor populations were of relatively equivalent size. The existence of two classes of insulin receptors in mammary tissue was first reported by Inagaki and Komoto (20) using mammary tissue from the mouse. They indicated that changes in the population of high affinity receptors parallel changes in DNA synthetic

Table 2 Comparison of binding sites and affinities for insulin, IGF-I and IGF-II in lactating mammary tissue from the cow and rat

	Rat			Bovine		
	Insulin	IGF-I	IGF-II	Insulin	IGF-I	IGF-II
KI (M-1)	$3.5x10^9$	$2.4x10^9$	$2.8x10^8$	$1.8x10^9$	$2.23x10^9$	$2.1x10^9$
K2 (M-1)	$14x10^8$	0	0	$2.5x10^7$	0	0
RI f mol	21.8	18.9	540	26.2	28.1	22.8
R2 f mol	98	0	0	29.5	0	0

(M-1) = litres/mol
f mol = femtomol

Mammary tissue was prepared as in methods Table 1. Bovine mammary tissue was obtained from a mid-lactation cow.

---

rate in this species. In early lactation, they reported a large increase in the low affinity receptor population and a smaller increase in the high affinity receptor population.

Collectively, these results suggest that the relative population of high and low affinity sites for insulin binding is associated with changes in the metabolic state of the mammary gland. In vitro studies suggest that IGF-I is more potent than insulin in stimulating mammary growth (3, 19). However, insulin is more potent than IGF-I in lactogenesis (28). Several studies have reported low or undetectable insulin requirements by mammary tissue during lactation (6, 22). In cattle, mammary uptake of insulin and concentration in mammary secretions is high prepartum and declines with the onset of lactation (6, 23). Prepartum milking of cows to induce premature lactogenesis is associated with a decline in insulin concentration in prepartum secretions (23).

These results also indicate that mammary tissue of ruminants and non-ruminants contains two populations of insulin receptors. Either or both of these receptor populations may display negative cooperativity (20). However, the relationship between these receptor populations and the low insulin requirement by mammary tissue during lactation (18, 22) and negative relationship between circulating insulin and milk yield (21) remains unresolved. The large gradient of glucose concentration across the mammary epithelial cell suggests that glucose uptake by lactating mammary gland requires a transport system that must be under endocrine regulation. In agreement with this hypothesis is the finding that the number of glucose transporters in mammary tissue is influenced by the hormonal environment in vitro (27).

INSULIN-LIKE GROWTH FACTOR I

Somatomedins are present in high concentration in mammary secretions during pregnancy and in colostrum (9, 15, 16, 23). They are also present in milk but at much lower concentrations (8, 33). The decline in concentration of milk insulin, IGF-I and IGF-II during the early postpartum period is not due to a dilution effect as milk yield increases. As shown in Fig. 1, the total output of IGF-I into bovine milk declines

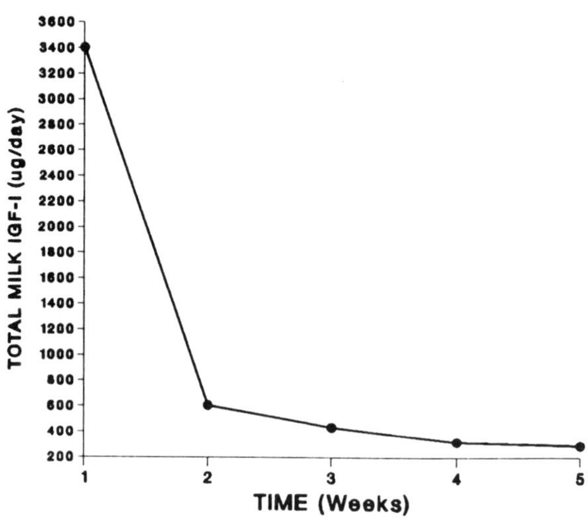

Fig. 1 IGF-I concentrations in weekly milk samples obtained from 10 first-lactation Holstein heifers.

IGF-I concentrations were determined using a radioimmunoassay. Total secreted IGF-I was estimated from the concentration in each sample multiplied by the average weekly milk production of the sample period.

---

across the first five weeks of lactation. This may be associated with a decline in receptor numbers (13) although initial reports indicated an increase in receptors for IGF-I and IGF-II in the immediate postpartum period (12). In dairy cows, the numbers of receptors for IGF-II are higher than those for IGF-I in late pregnancy and early lactation and this difference in receptor population was reflected in the concentration of these hormones in milk (23). However, as shown in Table 2 in mid-lactation, mammary tissue of a cow contained equivalent populations of type I and II somatomedin receptors. This was also reflected in milk concentrations of these hormones (33).

In both the rat and cow (Tables 1 and 2) it is apparent that there appears to be a single population of type I somatomedin receptors in contrast to the two insulin receptor populations. In rat mammary tissue the Type I receptor population was highest during pregnancy and may be associated with the mammary growth that is occurring during this period.

Recently, a truncated form of IGF-I was isolated from colostrum where it occurs in high concentration (1). This modified IGF-I, lacking the N-terminal tripeptide Gly-Pro-Glu, possesses enhanced biological activity compared with native IGF-I. We evaluated a recombinant form of this variant against bovine IGF-II and IGF-I in inducing DNA synthesis in bovine mammary tissue in vitro. As shown in Fig. 2, the -3N-IGF-I is approximately 4-fold more potent than native IGF-I in stimulating DNA synthesis in bovine mammary tissue in collagen gels. This difference in potency is not due to a greater affinity of the IGF-I receptor for the -3N-IGF-I.

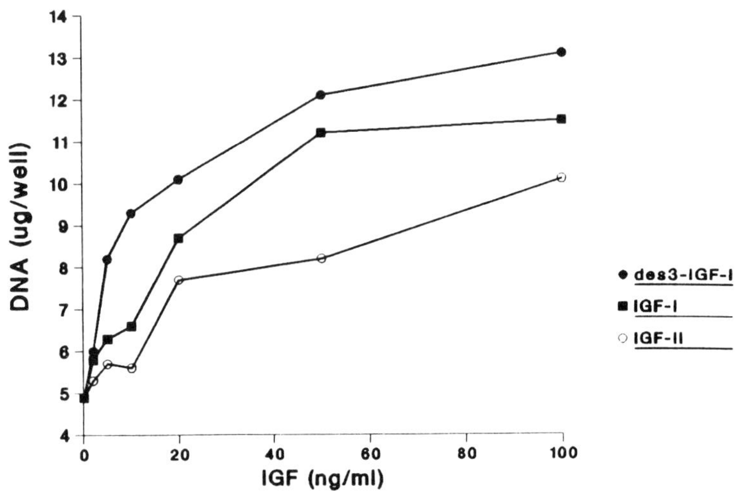

Fig. 2 Induction of DNA synthesis in mammary epithelial cells by IGF-I, IGF-II and a truncated form of IGF-I.

Epithelial cells were isolated from mammary tissue of pregnant, non-lactating heifers and cultured for 10 days within a collagen matrix. The basal medium (DME/F-12 + 3% fetal bovine serum) was supplemented with various concentrations of either IGF-I, IGF-II, or DES-3 IGF-I. After the incubation period, the cultures were terminated and total DNA was determined.

Using bovine mammary tissue as shown in Fig. 3, there was no difference between native and truncated forms of IGF-I for the IGF-I receptor. Low molecular weight binding proteins are secreted by bovine mammary tissue in culture (D. Clemmons personal communication). These low molecular weight binding proteins may require the N-terminal tripeptide missing in the truncated form to bind IGF-I as has been recently shown for binding proteins of similar molecular weight produced by bovine kidney (MDBK) cells in culture (31). Although not yet demonstrated, such a difference in ability of binding proteins to recognize the truncated molecule may result in greater availability of this molecule for the receptor on the mammary epithelial cell and hence its greater biological activity.

As reported earlier the uptake of IGF-I by the mammary gland appears to decline with onset of lactation as measured by falling concentrations in milk and reduced total IGF-I in milk (23, 33, Fig. 1). The high concentrations of IGF-I in mammary secretions during pregnancy and their potency in inducing cell division in mammary epithelial cells in vitro suggest that this somatomedin is involved in regulation of mammary growth. Although the population of type II receptors is greater than type I receptor population in mammary tissue during pregnancy, the potency of IGF-II in inducing cell divisions is much lower than IGF-I (Fig. 2). Furthermore, the addition of IGF-I and IGF-II to culture media does not produce additive effects on bovine mammary tissue growth in vitro. Likewise, the addition of insulin to cultures containing IGF-I does not result in additive growth (M. McGrath, personal communication). Epidermal growth factor (EGF)

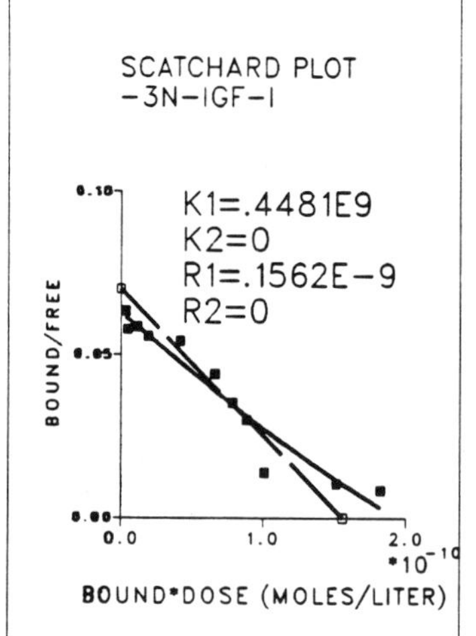

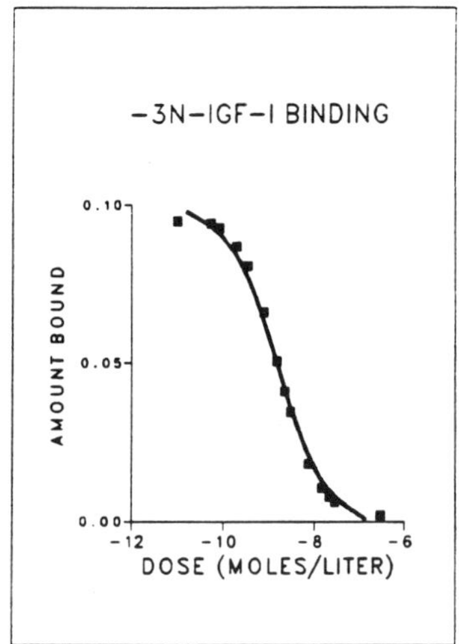

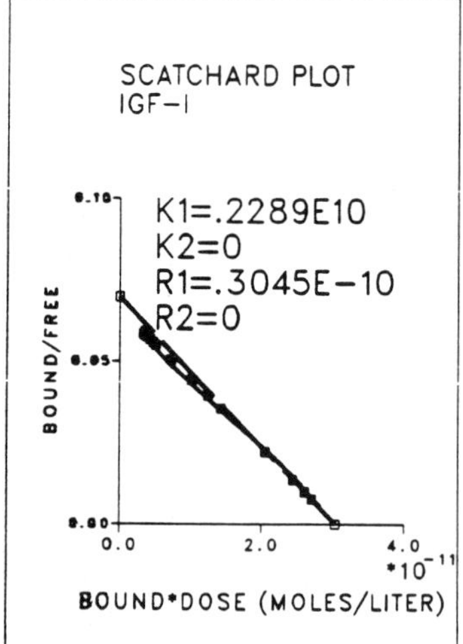

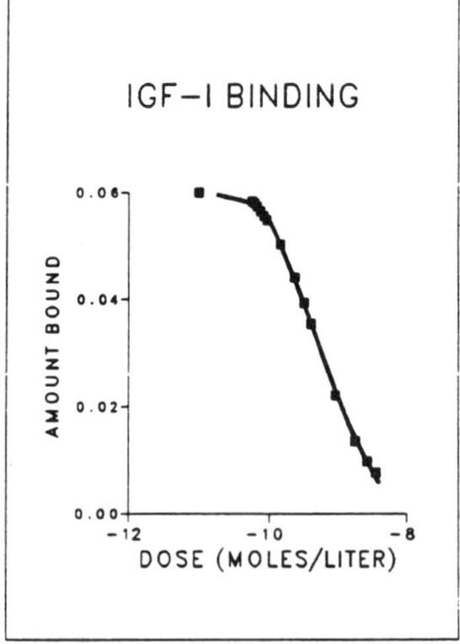

Fig. 3 Binding of IGF-I and -3N-IGF-I by mammary microsomal membrane.

Microsomal membranes were prepared from a homogenate of mammary tissue from a lactating cow by differential centrifugation in 0.3 M sucrose/0.025 M Tris buffer. Radiolabelled IGF-I and -3N-IGF-I were bound to mammary membrane by incubating radiolabelled ligand and 480 μg of membrane in the presence of various levels of unlabelled ligand (0-100 ng/tube) in a total volume of 450 μl of 25 mM Tris buffer with 0.1% BSA. Incubation was carried out for 18 hours at 4°C. The membranes were separated by centrifugation and counted in a gamma-counter. The binding data and the displacement curves were analyzed by Scatchard analysis. Data are representative examples of analyses from tissue obtained from several animals.

K1 = Affinity constant (high affinity sites)
K2 = Affinity constant (low affinity sites)
R1 = High affinity receptors
R2 = Low affinity receptors

also induces cell division in mammary tissue in vitro and the population of EGF receptors is higher in bovine mammary tissue during pregnancy than during lactation (30). Addition of these two growth factors to culture media containing bovine mammary tissue results in synergistic growth responses (25). Exogenous administration of these two growth factors by intramammary infusion during late pregnancy in ewes resulted in increased mammary growth (25).

The role of IGF-I in regulation of mammary metabolism has only recently received attention due to increased availability of recombinant forms of this molecule. The population of IGF-I receptors during established lactation appears to be lower when expressed on a membrane protein basis (Table 1, Ref. 13). Although IGF-I is not as potent as insulin in inducing casein synthesis from mid-pregnant mouse mammary tissue in vitro the effects of these two hormones were additive (28). Exogenous treatment of lactating cows with somatotropin results in increased plasma and milk IGF-I concentrations. However, the concentration of IGF-I in milk of bST treated animals is well within the range of milk somatomedin concentrations detected across lactation (8, 33). In fact, season and parity had as great, or even greater effects on milk IGF-I concentrations as treatment of cattle with somatotropins (8, 33).

Conclusive studies delineating the role of IGF-I in lactating mammary tissue are not yet available. Contradictory results have been published using in vitro approaches (3, 27, 28, 29). Clearly, additional work is required to reconcile these differences to establish a functional role of IGF-I in mammary metabolism during lactation.

INSULIN-LIKE GROWTH FACTOR II

Insulin-like growth factor II binds to the type I and type II receptor (2). Neither insulin nor IGF-I competes with IGF-II for the type II receptor across physiological concentrations of these molecules when recombinant molecules are used as the ligands. Previous reports of IGF-I cross-reactivity with the type II receptor were probably due to IGF-II contamination of the serum purified IGF-I preparations (2, Fig. 4). In contrast IGF-II does compete for the type I somatomedin receptor (Fig. 5).

Availability of this molecule has been extremely limited to date resulting in very little published information. Type II receptors are the dominant form of somatomedin receptor during pregnancy and early lactation (12, 13). This is reflected in milk concentrations of this hormone (23). Concentrations of IGF-II in mammary secretions are also elevated during pregnancy and then fall with onset of lactation (23). This may be associated with a lower population of type II receptors during established lactation (13, Table 2).

Mammary tissue from rats was examined at various time points during gestation and lactation (Table 1). Scatchard analysis indicate the presence of only one type of IGF-II receptor with affinity constants that changed little across pregnancy and lactation. Both affinity and number of IGF-II receptors was found to be lower in virgin than pregnant or lactating animals (Table 1). Furthermore, there was very little change in the receptor density in comparing mammary tissue from pregnant and lactating animals. In mid-lactation (Table 2) the available type II receptor population is essentially equivalent to the size of the type I receptor population in bovine mammary tissue examined. Since high affinity type I and II receptors are present during lactation the possibility exists that both IGF-I and IGF-II have biological effects on lactating bovine mammary tissue.

Exogenous administration of somatotropin to lactating cattle results in increases in plasma but not milk IGF-II concentrations (8). To date, there are no published reports of biological responses of lactating mammary tissue to IGF-II. Since IGF-II will bind

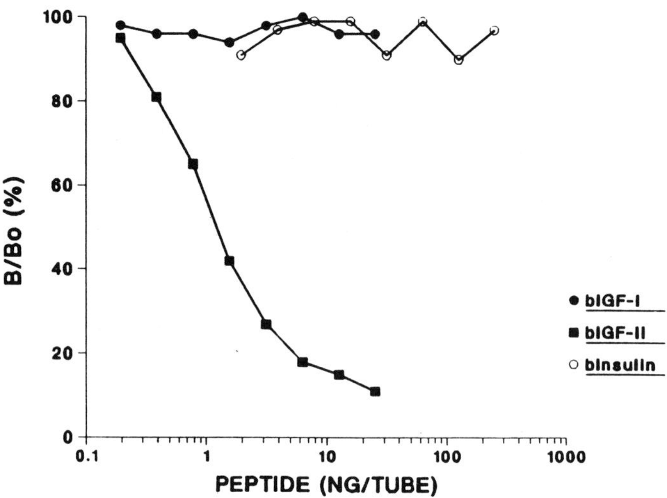

Fig. 4 Somatomedin type II receptor competition curve.

Recombinant bovine IGF-I and bovine insulin were not capable of displacing IGF-II at concentrations which were equivalent or 100-fold in excess respectively.

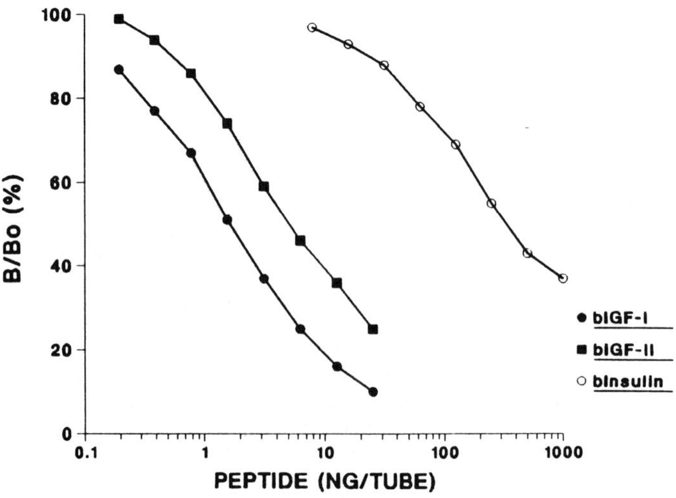

Fig. 5 Somatomedin type I receptor competition curve.

Recombinant bovine IGF-II and bovine insulin were 29.2% and 0.6% as effective as recombinant bovine IGF-I in displacing IGF-I from the type I receptor.

to the IGF-I receptor at relatively high concentrations (Fig. 4) it is not surprising that it will stimulate DNA synthesis in non-lactating mammary tissue (Fig. 2).

REFERENCES

1 Ballard, F.J., Francis, G.L., Ross, M., Bagley, C.J., May, B. and Wallace, J.C. (1987). Natural and synthetic forms of insulin-like growth factor-I (IGF-I) and the potent derivative, destripeptide IGF-I: biological activities and receptor binding. Biochemical and Biophysical Research Communications **149,** 398-404.

2 Barenton, B., Guyda, H.J., Goodyer, C.G., Polychronakos, C. and Posner, B.I. (1987). Specificity of insulin-like growth factor binding to type-II IGF receptors in rabbit mammary gland and hypophysectomized rat liver. Biochemical and Biophysical Research Communications **149,** 555-561.

3 Baumrucker, C.R. (1986). Insulin-like growth factor I (IGF-I) and insulin stimulates lactating bovine mammary tissue DNA synthesis and milk production in vitro. Journal of Dairy Science **69** (Suppl. 1), 120.

4 Baxter, R.C., Zaltesman, Z. and Turtle, J.R. (1984). Immunoreactive somatomedin-C/insulin-like growth factor I and its binding protein in human milk. Journal of Clinical Endocrinology and Metabolism **58,** 955-1000.

5 Bickerstaffe, R., Annison, E.F. and Linzell, J.L. (1974). The metabolism of glucose, acetate, lipids and amino acids in lactating dairy cows. Journal of Agricultural Science, Cambridge **82,** 71-89.

6 Campbell, P.G. and Baumrucker, C.R. (1986). Changes in [^{125}I] insulin binding to bovine mammary gland microsomes during pregnancy and lactation. Journal of Dairy Science **69** (Suppl. 1), 167.

7 Collier, R.J., McNamara, J.P., Wallace, C.R. and Dehoff, M.H. (1984). A review of endocrine regulation of metabolism during lactation. Journal of Animal Science **59,** 498-510.

8 Collier, R.J., Li, R., Johnson, H.D., Becker, B.A., Buonomo, F.C. and Spencer, K.J. (1988). Effect of sometribove, (methionyl bovine somatotropin BST) on plasma insulin-like growth factor I (IGF-I) and II (IGF-II) in cattle exposed to heat and cold stress. Journal of Dairy Science **71** (Suppl. 1), 228.

9 Corps, A.N., Brown, K.D., Rees, L.H., Carr, J. and Prosser, C.G. (1988). The insulin-like growth factor I content in human milk increases between early and full lactation. Journal of Clinical Endocrinology and Metabolism **67,** 25-29.

10 Davis, S.R. and Collier, R.J. (1985). Mammary blood flow and the regulation of substrate supply for milk synthesis. Journal of Dairy Science **68,** 1041-1058.

11 Davis, S.R., Gluckman, P.D., Hart, I.C. and Henderson, H.V. (1987). Effects of injecting growth hormone or thyroxine on milk production and blood plasma concentrations of insulin-like growth factors I and II in dairy cows. Journal of Endocrinology **114,** 17-24.

12 Dehoff, M.H., Elgin, R.G., Collier, R.J. and Clemmons, D.R. (1988). Both type I and II insulin-like growth factor receptor binding increase during lactogenesis in bovine mammary tissue. Endocrinology **122,** 2412-2417.

13 Disenhaus, C., Belair, L. and Djiane, J. (1988). Caractérisation et évolution physioloquie des recepteurs pour les < < insulin-like growth factors > > I et II (IGF's) dans la glande mammaire de brebis. Reproduction Nutrition Dévéloppment **28,** 241-252.

14 Faulkner, A., Chaiyabutr, N., Peaker, M., Carrick, D.T. and Kuhn, N.J. (1981). Metabolic significance of milk glucose. Journal of Dairy Research **48,** 51-57.

15 Francis, G.M., Read, L.C., Ballard, F.J., Bagley, C.J., Upton, F.M., Grovestock, P.M. and Wallace, J.C. (1986). Purification and partial sequence analysis of insulin-like growth factor-I from bovine colostrum. Biochemical Journal **233,** 207-213.

16 Francis, G., Upton, F.M., Ballard, F.J., McNeil, K.A. and Wallace, J.C. (1988). Insulin-like growth factors I and II in bovine colostrum. Biochemistry Journal **251,** 95-103.

17 Gertler, A., Cohen, N. and Maoz, A. (1983). Human growth hormone but not ovine or bovine growth hormones exhibits a galactopoietic prolactin-like activity in organ culture from bovine lactating mammary gland. Molecular and Cellular Endocrinology **35,** 51-55.

18 Hove, K. (1978). Effects of hyperinsulinemia on lactose secretion and glucose uptake by the goat mammary gland. Acta Physiologica Scandinavia **104,** 422-431.

19 Imagawa, W., Spencer, E.M., Larsen, L. and Nandi, S. (1986). Somatomedin-C substitutes for insulin for the growth of mammary epithelial cells from normal virgin mice in serum-free collagen gel culture. Endocrinology **119,** 2695-2699.

20 Inagaki, Y. and Kohmoto, Y. (1982). Changes in Scatchard plots for insulin binding to mammary epithelial cells from cycling, pregnant and lactating mice. Endocrinology **110,** 176-181.

21 Koprowski, J.A. and Tucker, H.A. (1973). Bovine serum growth hormone, corticoids and insulin during lactation. Endocrinology **93,** 645-658.

22 Laarveld, B., Christensen, D.A. and Brockman, R.P. (1981). The effect of insulin on net metabolism of glucose and amino acids by the bovine mammary gland. Endocrinology **108,** 2217-2223.

23 Malven, P.V., Head, H.H., Collier, R.J. and Buonomo, F.C. (1987). Periparturient changes in secretion and mammary uptake of insulin and in concentrations of insulin and insulin-like growth factors in milk of dairy cows. Journal of Dairy Science **70,** 2254-2265.

24 McGrath, M.F. (1987). A novel system for mammary epithelial cell culture. Journal of Dairy Science **70**, 1967-1980.

25 McGrath, M.F. and Collier, R.J. (1988). Effect of epidermal growth factor (EGF) and insulin-like growth factor I (IGF-I) in ruminant mammary development. Journal of Dairy Science **71** (Suppl. 1), 229.

26 Minuto, F., Del Monte, P., Barreca, A., Nicolin, A. and Giordano, G. (1987). Partial characterization of somatomedin-C-like immunoreactivity secreted by breast cancer cells in vitro. Molecular and Cell Endocrinology **54,** 179.

27 Prosser, C.G. and Topper, Y.J. (1986). Changes in the rate of carrier-mediated glucose transport by mouse mammary epithelial cells during ontogeny: hormone dependence delineated in vitro. Endocrinology **119,** 91-96.

28 Prosser, C.G., Sankaran, L., Henninghousen, L. and Topper, Y.J. (1987). Comparison of the roles of insulin and insulin-like growth factor I in casein gene expression and in the development of α-lactalbumin and glucose transport activities in the mouse mammary epithelial cell. Endocrinology **120,** 1411.

29 Shamay, A., Cohen, N., Niwa, M. and Gertler, A. (1988). Effect of insulin-like growth factor I on deoxyribonucleic acid synthesis and galactopoiesis in bovine undifferentiated and lactating mammary tissue in vitro. Endocrinology **123,** 804-809.

30 Spitzer, E. and Grosse, R. (1987). EGF receptors on plasma membranes purified from bovine mammary gland of lactating and pregnant animals. Biochemistry International **14,** 581-588.

31 Szabo, L., Mottershead, D.G., Ballard, F.J. and Wallace, J.C. (1988). The bovine insulin-like growth factor (IGF) binding protein purified from conditioned medium requires the N-terminal tripeptide in IGF-I for binding. Biochemical Biophysical Research Communications **151,** 207-214.

32 Topper, Y.J., Nicholas, K.R., Sankaran, L. and Kulski, J.K. (1984). Insulin biology from the perspective of studies on mammary gland development. In Biochemical Actions of Hormones (Litwack, G. ed.), p 163. Academic Press, New York and London.

33 Torkelson, A.R., Lanza, G.M., Birmingham, B.K., Vicini, J.L., White, T.C., Dyer, S.E., Madsen, K.S. and Collier, R.J. (1988). Concentrations of insulin-like growth factor (IGF-I) in bovine milk: effect of herd, stage of lactation and sometribove, USAN (recombinant methionyl bovine somatotropin). Journal of Dairy Science **71** (Suppl. 1), p. 169.

IMMUNOLOGICAL ENHANCEMENT

ANTIGEN-ANTIBODY COMPLEXES THAT ENHANCE GROWTH

A.T. Holder[1] and R. Aston[2]

[1] Department of Endocrinology and Animal Physiology, AFRC Institute for Grassland and Animal Production, Hurley, Maidenhead, Berkshire SL6 5LR, U.K.
[2] Peptide Technology Ltd., P.O. Box 444, Dee Why, N.S.W. 2099, Australia

INTRODUCTION

The use of exogenous bovine growth hormone (bGH) as a means of promoting growth and lactation in sheep and cattle has been the subject of much research (11, 18, 22) and public comment. However, the recent EEC ban on the use of hormone-based growth promoters has necessarily focused attention on alternative methods of improving animal production. One such approach, outlined here, involves immunological manipulation of the endocrine system using either monoclonal antibodies (MAbs) or antisera of restricted specificity.

MONOCLONAL ANTIBODIES

Much of this work has involved two panels of MAbs prepared against oGH or hGH (Fig. 1). Competition assays between different MAbs for binding to hGH or bGH have defined at least four non-overlapping antigenic determinants on each hormone (2, 4, 16). For hGH, two of these determinants are shared with the closely related hormone human chorionic somatomammotrophin (hCS) but none are shared with bGH. MAbs prepared against oGH bound to bGH with equal efficacy but did not cross react with hGH. In this paper doses of MAb have generally been expressed as ABT_{50} values which correspond to the dilution (titre) of MAb required to bind 50% of the tracer in a liquid phase RIA.

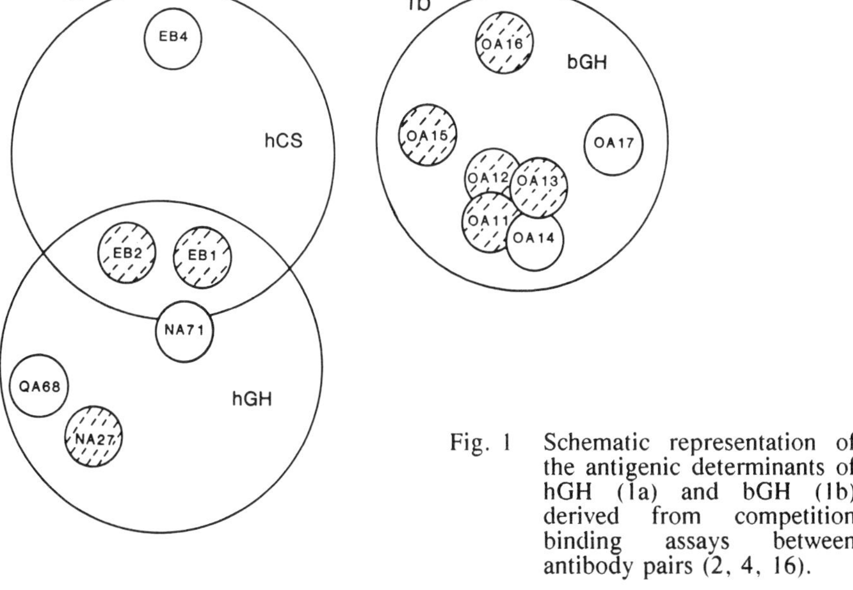

Fig. 1 Schematic representation of the antigenic determinants of hGH (1a) and bGH (1b) derived from competition binding assays between antibody pairs (2, 4, 16).

Investigations into the structure/function relationships of GH using these panels of MAbs as specific structural probes have shown (a) that MAbs can differentially affect the binding of GH to various tissues and (b) that certain MAbs can enhance the biological actions of GH in vivo (2, 3, 6, 12, 14). The former was not unexpected, however, the latter runs counter to the general concept that antibodies against hormones abrogate hormone activity.

GH/MAb COMPLEXES AND GROWTH

Hypopituitary Snell dwarf mice were used as the animal model to screen MAbs for their ability to enhance the growth-promoting activity of GH. Enhancement was monitored by measuring increases in body weight and/or cartilage metabolism, as assessed by increased incorporation of [^{35}S]-labelled sulphate (35S) into costal cartilage in vivo (15).

In short term experiments (72 h treatment periods) most MAbs were found to enhance the somatogenic actions of either hGH or bGH (2, 3, 12) and none were found to be inhibitory (Table 2). For hGH, enhancement was most marked for those MAbs which recognized antigenic determinants common to both hGH and hCS (EB1 and EB2) while those specific for hGH (NA71 and QA68), were only mildly enhancing at relatively high concentrations. With a constant dose of hGH enhancement was found to be dose dependent for MAb concentration and vice versa (Fig. 2). At optimum levels hGH/MAb-EB1 complexes gave a 400-600% increase in both 35S uptake and weight gain over that observed for hGH alone (3, 12).

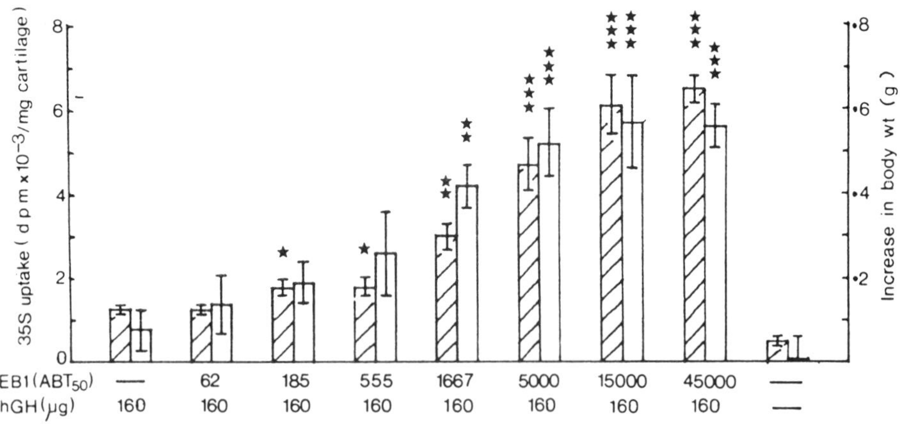

Fig. 2 The effect of various doses of MAB EB1 (62-45000 ABT$_{50}$/day for 2 days) complexed with a constant dose of hGH (160 μg) on uptake of labelled sulphate (35S) into dwarf mouse costal cartilage in vivo (hatched bars) and increase in body weight (open bars). Control animals were injected with PBS. Values are means ± SEM, *$P < 0.01$, *** $P < 0.001$ compared with hGH alone (Student's T-test) (12)

In long-term treatment experiments (42 days), mice receiving hGH/MAb-EB1 showed a five-fold increase in weight gain compared to groups receiving hGH alone. The data suggested that had treatment been continued then the final weight of the mice receiving hGH/MAb-EB1 complexes would not have differed from those treated with hGH alone; they would have simply achieved this 'target' weight more quickly (12).

The observation that MAb-EB1 cross-reacted with marmoset GH offered the opportunity to look at the effects of this MAb on the action of endogenous GH in vivo. Three groups of marmosets were injected with either MAb-EB1 alone (48000 ABT_{50} /injection), hGH (400 μg/injection) or saline 3 times per week for a period of 7 weeks. The growth rate and final body weight of marmosets treated with MAb-EB1 alone were significantly greater than that of either saline or hGH treated-animals (Fig. 3). Thus it is possible to enhance the activity of physiological levels of endogenous GH in normally growing marmosets (12).

A similar picture of enhancement was observed for bGH. Again the degree of enhancement varied according to epitope specificity; however, MAbs which bound to overlapping determinants of bGH (OA11, OA12, OA13 and OA14; Fig. 1b) were found to exhibit different degrees of enhancement (2). There was no apparent relationship between the degree of enhancement and the affinity of various monoclonals for bGH. In short-term experiments bGH/MAb-OA11 complexes promoted a 250% increase in 35S uptake compared to that for bGH alone, an enhancement similar to that observed for hGH/MAb-EB1. This response was dose dependent for both bGH and MAb-OA11 (2).

Many of the somatogenic actions of GH are thought to be mediated by insulin-like growth factor I (IGF-I/somatomedin-C). Thus the observation that MAb-mediated enhancement of the growth-promoting activity of bGH in hypophysectomized rats is associated with increased levels of serum IGF-I supports this hypothesis and also suggests that enhancement occurs via normal growth pathways (30).

Fig. 3 Effects of treatment with hGH (400 μg/injection; ■) MAB-EB1 alone (48000 ABT_{50} /injection; ▲) or phosphate buffered saline (●) on weight gain in normal marmosets injected three times per week for 7 weeks (12)

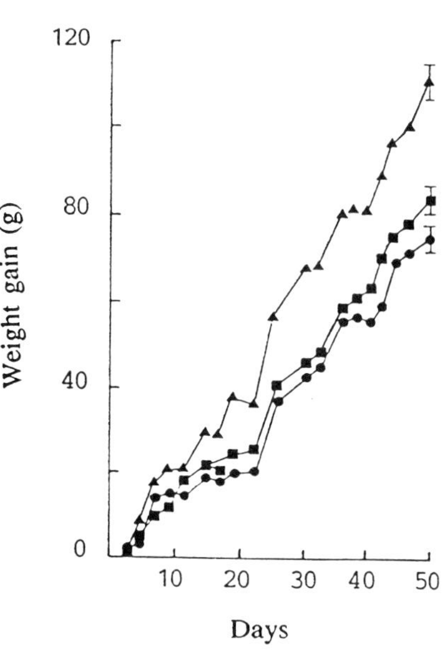

GH/MAb COMPLEXES AND BODY COMPOSITION

In addition to its somatogenic and lactogenic effects GH is also known to influence body composition via its effects on intermediary metabolism, particularly fat deposition (9, 11, 17) and muscle protein synthesis (1). A set of experiments were designed to compare the effects of treatment with phosphate buffered saline (PBS), hGH and hGH/MAb-EB1 on body composition in dwarf mice over a 10 day treatment period. Within each of these different treatment groups there were 3 subgroups which received 50, 75 or 100% of an ad libitum diet (14).

PBS-treated groups lost more or gained less weight than the equivalent groups treated with hGH, which in turn lost more or gained less weight than the equivalent groups treated with hGH/MAb-EB1. A similar relationship was observed when weight gain or loss was expressed as g/g food eaten; thus, MAb-EB1 enhanced the increased efficiency of food utilization observed for hGH alone. The ability of hGH to decrease fat content (when expressed as percent of body weight) was also enhanced when complexed with MAb-EB1. In addition, groups treated with hGH/MAb-EB1 had a significantly greater protein content than the equivalent group treated with hGH alone, which resulted in a significantly higher protein:fat ratio (Table 1). Throughout all groups there was an inverse relationship between water content and fat content when expressed as a percentage of body weight. These results suggested that the ability of hGH to repartition nutrients in favour of accretion of protein rather than fat is greatly enhanced when it is complexed with MAb-EB1. These conclusions are supported by the observation that MAb-EB1 can enhance the anabolic action of hGH on muscle protein synthesis in dwarf mice (6).

Table 1 Fat, protein and protein/fat ratio in dwarf mice treated for 10 days with PBS; hGH (40 mU/day) alone or complexed with MAb-EB1 (15000 ABT_{50}). Mice received 50 or 75% of the food consumed by the ad libitum group (100%) on the same treatment. Values are means ± SEM (Ref. 14).

Treatment	Fat (% body wt)	Protein (% body wt)	Protein/Fat ratio
PBS			
50% diet	14.0 ± 2.0	15.7 ± 0.6	1.29 ± 0.31
75% diet	18.5 ± 0.7	14.4 ± 0.2	0.78 ± 0.04
100% diet	17.4 ± 0.9	14.1 ± 0.1	0.80 ± 0.05
hGH			
50% diet	9.7 ± 1.3	16.3 ± 0.3	1.89 ± 0.33
75% diet	12.8 ± 0.4***	14.6 ± 0.2	1.15 ± 0.06***
100% diet	14.0 ± 0.9*	14.3 ± 0.2	1.05 ± 0.08*
hGH/MAb-EB1			
50% diet	5.6 ± 0.4**$^{+}$	17.0 ± 0.2*$^{+}$	3.15 ± 0.26***$^{+++}$
75% diet	7.4 ± 0.6***$^{+++}$	16.5 ± 0.3***$^{+++}$	2.34 ± 0.23***$^{+++}$
100% diet	11.4 ± 1.4**	15.1 ± 0.1***$^{++}$	1.52 ± 0.17**$^{+}$

*P < 0.05, **P < 0.01, ***P < 0.001 compared with PBS control group on the same diet; $^{+}$P < 0.05, $^{++}$P < 0.01, $^{+++}$P < 0.001 compared with hGH alone group on the same diet (Student's t-test).

GH/MAb COMPLEXES AND LACTOGENIC ACTIVITY

MAb mediated enhancement of the somatogenic actions of GH suggested that it might also be possible for MAbs to enhance the lactogenic actions of this hormone. The panel of MAbs to hGH (Fig. 1) were investigated using the pigeon crop sac assay (21) for their ability to enhance the lactogenic activity of hGH. MAbs which enhanced the somatogenic activity of hGH (EB1 and EB2) were also effective enhancers of lactogenic activity, increasing crop sac mucosa weight by up to 100% compared to hGH alone. Similarly MAbs NA71 and QA68, which were relatively poor enhancers of growth, either provoked no response (NA71) or inhibited (QA68) the lactogenic activity of hGH (Table 2). Using MAb-EB1 which cross-reacts with hCS it was also possible to enhance the lactogenic activity of this hormone (3).

Enhancement of the lactogenic activity of bGH was examined in normal adult lactating ewes using MAb-OA11 (23), which was known to enhance somatogenic activity. Sheep were treated with either bGH, MAb-OA11 or bGH/MAb-OA11 complexes for 21 days starting 8 weeks post lambing. Milk yield was expressed as a percent of the mean daily yields for 6 days immediately preceding treatment. In groups treated with bGH or bGH/MAb-OA11, milk yield had increased dramatically by day 4 reaching maximal levels around day 10 (Fig. 4); milk yield was significantly greater in animals receiving bGH/MAb-OA11 than those receiving bGH alone. MAb-OA11 alone had little effect on the normal decline in milk yield expected over this period and thus did not appear to enhance endogenous GH. Serum IGF-I levels were also measured before and during (day 16) treatment and were found to reflect lactational performance, being greater in animals treated with bGH/MAb-OA11 than in those treated with bGH alone.

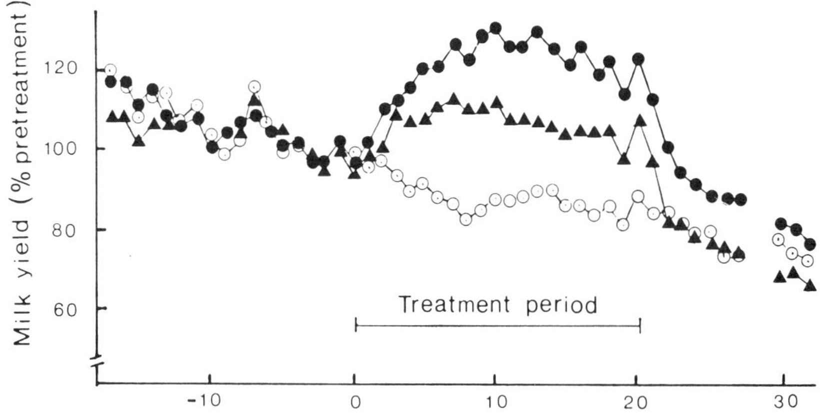

Fig. 4 Mean milk yield for ewes treated daily with MAB-OA11 (2 mg/injection, 0), bGH (10 mg/injection, Δ) or these doses of MAB-OA11 and bGH combined (0). Milk yield was expressed as a percentage of pretreatment yields for days -6 to 0 (23)

GH/MAb COMPLEXES AND RECEPTOR BINDING

MAbs have been shown to permit the binding of GH to certain classes of receptor yet not others. However there does not appear to be a coherent theme linking the effects of MAbs on receptor binding to their ability to enhance hormone action. For example, MAbs which can enhance the somatogenic and lactogenic actions of hGH in vivo (EB1 and EB2) will permit the binding of hGH to pregnant rabbit and rat liver microsomes yet inhibit the binding of hGH to lymphoid tissue. Conversely, MAbs which have little enhancing potential (NA71 and QA68) are less effective at inhibiting the binding of hGH to lymphoid tissue, and more effective at inhibiting binding to liver microsomes (3, 4, 29). It would be convenient if this theme were to be repeated for bGH. However MAb-OA16, which enhanced the biological activity of bGH in vivo, markedly inhibited the binding of bGH to pregnant rabbit liver microsomes. Furthermore MAb-OA15 which appeared to increase the binding of bGH to rabbit liver microsomes in vitro, showed low enhancing potential in vivo (2). These data are given in Table 2.

The discrepancy between the ability of certain MAbs to enhance the biological activity of GH and their effects on receptor binding in vitro is likely to result from a variety of factors which are present in vivo yet absent from the more simple in vitro binding systems used. In addition some of the confusion may be a result of the use of heterologous systems.

Table 2 Properties of MAbs to hGH and oGH/bGH (subjective assessment based on data in Refs. 2, 3, 4, 12, 23, 29)

MAB	Affinity for GH 1/nmol)	Potentiation or inhibition of GH activity		Inhibition or enhancement of receptor binding	
		Somatogenic	Lactogenic*	IM9 cells	RLM
hGH					
EB1	0.43	+ + +	+ + +	- - -	0
EB2	1.00	+ + +	+ +	- -	-
QA68	1.00	+	- -	-	- -
NA71	5.2	+	0	- -	- - -
bGH					
OA11	0.43	+ +	+ + +	NA	+
OA12	0.27	+ +	?	NA	- -
OA13	0.31	+ + +	?	NA	- -
OA14	-	0	?	NA	- - -
OA15	0.73	+	?	NA	+ +
OA16	0.11	+	?	NA	- - -
OA17	0.36	0	?	NA	-

+ + +, 'Maximum' enhancement; - - -, 'maximum' inhibition; 0, no effect; ?, not tested; NA, not applicable; *, for hGH and bGH lactogenic activity was assessed by the pigeon crop sac assay and increased milk yield in ewes, respectively; RLM, pregnant rabbit liver microsomes.

COVALENT LINKING STUDIES

In order to understand the phenomenon of enhancement more clearly it was necessary to establish whether enhancement was a distinct property of a GH/MAb complex, or whether this was a more general result of a specific association between GH and any large protein. Thus covalently linked complexes of hGH/ovalbumin and hGH/bovine serum albumin (BSA) were examined for their effect on 35S uptake into costal cartilage of hypopituitary dwarf mice in vivo in short-term experiments (72 h). Both ovalbumin and BSA were cross linked to hGH using dimethylsuberimidate (DMS) or glutaraldehyde using standard techniques. Levels of BSA and ovalbumin were arranged to be approximately equimolar with hGH in the reaction mixtures such that the final concentration of hGH was 100 mU/0.1 ml (the injection volume). The same concentration of GH was also cross linked in the absence of ovalbumin or BSA.

The results illustrated in Fig. 5 clearly show that the biological potency of hGH was markedly enhanced when cross linked to either ovalbumin or BSA using DMS or glutaraldehyde. The effect of injecting hGH with added but not cross linked ovalbumin or BSA did not differ from those of hGH alone. Furthermore hGH when cross linked to itself promoted a significantly higher uptake of 35S than the same dose of unlinked hGH. This last observation is of interest since hGH readily forms dimers which are not noted for enhanced activity (28), however the nature of this association is likely to be different from that between cross linked hGH molecules.

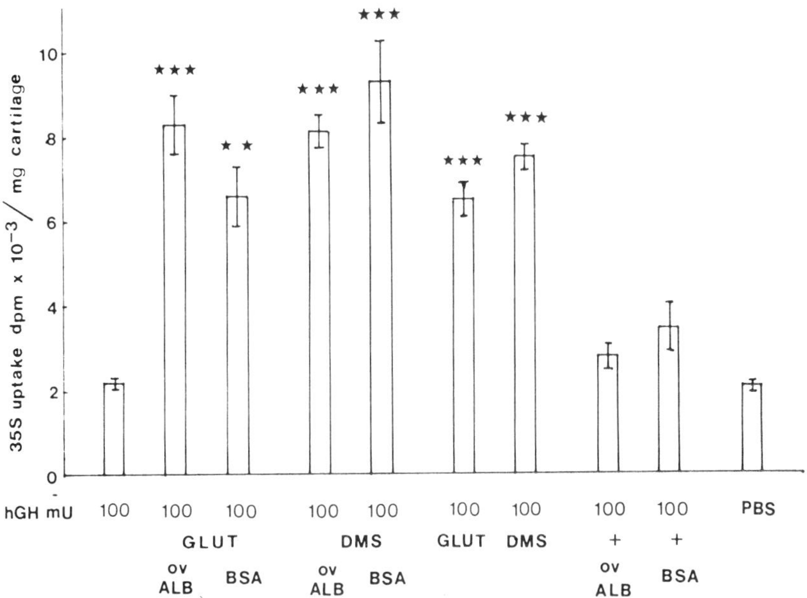

Fig. 5 Effects of covalently linked hGH/BSA, hGH/ovalbumin and hGH/hGH complexes on uptake of 35S into costal cartilage in vivo. Controls were PBS, hGH and hGH with added BSA or ovalbumin. Values are means ± SEM; $^{**}P < 0.01$, $^{***}P < 0.001$ compared with control groups (Student's t-test)

Although there is no direct evidence, it seems likely that the mechanism of enhancement seen here is the same as that for MAb-mediated enhancement of hormone activity. Cross linking agents act through lysine residues which, being hydrophilic, are highly represented on the surface of proteins. This suggests that enhancement may not be dependent upon the interaction of a MAb with a specific antigenic determinant but is rather a more general surface phenomenon. The relevance of these results to MAb-mediated enhancement is discussed further in the next section.

MECHANISM OF ENHANCEMENT

Although the mechanism of MAb-mediated enhancement of hormone action is unclear, it is possible to reduce the number of potential candidates. The following comments presuppose, perhaps rather naively, that enhancement of the various actions of GH is mediated via a common mechanism(s). Thus general conclusions have been drawn from data concerned with the enhancement of different biological actions of hGH or bGH in different species.

Antibody facilitated receptor cross linking has been implicated in the enhancement of the biological actions of epidermal growth factor (EGF) and insulin in vitro (20, 26, 27). However, this mechanism is dependent upon antibody bivalency. Since univalent Fab fragments of enhancing MAbs are as effective as the bivalent antibody in enhancing GH action (2, 3), this mechanism seems unlikely.

Antibody induced conformational change in an antigen has been shown to enhance the binding of a second antibody to the antigen, or even the biological activity of the antigen itself (7, 19). For such a mechanism to apply here, each enhancing MAb would have to induce a 'beneficial' structural modification leading to enhanced activity, even though they may bind to quite distinct sites on GH. In addition MAb-EB1 would have to achieve this with regard to both somatogenic and lactogenic receptors, which are thought to be structurally distinct (5, 29), by binding to a single epitope.

Prolongation of the systemic half life of exogenous insulin is known to occur in certain diabetic patients presenting with antibodies to insulin, and leads to enhanced biological activity. The antibodies are thought to act as a buffering system, slowly releasing free insulin (10, 24). These results raise two issues: firstly, can prolongation of systemic half life account for MAb-mediated enhancement of hormone action. This seems unlikely since enhancement of the lactogenic actions of hGH in the pigeon crop sac assay is independent of systemic distribution of the hormone. However, this does not exclude prolongation of hGH half-life within the crop sac. The second question is whether it is the free hormone or GH/MAb complex which interacts with the receptors. If the MAbs merely act as a buffering system, slowly releasing free hormone, then there should be some relationship between antibody affinity and enhancement; none was observed. Furthermore, enhancement occurred when hGH was covalently linked to ovalbumin or BSA. Thus although receptors may strip GH from the antibody, perhaps due to their greater affinity for hGH, dissociation of the complex does not appear to be necessary in order to obtain enhancement.

The ability of MAbs to inhibit the binding of GH to certain receptor subtypes but not others (2, 3, 4, 29) suggests that enhancement may be achieved by restricting GH binding to receptors associated with enhanced activities; in effect increasing GH availability to certain receptor subtypes at the expense of others.

Enhancement could also be due to changes in receptor processing. The interaction of GH with its receptor triggers intracellular events which mediate hormone action and is followed by internalization of GH/receptor complexes and subsequent processing of GH. In adipocytes GH is processed by two independently regulated pathways, lysosomal degradation or release of the intact hormone (25). It is possible that enhancing MAbs inhibit internalization leading to protracted receptor activation and increased stimulation of intracellular events. Alternatively, MAbs may inhibit lysosomal degradation or encourage GH processing via the pathway leading to release of intact GH or even GH/MAb complexes.

ENHANCEMENT OF GH ACTIVITY BY ACTIVE IMMUNIZATION

The ability of MAbs to enhance various biological actions of GH has been demonstrated unequivocally and the advantages of this in terms of animal production are obvious. In order to obtain these effects within the context of current farming practice it would be necessary to manipulate the animal's own immune system in order to produce antibodies of appropriate specificity.

Perhaps the simplest way of achieving this is to locate the antigenic determinants which are associated with enhancing MABs, and to immunize animals with peptides resembling these regions. The antibody response, although polyclonal, would be of restricted specificity and recognition of the endogenous hormone by these antibodies would lead to enhancement. It has been shown in marmosets that MAbs will enhance the biological actions of endogenous GH (12) and the ability of antibodies raised against peptide fragments to recognize the complete protein is also established (8). The problem of animals recognizing the peptides as self may be circumvented by making subtle modifications to the structure of these peptides which will increase their antigenicity without loss of specificity. In practice it has been possible to raise antisera to peptides of bGH which will enhance the biological activity of bGH in hypopituitary Snell dwarf mice and sheep (data not given).

CONCLUSIONS

Although much of this work has been concerned with MAb-mediated enhancement of the actions of GH and hCS, the wider applications of this phenomenon to other hormones such as thyrotrophin (13), should not be ignored.

This approach to the manipulation of animal production has a number of advantages. Active immunization techniques require only transitory exposure of the animal to minute amounts of antigen in order to achieve relatively long-lasting effects. Furthermore the active components, ie antibody and hormone are endogenous, being part of the normal genetic repertoire of the animal. Immunogens which influence growth, carcass composition, lactation and reproduction, may play an important role in the future of animal production.

REFERENCES

1 Albertson-Wikland, K., Eden, S. and Isaksson, O. (1980). Analysis of early responses to growth hormone on amino acid transport and protein synthesis in diaphragms of young normal rats. Endocrinology **106**, 291-297.

2 Aston, R., Holder, A.T., Ivanyi, J. and Bomford, R. (1987). Enhancement of bovine growth hormone activity in vivo by monoclonal antibodies. Molecular Immunology **24**, 143-150.

3 Aston, R., Holder, A.T., Preece, M.A. and Ivanyi, J. (1986). Potentiation of the somatogenic and lactogenic activity of human growth hormone with monoclonal antibodies. Journal of Endocrinology **110**, 381-388.

4 Aston, R. and Ivanyi, J. (1985). Monoclonal antibodies to growth hormone and prolactin. Pharmacology and Therapeutics **27**, 403-424.

5 Barnard, R., Bundesen, P.G., Rylatt, D.B. and Waters, M.J. (1985). Evidence from the use of monoclonal antibody probes for the structural heterogeneity of the growth hormone receptors. Biochemical Journal **231**, 459-468.

6 Bates, P.C., Holder, A.T. and Aston, R. (1987). Effect of growth hormone and monoclonal antibody on growth and muscle protein synthesis in Snell dwarf mice. Journal of Endocrinology **112** (Suppl.), 60.

7 Duncan, J.S., Hewitt, J. and Weston, P.D. (1982). Activation of beta-galactosidase by monoclonal antibodies. Biochemical Journal **205,** 219-224.

8 Geysen, H.M., Barteling, S.J. and Meloen, R.H. (1985). Small peptides induce antibodies with a sequence and structural requirements for binding antigen comparable to antibodies raised against the native protein. Proceedings of the National Academy of Sciences, U.S.A. **82,** 178-182.

9 Goodman, H.M. and Grichting, G. (1983). Growth hormone and lipolysis: a re-evaluation. Endocrinology **113,** 1697-1702.

10 Gray, R.S., Cowan, P., di Mario, U., Elton, R.A., Clarke, B.F. and Duncan, L.J.P. (1985). Influence of insulin antibodies on pharmacokinetics and bioavailability of recombinant human and highly purified beef insulins in insulin dependent diabetics. British Medical Journal **290,** 1687-1691.

11 Hart, I.C. and Johnsson, I.D. (1986). Growth hormone and growth in meat producing animals. In Control and Manipulation of Animal Growth (P.J. Buttery, N.B. Haynes and D.B. Lindsay, eds.), pp 135-159. Butterworths, London.

12 Holder, A.T., Aston, R., Preece, M.A. and Ivanyi, J. (1985). Monoclonal antibody-mediated enhancement of growth hormone activity in vivo. Journal of Endocrinology **107,** R9-R12.

13 Holder, A.T., Aston, R., Rest, J.R., Hill, D.J., Patel, N. and Ivanyi, J. (1987). Monoclonal antibodies can enhance the biological activity of thyrotrophin. Endocrinology **120,** 567-573.

14 Holder, A.T., Blows, J.A., Aston, R. and Bates, P.C. (1988). Monoclonal antibody enhancement of the effects of human growth hormone on growth and body composition in mice. Journal of Endocrinology **117,** 85-90.

15 Holder, A.T.., Wallis, M., Biggs, P. and Preece, M.A. (1980). Effects of growth hormone, prolactin and thyroxine on body weight, somatomedin-like activity and in vivo sulphation of cartilage in hypopituitary dwarf mice. Journal of Endocrinology **85,** 35-47.

16 Ivanyi, J. (1982). Study of antigenic structure and inhibition of activity of human growth hormone and chorionic somato-mammotrophin by monoclonal antibodies. Molecular Immunology 19, 1611-1618.

17 Jagannadha, A. and Ramachandran, J. (1978). Growth hormone and the regulation of lipolysis. In Hormonal Proteins and Peptides, vol.4 (C.H. Li, ed), pp 43-60. Academic Press, New York.

18 Johnsson, I.D. and Hart, I.C. (1986). Manipulation of milk yield with growth hormone. In Recent Advances in Animal Nutrition. Studies in the Agricultural and Food Sciences (W. Haresign and D.J.A. Cole, eds.), pp 105-123. Butterworths, London.

19 Lubeck, M. and Gerhard, W. (1982). Conformational changes at lopologically distinct antigenic sites on the influenza A/PR/8/34 virus HA molecule are induced by the binding of monoclonal antibodies. Virology **118,** 1-7.

20 Lyen, K.R., Smith, R.M. and Jarrett, L. (1983). Differences in the ability of anti-insulin antibody to aggregate monomeric ferritin-insulin occupied receptor sites on liver and adipocyte plasma membranes. Diabetes **32,** 648-653.

21 Nicoll, C.S. (1967). Bio-assay of prolactin. Analysis of the pigeon crop-sac response to local prolactin injection by an objective and quantitative method. Endocrinology **80,** 641-655.

22 Pell, J.M. and Bates, P.C. (1987). Collagen and non-collagen protein turnover in skeletal muscle of growth hormone treated lambs. Journal of Endocrinology **115,** R1-R4.

23 Pell, J.M., Johnsson, I.D., Pullar, R., Morrell, D.J., Hart, I.C., Holder, A.T. and Aston, R. (1988). Potentiation of growth hormone activity in sheep using monoclonal antibodies. Journal of Endocrinology (in press).

24 Reeves, W.G. (1986). The immune response to insulin: characterisation and clinical consequences. In The Diabetes Annual/2 (K.G.M.M. Alberti and L.P. Krall, eds.), pp 81-93. Elsevier, B.U.

25 Roupas, P. and Herington, A.C. (1987). Processing growth hormone by rat adipocytes in primary culture: differentiation between release of intact hormone and degradative processing. Endocrinology **121,** 1521-1530.

26 Schechter, Y., Chang, K-J., Jacobs, S. and Cuatrecasas, P. (1979). Modulation of binding and bioactivity of insulin by anti-insulin: relation to possible role of receptor self-aggregation in hormone action. Proceedings of the National Academy of Sciences, U.S.A. **76,** 2720-2724.

27 Schechter, Y., Hernaez, L., Schlessinger, J. and Cuatrecasas, P. (1979). Local aggregation of hormone-receptor complexes is required for activation by epidermal growth factor. Nature **278,** 835-838.

28 Tanaka, T., Aiba, T. and Shishiba, Y. (1984). Discrepancy between immunoactivity and bioactivity of big-big and big human growth hormones in acromegaly. Endocrinologia Japonica **31,** 133-140.

29 Thomas, H., Green, I.C., Wallis, M. and Aston, R. (1987). Heterogeneity of growth-hormone receptors detected with monoclonal antibodies to human growth hormone. Biochemical Journal **243,** 365-372.

30 Wallis, M., Daniels, M., Ray, K.P., Cottingham, J.D. and Aston, R. (1987). Monoclonal antibodies to bovine growth hormone potentiate effects of the hormone on somatomedin C levels and growth of hypophysectomised rats. Biochemical and Biophysical Research Communications **149,** 187-193.

TRANSGENICS

INSERTION OF GROWTH HORMONE GENES INTO PIG EMBRYOS

V.G. Pursel[1], K.F. Miller[1], D.J. Bolt[1], C.A. Pinkert[2], R.E. Hammer[2], R.D. Palmiter[3] and R.L. Brinster[2]

[1] U.S. Department of Agriculture, Agricultural Research Service, Beltsville, Maryland, U.S.A.
[2] University of Pennsylvania, School of Veterinary Medicine, Philadelphia, Pennsylvania, U.S.A.
[3] University of Washington, Howard Hughes Medical Institute, Seattle, Washington, U.S.A

Recent research has clearly demonstrated that genes coding for growth hormone (GH) and growth hormone releasing factor (GRF) can be integrated into the genome of domestic swine by microinjection of the cloned genes into a pronucleus or nucleus of fertilized pig ova. The percentage of injected ova that developed into transgenic pigs varied from 0.30% to 1.69%. The proportion of transgenic pigs that expressed the gene varied from 17% to 100%, depending upon the composition of the fusion gene. The elevation of foreign GH in expressing transgenics stimulated the expected elevation in insulin-like growth factor (IGF-I). Enhanced growth rate has been reported for some transgenic GH pigs but not for others, with the difference possibly due primarily to dietary constraints and general health of the pigs. Expression of the GH genes has markedly reduced subcutaneous fat and improved the efficiency of converting feed into meat. The persistent excess GH in transgenic pigs was detrimental to general health; lameness, lethargy, and gastric ulcers were the most prevalent problems. Females that expressed foreign GH genes were anestrus. Most of the transgenic pigs that were reproductively sound transmitted the gene to a portion of their progeny.

INTRODUCTION

The technology for introducing cloned genes into animals has been available only since 1980. Animals that have integrated a cloned gene into the genome are called transgenic (6).

Insertion of genes into mice has become extraordinarily useful for studies of gene function, developmental biology, physiology and immunology. The potential that gene transfer offers was most dramatically demonstrated by the production of the transgenic "super mouse", which was the consequence of a high concentration of rat growth hormone (rGH) being secreted. Palmiter and coworkers (11) used the regulatory (promoter) sequences of a mouse metallothionein (MT) gene to direct the transcription of the rGH structural gene and thereby avoid the usual hormonal feedback mechanism that regulates GH.

Stimulation of growth rate is not the only characteristic of GH that might be of considerable economic value to livestock production. Rather, increased feed efficiency, lean muscle mass, and milk production, along with reduced subcutaneous fat, are all attributes resulting from elevation of GH. For these reasons, interest in the production of transgenic animals with GH fusion genes has been intense. The purpose of this report is to review the progress of the efforts by several research groups to produce transgenic pigs with several GH fusion genes.

INTEGRATION OF FUSION GENES

Although transgenic animals can be produced by use of retrovirus infection or introduction of transformed embryonic stem cells into the blastocyst, to date only microinjection of DNA into a pronucleus or nucleus has been successfully used in pigs. The microinjection process is similar to that used routinely for mice except that the opacity of the cytoplasm of pig ova makes visualization of the pronuclei extremely difficult. Centrifugation of pig ova at 7000 to 15000 x G for 3 to 5 min stratifies the cytoplasm and results in the pronuclei of 1-cell ova or nuclei of 2-cell ova becoming discernible in the equatorial segment with the aid of interference contrast microscopy. The centrifugation technique developed by Wall et al. (20) has been used in all gene transfers reported to date for swine.

The percentage of injected pig ova that survive microinjection and are born has varied from 4% to 25% (Table 1). A number of factors influence this. Vize et al. (19) reported that no pregnancies survived to term when an average of only 13 injected ova were transferred to recipients; therefore, they subsequently transferred 30 injected ova. We found that injected 2-cell ova survived to term better than injected 1-cell ova (15). Other factors one would expect to influence survival are the skill of the micromanipulator, the duration of in vitro culture, the concentration and form of the DNA, and the synchrony of donor and recipient at the time of embryo transfer.

A number of fusion genes containing the coding region for growth hormones have now been successfully integrated into the pig genome (Table 1). Human growth hormone (hGH) structural gene driven by a murine metallothionein-I (MT) promoter was used to produce the first transgenic pigs (1, 9). Subsequently, the MT promoter fused to bovine growth hormone (bGH) structural gene was integrated into pigs (16); and most recently, the phosphoenolpyruvate carboxykinase (PEPCK) promoter fused to bGH has been integrated into pigs (21). While most of the genes that have been transferred into pigs contained genomic sequences that included introns, the Moloney murine leukemia virus (MLV)-rGH fusion gene contained only the cDNA for rGH (4). The fusion gene transferred into pigs by Vize et al. (19) was composed of a human MT-IIA promoter ligated to the pGH coding region plus the last exon and 3 noncoding sequences of pig genomic DNA. In addition, the MT promoter and mouse albumin promoter (Alb) were fused to a human growth hormone releasing factor (hGRF) minigene composed of a mixture of cDNA and genomic DNA and transferred into pigs (14).

The efficiency of producing transgenic pigs, expressed as the percentage of injected ova, varied from 0.30 to 1.69 (Table 1). When several of the same fusion genes were microinjected into mice, the efficiency was about 3%.

The number of transgene copies per cell that integrated into pigs varied from 1 to 490 for MT-hGH (9), < 1 to 28 for MT-bGH (10), < 1 to 15 for MT-pGH (19), and 8 for MLV-rGH (4), with most integrations probably at a single site. Southern blot analysis showed that transgenic pigs with MT-hGH contained intact copies of the fusion gene with many oriented in tandem head-to-tail arrays and some in a head-to-head configuration (9). Only one of four MT-pGH transgenic pigs studied contained the transgene organized in the head-to-tail array (19).

EXPRESSION OF INTEGRATED GENES

Immunodetectable concentrations of hGH, bGH and rGH and elevated pGH and GRF were present in the blood plasma of many of the transgenic pigs (Table 1). The proportion of transgenic pigs that expressed the inserted gene appeared to vary considerably. Only 1 of 6 pigs expressed the MT-pGH gene (19) and only 2 of 8 pigs expressed the MT-hGRF gene (14), while 11 of 18 and 8 of 11 pigs expressed the MT-hGH and MT-bGH genes, respectively (9, 16).

Table 1 Efficiency of transferring growth-related genes into pigs

Fusion gene	Citation	Ova injected (no.)	Offspring (no.)	Offspring (%)	Transgenic[a] (no.)	Transgenic[a] (%)	Expressing[b] (no.)	Expressing[b] (%)
MT-hGH	9	2035	192	9.4	20	0.98[a]	11/18	61[b]
MT-hGH	1	268	15	5.6	1	0.37	c	c
MT-bGH	16	2198	149	6.8	11	0.50	8/11	73
MLV-rGH	4	59	15	25.4	1	1.69	1/1	100
MT-pGH	19	423	17	4.0	6	1.42	1/6	17
MT-hGRF	14	2627	238	8.9	8	0.30	2/8	25
Alb-hGRF	14	968	132	13.6	5	0.52	3/3	100
PEPCK-bGH	21	1057	c	c	c	c	2	c

[a] Percentage of injected ova resulting in a pig with gene integration
[b] Percentage of transgenic pigs expressing the fusion gene
[c] Data incomplete

Possibly the lack of introns in the MT-pGH construct was responsible for the low proportion of pigs expressing the gene. Brinster et al. (2) reported that a large number of cDNA-based constructs were either not expressed or were expressed poorly when integrated into mice.

In regard to the MT-hGRF gene, only 25% of transgenic pigs and 14% of the transgenic lambs (17) expressed the gene. This finding was not anticipated because 11 of 14 transgenic mice expressed the gene (8). A similar disparity in expression of MT-pGH by transgenic pigs (1 of 6) and mice (13 of 18) was reported (18, 19). Integration of the Alb-hGRF construct in pigs resulted in 3 of 3 expressing the gene even though the same hGRF minigene with one intron was involved in both the MT-hGRF and Alb-hGRF constructs (14). Apparently, the albumin promoter/regulatory sequences provided something that overcame a deficiency in the MT-hGRF construct.

The proportion of transgenic pigs producing hGH was similar over a wide range of gene copies per cell (Table 2). Therefore, whether the MT-hGH gene was expressed or not was unrelated to the number of gene copies integrated into the genome.

Table 2 Relationship of integrated copies of gene per cell and expression of MT-hGH gene in transgenic pigs (9)

Gene copies per cell	Number of pigs: Integrated gene	Number of pigs: Expressed gene
1	4	2
2-10	7	5
11-100	4	3
>100	3	2

Plasma GH concentrations at birth ranged from 3 to 949 ng/ml for hGH (9) and 5 to 944 ng/ml for bGH (10). The single transgenic pig expressing the MLV-rGH gene had 500 to 1300 ng rGH/ml plasma (4), and one pig expressing MT-pGH had 28 ng pGH/ml plasma (19). The number of gene copies per cell and concentration of plasma hGH were not related in MT-hGH transgenic pigs (9). However, the concentration of plasma bGH was positively correlated with the number of gene copies per cell in MT-bGH transgenic pigs (10).

In pigs expressing MT-hGH or MT-bGH, the concentrations of foreign GH during the first 180 days varied from 3 to 8710 ng/ml (10). The most likely cause of variability is that the site of integration and consequently the genomic sequences on both sides of the integration site differ for each transgenic pig and influence the level of gene expression. However, each individual pig tended to maintain a characteristic level of expression. Variation within pigs did not seem to correspond to age or body weight, since plasma GH in some pigs increased with time while others decreased with time (10).

Considerable evidence has been collected to support the conclusion that the foreign GH produced in transgenic pigs is biologically active. Transgenic pigs expressing the hGH gene rarely had detectable concentrations of plasma pGH (10), which indicates the negative feedback mechanism was functioning. Furthermore, insulin-like growth factor-I (IGF-I) concentrations were 2-fold to 7-fold higher in pigs transgenic with hGH, bGH and rGH than in control pigs or non-expressing transgenic pigs (4, 10), which indicates foreign GH was able to bind to GH receptors of hepatocytes to stimulate IGF-I synthesis.

The concentrations of GRF in plasma of transgenic pigs with MT-hGRF and Alb-hGRF were 130 to 380 pg/ml and 400 to 8000 pg/ml, respectively (14). These values are 10- to 500-fold higher than plasma GRF concentrations found in littermate control pigs. However, most of the plasma GRF is the 3-44 metabolite (L.A. Frohman, unpublished data), which is consistent with the lack of a detectable increase in concentration of plasma pGH in the transgenics when compared to the littermate controls (14).

The MT promoter appears to provide appropriate tissue-specific expression of the transgenes in pigs. MT-hGH and MT-bGH genes produced messenger RNA in liver, kidney, testis, adrenal and several other tissues (14), while rGH mRNA was highest in spleen, lung, colon and jejunum with lesser amounts in the kidney, lymph nodes and bone marrow (4).

Concentrations of foreign GH in plasma and GH mRNA in tissues were considerably lower in transgenic pigs harbouring the MT-hGH and MT-bGH genes than in transgenic mice harbouring the same fusion genes. In mice, concentrations of hGH and bGH in plasma were frequently elevated more than 10-fold after zinc stimulation (7, 12). In contrast, addition of 1000 to 3000 ppm zinc to the feed resulted in little more than a doubling of the bGH concentration in plasma of transgenic pigs (14).

PERFORMANCE AND PHYSIOLOGICAL CHARACTERISTICS

The enhanced growth rate and body size of transgenic mice that expressed foreign GH genes brought the expectation that integration of similar fusion genes into the pig genome might produce comparable results. This expectation was not realized in the founder population of MT-hGH and MT-bGH transgenic pigs (9, 16). However, in a subsequent study (13), daily weight gains were 16.5% faster for MT-bGH transgenic pigs than for littermate control pigs when dietary protein was not limiting during the 30 to 90 kg growth period (Table 3).

Table 3 Growth performance of MT-bGH transgenic pigs and control pigs from 30 to 90 kg body weight (13)

		Daily gain (g/day)		Feed efficiency (kg feed/kg gain)	
Pigs	N	Mean	SE	Mean	SE
Control	14	774	21	2.88	0.10
Transgenic	6	902	35	2.38	0.16
P value		0.006		0.018	

Diet contained 18% crude protein plus 0.25% lysine

In addition, a transgenic pig expressing MT-pGH gained 492 g per day faster than littermate control pigs during the 20 to 90 kg growth period (19). Based on several recent studies of pigs treated with exogenous pGH, it seems likely that the growth rate of pigs will not respond to the extent found in transgenic mice because appetite depression accompanies elevated GH in pigs (3, 5). In comparison to littermate controls, feed intake was depressed 20% in one group of MT-bGH transgenic pigs (14) and 17% in pigs fed ad libitum in the data shown in Table 3, which are comparable to a 14% and 17% depression in feed intake reported for pigs injected with pGH (3, 5).

The founder population of MT-bGH transgenic pigs were 16% more efficient in converting feed into body weight gains than littermate controls (14). A subsequent MT-bGH generation of transgenics was 17.4% more efficient than their sibling control pigs (Table 3). Similar improved feed efficiencies of 23% (3) and 25% (5) were reported for pigs injected with exogenous pGH in comparison to littermate controls.

The elevation of foreign GH in transgenic pigs with MT-hGH and MT-bGH have resulted in marked repartitioning of nutrients from subcutaneous fat into other carcass components, including muscle, skin, bone and certain organs. Ultrasonic estimates or slaughter measurements of backfat thickness at the tenth rib of hGH and bGH transgenic pigs at about 90 kg body weight averaged 7.0 mm and 7.9 mm, respectively, while littermate control pigs averaged 18.5 and 20.5 mm, respectively ($P<0.01$ in each case, 14). Additionally, the backfat measurements do not adequately reflect the lack of subcutaneous fat in the transgenic pigs because the skin over the tenth rib was about 1 mm thicker for transgenic pigs than for control pigs (V.G. Pursel and M.B. Solomon, unpublished data).

Ebert et al. (4) reported that by 9 months of age, a transgenic boar with MLV-rGH was 26% heavier and linear bone growth of fore and hind limbs was greater than for a littermate control boar. In contrast, MT-hGH and MT-bGH transgenic pigs have not grown to a larger body size, and the femur, tibia and humorus of MT-bGH transgenics at 8 and 10 months of age were not longer than for full sib controls (14). Additional investigation is required to determine whether this difference is due to structural differences of bGH and rGH that affect binding to GH receptors in epiphyseal chondrocytes, to some other physiological factor, or whether the single MLV-rGH transgenic boar represents a unique occurrence.

Transgenic pigs expressing MT-hGH and MT-bGH were moderately hyperglycemic, averaging 10 to 40 mg/dL above littermate control pigs, and insulin concentrations were elevated about 20-fold above littermates (14). A similar degree of hyperglycemia has been reported in pigs treated with exogenous pGH (3, 5), and concentrations of serum insulin increased 2- to 7-fold above control pigs. In contrast, a MLV-rGH transgenic pig was consistently hyperglycemic (serum glucose more than 3-fold higher than normal), and glucosurea was observed (4).

The MT-hGH and MT-bGH transgenic pigs exhibited a number of notable health problems, including lameness, susceptibility to stress, peptic ulcers, parakeratosis, lethargy, anestrus in gilts, and lack of libido in boars (14, 16). Pathology in joints, characteristic of osteochonditis dissecans, was also observed in the MLV-rGH transgenic pig (4) and in some groups of pigs treated with exogenous pGH for 57 days (5). The health problems that have been observed in pigs exposed to high concentrations of GH are all conditions that are prevalent in the pig population but at a much lower incidence and with less severity. Possibly these ailments would not be exhibited in pigs expressing GH fusion genes if the foundation stock was not predisposed to such conditions.

TRANSMISSION OF TRANSGENES

If transgenic livestock are produced that are proven to have economic importance to the livestock industry, it will be essential that the transgenes be transmitted to progeny in a consistent manner. Of 12 transgenic pigs tested to date, 10 pigs successfully transmitted the transgene (MT-hGH, MT-bGH, MT-pGH or MT-hGRF) to one or more progeny (14, 16, 19). One of six MT-hGH pigs (16) and one of three MT-bGH (14) transgenic boars that failed to transmit the transgene to their progeny were probably mosaic for the gene, with integration only in the somatic cells. In another MT-hGH boar, integration was evidently mosaic in the germline since the transgene was only transmitted to 1 of 33 progeny (16). Mosaicism in the germline is reported to occur in 25 to 36% of transgenic mice produced by microinjection (22).

Germline transmission has been obtained from both expressing and non-expressing pigs. All transgenic progeny with MT-hGH, MT-bGH, or MT-hGRF fusion genes have expressed the transgene if their sire also expressed the gene (14, 16). However, transgenic progeny did not express the MT-hGH or MT-pGH fusion gene if their parent was a non-expressing transgenic (14, 16, 19).

The results suggest that transgenes become stably integrated into the pig genome and can be expected to function in progeny in the same manner as they did in the founder transgenic pig.

CONCLUSIONS

Recent research has clearly demonstrated that GH fusion genes can be integrated into the pig genome. Many of the physiological effects of a transgene expression were similar to effects obtained when pGH was administered by daily injection. However, the over-production of GH in transgenic pigs negatively affects the health of the pigs. Production of transgenic pigs with only the positive consequences of elevated GH will require use of gene regulators that permit tight control over the timing and concentration of GH production.

REFERENCES

1 Brem, G., Brenig, B., Goodman, H.M., Selden, R.C., Graf, F., Kruff, B., Springman, K., Hondele, J., Meyer, J., Winnaker, E.-L. and Krausslich H. (1985). Production of transgenic mice, rabbits and pigs by microinjection into pronuclei. ZF Zuchthygiene **20**, 251-252.

2 Brinster, R.L., Allen, J.M., Behringer, R.R., Galinas, R.E. and Palmiter, R.D. (1988). Introns increase transcriptional efficiency in transgenic mice. Proceedings of the National Academy of Science, U.S.A. **85**, 836-840.

3 Campbell, R.G., Steele, N.C., Caperna, T.J., McMurtry, J.P., Solomon, M.B. and Mitchell, A.D. (1988). Interrelationships between energy intake and endogenous porcine growth hormone administration on the performance, body composition, and protein and energy metabolism of growing pigs weighing 25 to 55 kilograms live weight. Journal of Animal Science **66**, 1643-1655.

4 Ebert, K.M., Low, M.J., Overstrom, E.W., Buonomo, F.C., Baile, C.A., Roberts, T.M., Lee, A., Mandel, G. and Goodman, R.H. (1988). A Moloney MLV-rat somatotropin fusion gene produces biologically active somatotropin in a transgenic pig. Molecular Endocrinology **2**, 277-283.

5 Evock, C.M., Etherton, T.D., Chung, C.S. and Ivy, R.E. (1988). Pituitary porcine growth hormone (pGH) and a recombinant pGH analog stimulate pig growth performance in a similar manner. Journal of Animal Science **66**, 1928-1941.

6 Gordon, J.W. and Ruddle, F.H. (1981). Integration and stable germ line transmission of genes injected into mouse pronuclei. Science **214**, 1244-1246.

7 Hammer, R.E., Brinster, R.L. and Palmiter, R.D. (1985). Use of gene transfer to increase animal growth. Cold Spring Harbor Symposium in Quantitative Biology **50**, 379-387.

8 Hammer, R.E., Brinster, R.L., Rosenfeld, M.G., Evans, R.M. and Mayo, K.E. (1985). Expression of human growth hormone releasing factor in transgenic mice results in increased somatic growth. Nature **315**, 413-416.

9 Hammer, R.E., Pursel, V.G., Rexroad, C.E. Jr., Wall, R.J., Bolt, D.J., Ebert, K.M., Palmiter, R.D. and Brinster, R.L. (1985). Production of transgenic rabbits, sheep and pigs by microinjection. Nature **315**, 680-683.

10 Miller, K.F., Bolt, D.J., Pursel, V.G., Hammer, R.E., Pinkert, C.A., Palmiter, R.D. and Brinster, R.L. (1988). Expression of human or bovine growth hormone gene with a mouse metallothionein I promoter in transgenic swine alters the secretion of porcine growth hormone and insulin-like growth factor-1. Journal of Endocrinology (in press).

11 Palmiter, R.D., Brinster, R.L., Hammer, R.E., Trumbauer, M.E., Rosenfeld, M.G., Birnberg, N.C. and Evans, R.M. (1982). Dramatic growth of mice that develop from eggs microinjected with metallothionein-growth hormone fusion genes. Nature **300**, 611-615.

12 Palmiter, R.D., Norstedt, G., Gelinas, R.E, Hammer, R.E. and Brinster, R.L. (1983). Metallothionein-human GH fusion genes stimulate growth of mice. Science **222**, 809-814.

13 Pursel, V.G., Campbell, R.G., Miller, K.F., Behringer, R.R., Palmiter, R.D. and Brinster, R.L. (1988). Growth potential of transgenic pigs expressing a bovine growth hormone gene. Journal of Animal Science **66** (Suppl. 1), 267 (Abstract).

14 Pursel, V.G., Miller, K.F., Bolt, D.J., Campbell, R.G., Pinkert, C.A., Mayo, K.E., Frohman, L.A., Palmiter, R.D., Brinster, R.L. and Hammer, R.E. (1989). Manuscript in preparation.

15 Pursel, V.G., Miller, K.F., Pinkert, C.A., Palmiter, R.D. and Brinster, R.L. (1987). Effect of ovum cleavage stage at microinjection on embryonic survival and gene integration in pigs. 11th International Congress on Animal Reproduction and Artificial Insemination, Dublin **4,** 480.

16 Pursel, V.G., Rexroad, C.E. Jr., Bolt, D.J., Miller, K.F., Wall, R.J., Hammer, R.E., Pinkert, C.A., Palmiter, R.D. and Brinster, R.L. (1987). Progress on gene transfer in farm animals. Veterinary Immunology and Immunopathology **17,** 303-312.

17 Rexroad, C.E. Jr. and Pursel, V.G. (1988). Status of gene transfer in domestic animals. 13th International Congress on Animal Reproduction and Artificial Insemination, Dublin **5,** 29-35.

18 Vize, P.D. (1987). Expression of porcine growth hormone in bacteria and transgenic animals. PhD Thesis. University of Adelaide, Adelaide, South Australia.

19 Vize, P.D., Michalska, A.E., Ashman, R., Lloyd, B., Stone, B.A., Quinn, P., Wells, J.R.E. and Seamark, R.F. (1988). Introduction of a porcine growth hormone fusion gene into transgenic pigs promotes growth. Journal of Cell Science **90,** 295-300.

20 Wall, R.J., Pursel, V.G., Hammer, R.E. and Brinster, R.L. (1985). Development of porcine ova that were centrifuged to permit visualization of pronuclei and nuclei. Biology of Reproduction **32,** 645-651.

21 Wieghart, M., Hoover, J., Choe, S.H., McCrane, M.M., Rottman, F.M., Hanson, R.W. and Wagner, T.E. (1988). Genetic engineering of livestock - transgenic pigs containing a chimeric bovine growth hormone (PEPCK/bGH) gene. Journal of Animal Science **66** (Suppl. 1), 266.

22 Wilkie, T.M., Brinster, R.L. and Palmiter, R.D. (1986). Germline and somatic mosaicism in transgenic mice. Developmental Biology **118,** 9-18.

INDUCED EXPRESSION OF A BOVINE GROWTH HORMONE CONSTRUCT IN TRANSGENIC PIGS

E.J.C. Polge[1], S.C. Barton[2], M.A.H. Surani[2], J.R. Miller[2], T. Wagner[3], F. Rottman[3], S.A. Camper[3], K. Elsome[1], A.J. Davis[2], J.A. Goode[2], G.R. Foxcroft[4] and R.B. Heap[2]

[1] Animal Biotechnology Cambridge, 307 Huntingdon Road, Cambridge CB3 0JQ, U.K.
[2] AFRC Institute of Animal Physiology and Genetics Research, Babraham, Cambridge CB2 4AT, U.K.
[3] Edison Animal Biotechnology Center, Ohio University, Athens, Ohio 45701, U.S.A.
[4] Department of Physiology and Environmental Science, University of Nottingham, Sutton Bonington, Loughborough LE12 5RD, U.K.

A construct consisting of a bovine prolactin promoter that directs transcription of the bovine growth hormone gene (bPRL.bGH) has integrated into pigs following microinjection into the pronuclei of 289 fertilized ova. Five of 20 live young were transgenic; the overall transmission of the transgene from founder stock to progeny was 42%. The bGH concentrations in systemic blood did not increase above the normal physiological range (less than 20 ng/ml) and growth rate was normal in all transgenic pigs. Episodic release of bGH was induced by sulpiride, a dopamine antagonist, and thyrotrophin releasing hormone (TRH) showing that the foreign gene was regulated through its promoter by the normal feedback mechanisms that control prolactin secretion, and that activity could be enhanced in a controllable manner.

INTRODUCTION

Gene insertion into mammalian genomes is important for studies of developmental biology, physiology and immunology. Original studies with growth hormone genes in mice demonstrated that introduction of a rGH structural gene directed by a mouse metallothionein (MT) promoter had a striking effect on subsequent growth rate. Later studies in pigs and sheep have so far produced less convincing results. Comparisons of the effects of fusion genes consisting of a MT (or other) promoter to direct transcription of a hGH or bGH structural gene show that growth rate is not consistently increased nor feed efficiency enhanced (3, 6). One of the reasons seems to relate to the unregulated expression of GH which results in excessively high concentrations of the hormone in systemic circulation. Attendant problems of decreased appetite, joint pathology and reproductive disorders has led to the view that prolonged expression of the GH gene at a high level is detrimental to animal health and welfare even though there may be a considerable improvement in the reduced deposition of fat (3, 6).

An alternative approach to the manipulation of animal performance may be required in domesticated farm animals already selected over many generations for desirable quantitative traits such as high growth rate, feed efficiency and reproductive performance.

Whereas experiments in mice show that the expression of the fusion gene, mMT.rGH, may be regulated by the introduction of Zn^{2+} into the drinking water, similar constructs are apparently overexpressed in pigs and sheep on a normal dietary regimen. We have therefore utilised a different construct, a bovine prolactin promoter genes to direct transcription of the bovine growth hormone gene. There were three reasons why this strategy was of potential interest. Camper and colleagues (1) have demonstrated that a 1-kilobase DNA fragment containing the promoter of the bovine prolactin gene will direct transcription of the chloramphenicol acetyltransferase gene after transfection into GH_3 cells, a rat pituitary tumour cell line, but not into COS-1 or HeLa cells. Their results imply that the promoter is likely to be expressed in pituitary cells in vivo but perhaps without any species specificity. Secondly, suppression derived from indigenous prolactin-inhibitory factors such as dopamine would result in a low expression of the bGH gene under normal circumstances. Thirdly, the availability of a specific radioimmunoassay for bGH would enable the expression of the bGH gene to be distinguished from that of the indigenous gene and enable studies to be carried out with appropriate antagonists and agonists predicted to influence promoter activity.

PRODUCTION OF TRANSGENIC PIGS

Animals were obtained from the Institute's herd of Large White pigs. Fertilised eggs were recovered from the oviducts of donors 52-54 h after an injection of hCG, and about 24 h after artificial insemination. A total of 289 eggs at the pronuclear stage were injected after centrifugation with a bGH gene directed by bovine prolactin (PRL) gene promoter (bPRL.bGH; (1)). Injected embryos were transferred to 12 recipients and 4 sows gave birth to a total of 20 piglets of which 5 (20%) were deemed to be transgenic by Southern blot analysis.

INTEGRATION OF TRANSGENE

Identification of integration of the foreign gene was achieved with a probe consisting of an internal 1.2 kb PVU II fragment from the bovine growth hormone gene (Fig. 1). Genomic DNA was prepared from ear punches digested with PVU II and hybridised with the 1.2 kb fragment. Copy number ranged from less than one copy per cell to >10 copies per cell. The animal with the highest copy number died shortly after analysis. The remaining four animals (nos. 25, 28, 35 and 39) were outcrossed to produce a total of 66 offspring, of which 28 were transgenic, giving an overall transmission rate of 42%. Transmission rate in individual litters varied from 0 to 77% as indicated in Fig. 2.

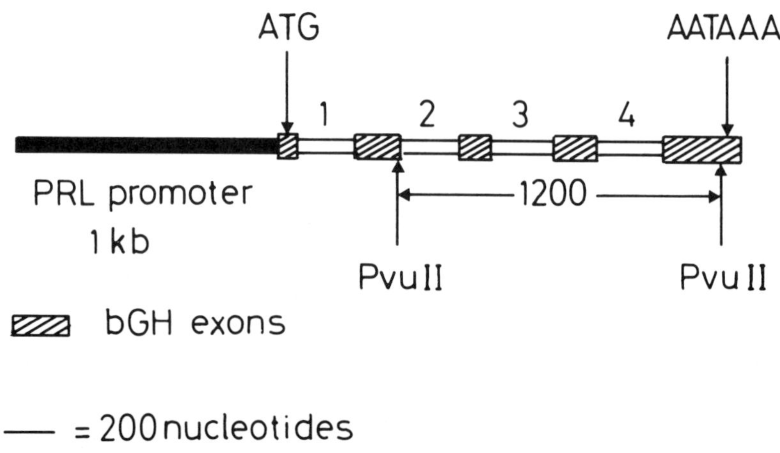

Fig. 1 Bovine prolactin promoter-bovine growth hormone construct showing the fragment used to detect integration (1.2 kb Pvu II fragment). The respective locations of the initiation codon and polyadenylation signal are indicated.

Family trees for each line show that male 25 sired two litters, and 50% of the progeny in each litter had inherited the transgene. In each case 80% (4 of 5) of the transgenic offspring were male. Female 28 had one litter and passed on the transgene to 40% of her progeny. Male 35 sired two litters with transmission frequencies of 16% and 25%, respectively. Male 39 sired two litters, the first of 5 piglets contained no transgenics but the second of 13 piglets had 10 transgenics, a transmission rate of 77% (Fig. 2). The overall frequency of transmission was equally distributed between males and females (14 males, 14 females), though in one litter all transgenic progeny were female (Fig. 2; 35 male x A46 female).

EXPRESSION OF TRANSGENE

Fig. 3 shows that the expression of the bPRL.bGH transgene was low as reflected in plasma concentrations of bGH in transgenic animals between days 73 and 280 after birth. The highest mean values were found in pigs 25 and 42 both of which showed positive results for bPRL.bGH integration. Concentrations of plasma bGH in other animals were indistinguishable from those in non-transgenic animals. Assay sensitivity (defined as 2 x SD of zero concentration of bGH and using guinea-pig anti-bGH antiserum, kindly provided by Dr I.C. Hart) was 0.5 ± 0.3 ng/ml. Values in non-transgenic animals derived either from non-specific cross-reactivity with compounds in pig plasma, or a low sustained level of bGH secretion in apparently non-transgenic animals. It is unlikely that they derived from cross-reactivity with endogenous pGH since the antiserum showed little recognition of this molecule (bGH, 100%; pGH, 6.4%). The data contrast markedly with those from animals in which a metallothionein promoter was used in which plasma bGH concentrations were high or excessively high.

Low plasma bGH concentrations were associated with growth rates that were unchanged by integration of the transgene (Fig. 4). During the growth phase animals were fed ad libitum on diets with standard protein composition (16%). No pathological effects were observed in the animals studied.

INDUCED EXPRESSION

The apparent lack of expression observed in the majority of transgenic pigs raised the question of whether the gene could be induced by appropriate stimuli. Sulpiride is a dopamine antagonist and produces hyperprolactinaemia, whereas thyrotrophin releasing hormone (TRH) stimulates prolactin release by activation of second messenger systems (cyclic AMP and phosphatidylinositides) and shows a strict requirement for calcium in vitro.

Animals were surgically prepared with indwelling catheters placed in a carotid artery and jugular vein. Catheters remained patent for up to 100 days when regularly flushed with sterile saline containing heparin (100 i.u./ml). Sulpiride (Dogmatil, Laboratories Delagrange, Paris) was infused intravenously at 0.8, 1.6 and 3.2 mg/min and arterial or venous blood samples were collected at frequent intervals. Bovine GH and porcine prolactin were measured by radioimmunoassay (2, 5).

Fig. 5 shows that there was a small, simultaneous increase in peripheral concentrations of bGH and pPRL 60-80 min after the start of infusion of 0.8 mg/min sulpiride. The response was much greater at 1.6 mg/min sulpiride. At 3.2 mg/min only pPRL showed a transient increase. Blood samples taken at 20 min intervals over a period of 340 min showed that there were no comparable episodes of bGH or pPRL release in the absence of antagonist infusion.

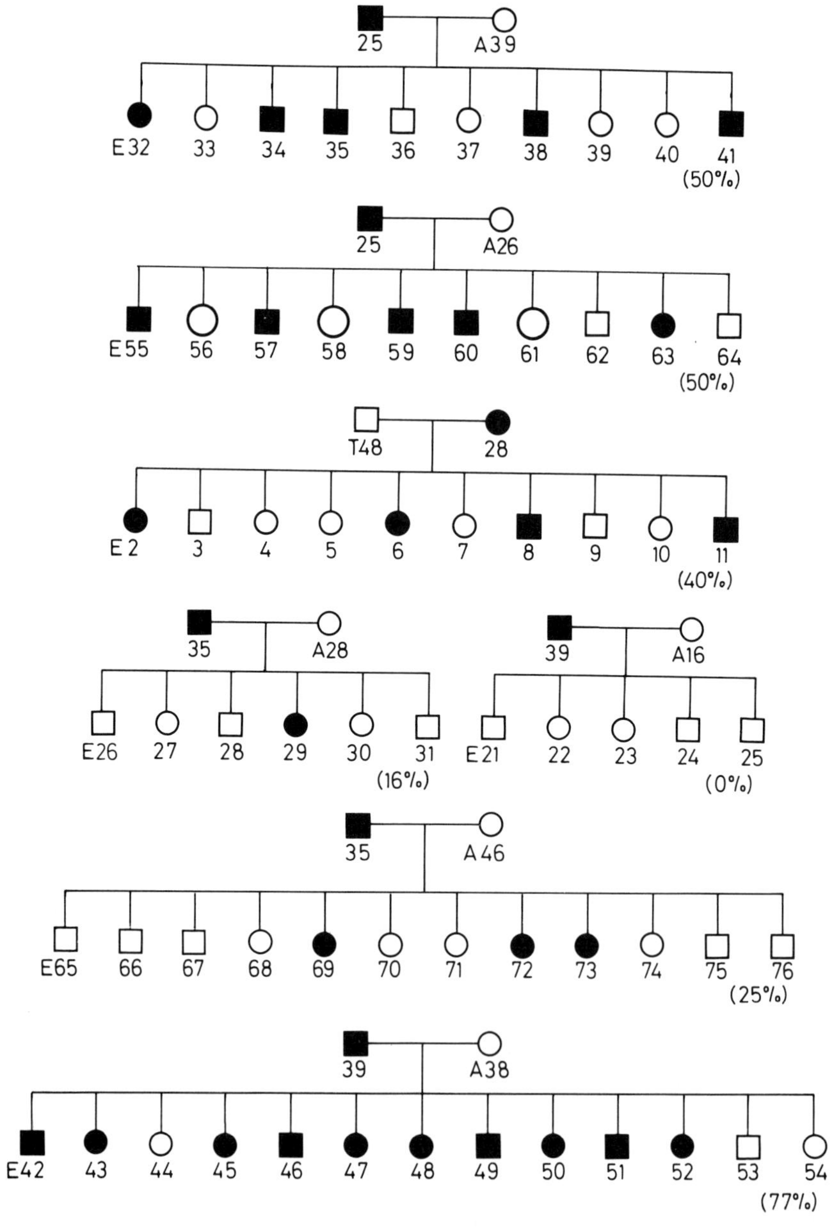

Fig. 2 Transmission of the transgene from the founder stock crossbred to non-transgenic pigs. Diagram shows transgenic animals by black symbols, non-transgenic by open symbols; square denotes male, circle denotes female. Percentage transmission rate given in brackets.

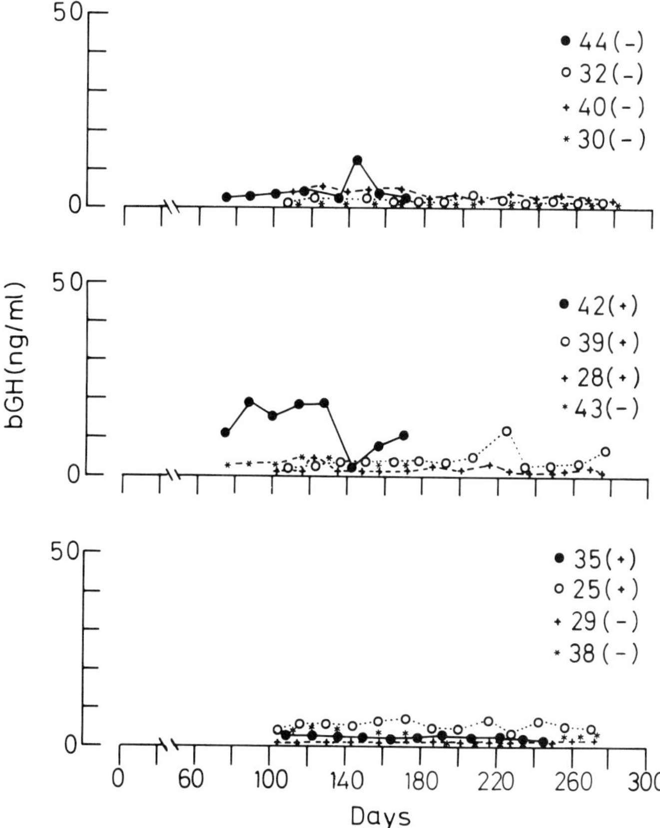

Fig. 3 Concentrations of bovine growth hormone in plasma of transgenic (+) and non-transgenic (-) pigs. Plasma samples were taken by venepuncture 1 to 2 h after feeding.

A pattern of refractoriness after an initial response was observed in other transgenic animals. Fig. 6 shows that induced secretion of bGH was variable with a clear response at 1.6 but not at 3.2 mg/min sulpiride infusion (intra-arterial) on day 1, and at 3.2 but not at 1.6 mg/min on day 2. Similarly, the repeated response in terms of indigenous pPRL secretion did not occur in every instance. The pattern of secretion induced by intra-arterial sulpiride infusion was more gradual in onset compared to that induced by a single intra-arterial injection of TRH. TRH produced a sharp dose-related increase in bGH release immediately after injection (Fig. 6). A similar episodic release was

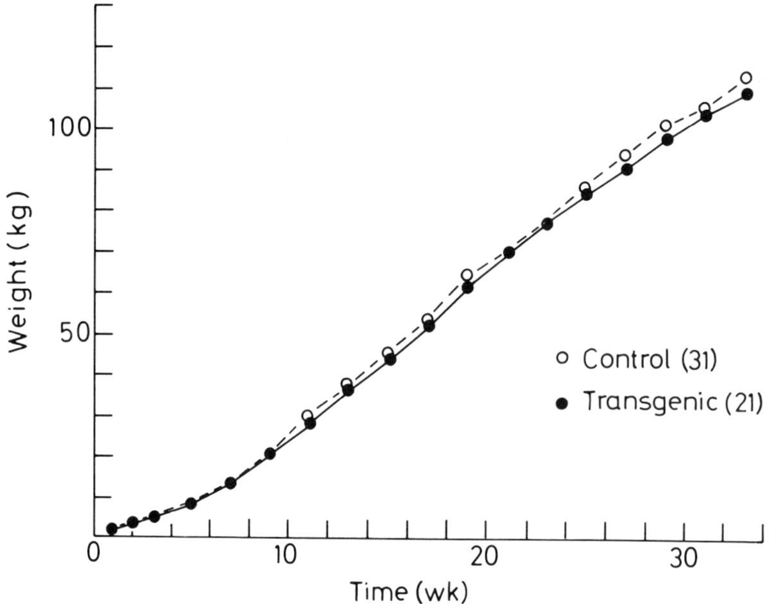

Fig. 4 Growth rates in transgenic and non-transgenic progeny; number of animals given in brackets.

---

observed in another transgenic pig treated with TRH (Fig. 7). There was no bGH response to sulpiride infusion in an animal in which there was no integration of the foreign gene (Fig. 7). The results obtained so far are summarised in Table 1.

COMMENT

The findings of these experiments show that a fusion gene consisting of a bovine PRL promoter to direct transcription of the bovine GH gene is integrated into the pig genome following microinjection of the DNA into the pronucleus of fertilised eggs. The efficiency of transgenic animals obtained was similar to that reported previously (3, 5) and the overall transmission rate of the transgene from founder stock to progeny was 42%, a figure close to that expected with a gene displaying Mendelian inheritance. There was no evidence for an increase in growth rate either of the founder stock or of the progeny. The lack of increase in growth rate was associated with a low circulating concentration of bGH and values greater than 10 ng/ml only occurred in one of five transgenic animals. This finding contrasts with those of others who used a construct consisting of a mouse metallothionein promoter and a bGH or hGH structural gene; values in excess of 1000 ng/ml foreign GH in systemic circulation were frequently obtained. However, it is also possible that bGH has reduced physiological effect in the pig and similar studies with pGH are needed to establish this possibility.

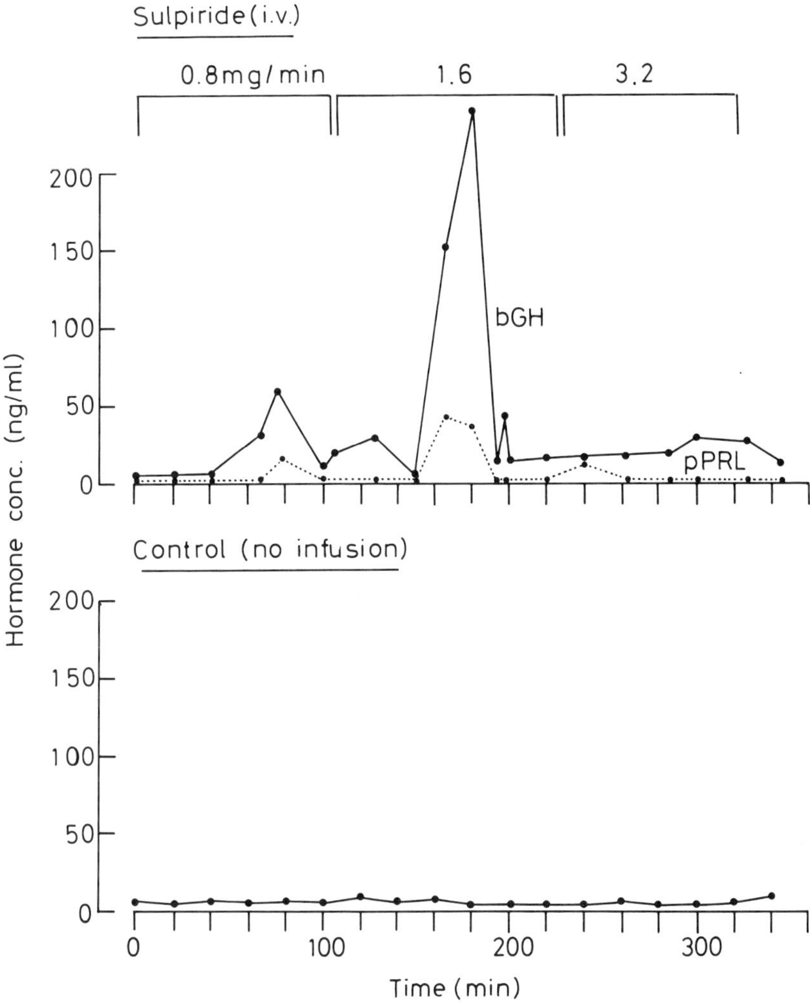

Fig. 5 Induced release of bovine growth hormone (bGH) and porcine prolactin (pPRL) in a transgenic male pig (No. 25). Top panel shows release during sulpiride infusion; lower panel shows results on a different day with no infusion.

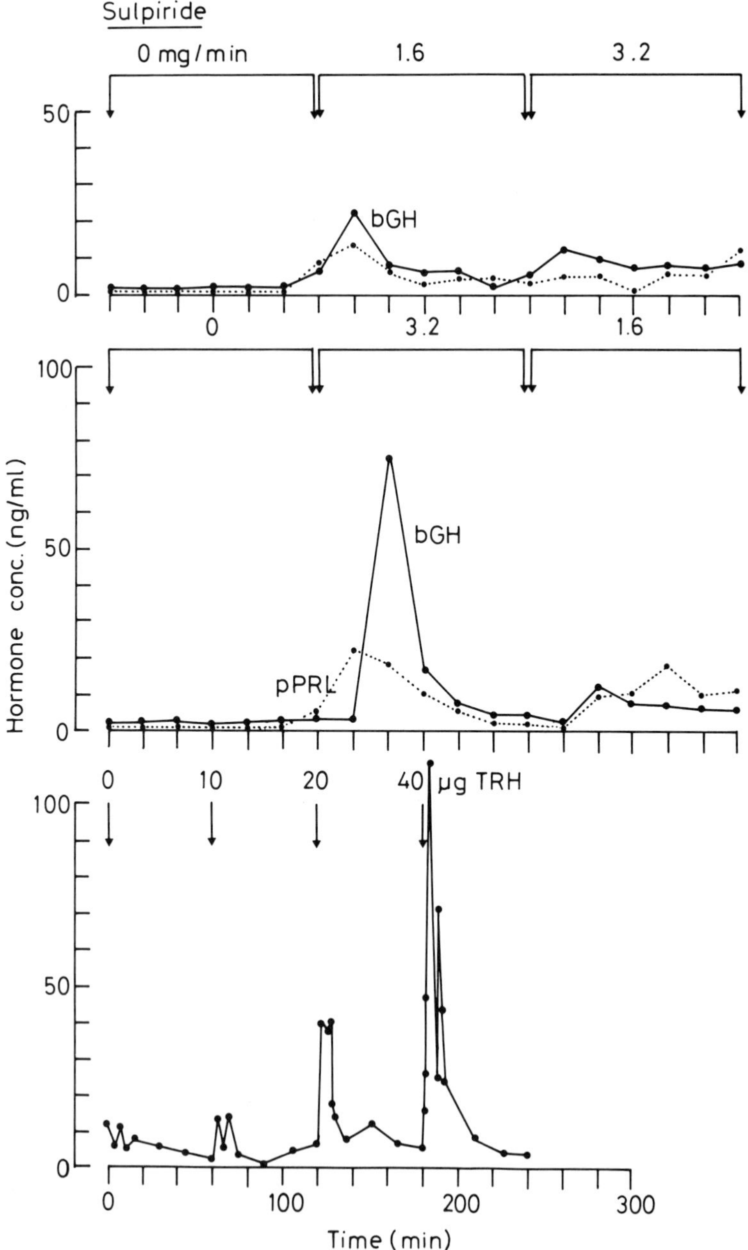

Fig. 6 Induced release of bovine growth hormone (bGH) and porcine prolactin (pPRL) in a transgenic male pig (E59). Top and middle panels show release during sulpiride infusions on two different days; lower panel shows effect of single injections of TRH.

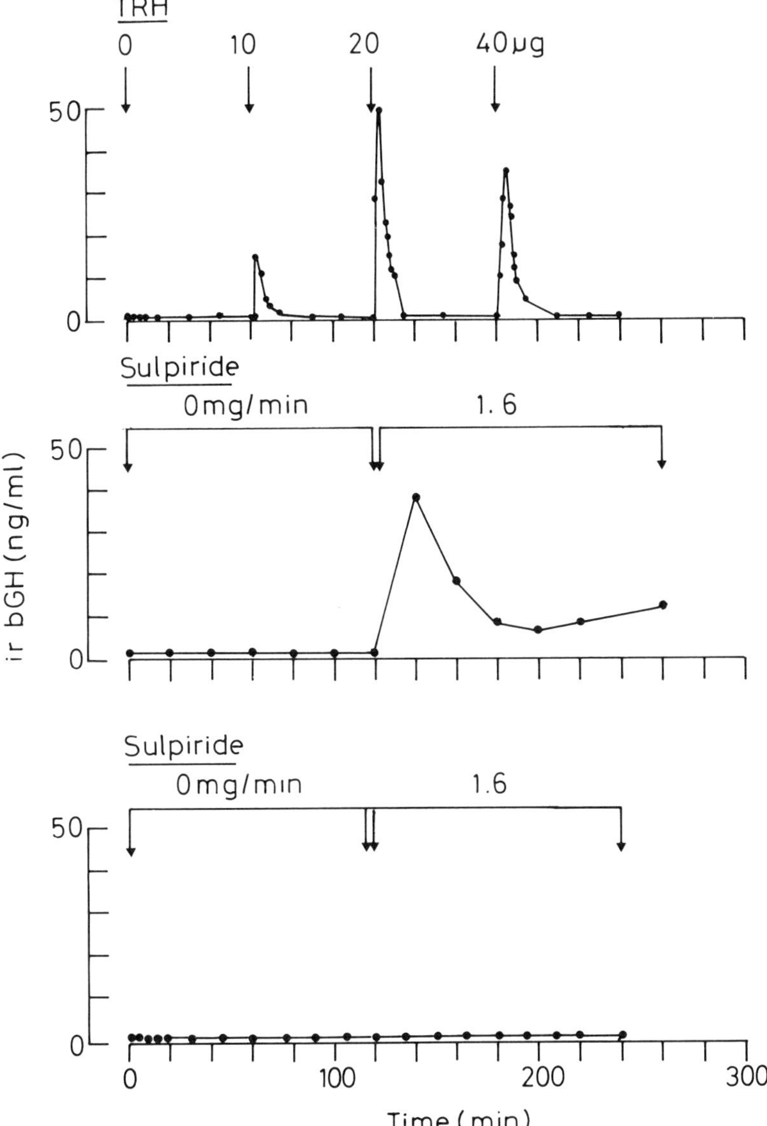

Fig. 7 Induced release of bovine growth hormone in a transgenic male pig (E51). Top panel shows effect of single injections of TRH; middle panel shows response to sulpiride infusion. Lower panel shows results in a non-transgenic male pig (E13).

Table 1 Induced expression of bovine growth hormone in transgenic pigs. Pig embryos were injected at pronuclear stage with fusion gene bPRL.bGH

No. of animal	Integration (Southern blot)	Treatment	
		Sulpiride	TRH
25	Yes	+++	nt
E59	Yes	++	+++
E63	Yes	+	++
E51	Yes	+	+++
E80	No	nt	-
E13	No	-	nt
E84	No	nt	-

nt = not tested
-, +, ++, +++; scoring represents concentration of bGH measured in circulation during continuous infusion of dopamine antagonist, sulpiride, or prolactin secretagogue, TRH.

From our present study it is notable that the bGH gene was induced by agents known to enhance prolactin secretion in other species. The fact that infusions of the dopamine antagonist, sulpiride, enhanced the secretion of bGH implies that the foreign gene was regulated by endogenous dopamine secretion, a well known inhibitor of prolactin release. The inability of sulpiride to cause a sustained release of bGH indicates a depletion of releasable stores, and this was confirmed by the lack of effect of a subsequent infusion on the same day. The close association between bGH and pPRL release during sulpiride infusions suggests that they are secreted by a similar tissue and it will be important to determine whether the integration of the bPRL.bGH construct shows any target cell specificity as was found for the bPRL-chloramphenicol acetyltransferase gene construct (1). The ability of a repeated infusion of sulpiride to enhance indigenous PRL secretion in the absence of a bGH response may reflect a higher level of production of releasable PRL than bGH.

Potentially the most promising finding was the ability of TRH to induce the pulsatile secretion of bGH. The patterned secretion of GH is clearly essential for optimal growth response in species such as the rat (4), and the prospect exists of regulating a foreign GH gene in the pig in an episodic manner during the exponential growth phase.

An intriguing question for the future will be the localisation of the foreign gene product. Is it distributed among lactotrophs, somatotrophs or mixed cells? The simultaneous release of GH and PRL is unusual under normal circumstances, but there are pathological cases involving pituitary tumours and acromegaly in man in which GH and PRL levels may both be raised.

Transgenic pigs with a bPRL.bGH construct show no signs of abnormality, pathology or reproductive disorder, plasma concentrations of bGH are within the normal range of endogenous PRL, and the prospect exists of manipulating the activity of the foreign gene in an episodic manner and at specific periods of the growth phase.

REFERENCES

1 Camper, S.A., Yao, Y.A.S. and Rottman, F.M. (1985). Hormonal regulation of the bovine prolactin promoter in rat pituitary tumor cells. Journal of Biological Chemistry **260**, 12246-12251.

2 Hart, I.C., Flux, D.S., Andrews, P. and McNeilly, A.S. (1975). Radioimmunoassay for ovine and caprine growth hormone: its application to the measurement of basal circulating levels of growth hormone in the goat. Hormone and Metabolic Research **7**, 35-40.

3 Pursel, V.G., Miller, K.F., Bolt, D.J., Pinkert, C.A., Hammer, R.E., Palmiter, R.D. and Brinster, R.L. (1989). Insertion of growth hormone genes into pig embryos. In Biotechnology of Growth Regulation (Heap, R.B., Prosser, C.G. and Lamming, G.E., eds.) p 181. Butterworths, London.

4 Robinson, I.C.A.F. and Clark, R.G. (1988). Growth promotion by growth hormone and insulin-like growth factor-I in the rat. In Biotechnology in Growth Regulation (Heap, R.B., Prosser, C.G. and Lamming, G.E., eds.) p 129. Butterworths, London.

5 Shaw, H.J. and Foxcroft, G.R. (1985). Relationships between LH, FSH and prolactin and reproductive activity in the weaned sow. Journal of Reproduction and Fertility **75**, 17-28.

6 Vize, P.D., Michalska, A.E., Ashman, R., Lloyd, B., Stone, B.A., Quinn, P., Wells, J.R.E. and Seamark, R.F. (1988). Introduction of a porcine growth hormone fusion gene into transgenic pigs promotes growth. Journal of Cell Science **90**, 295-300.

ACCEPTABILITY OF BIOTECHNOLOGY

CRITERIA FOR THE PUBLIC ACCEPTABILITY OF BIOTECHNOLOGICAL INNOVATIONS IN ANIMAL PRODUCTION

T.B. Mepham

Department of Physiology and Environmental Science, University of Nottingham Faculty of Agricultural and Food Sciences, Sutton Bonington, Loughborough, Leicestershire LE12 5RD, U.K.

> "How ... can those who understand the techniques and benefits of biotechnology promote its understanding and acceptance, under the constraints and opportunities of democratic societies?" (10)

INTRODUCTION

The major premise of this paper is that implementation of biotechnological innovations in agriculture will be highly dependent on their public acceptability. The thesis is that this can only be achieved by full and open discussion on risks and benefits, and the aim of the paper is to identify the main areas of concern and to propose a set of criteria to facilitate objective assessment. The social ramifications of biotechnology are extremely wide and their academic implications thus multidisciplinary. Ideally, biotechnological assessment needs to be interdisciplinary (33) and decisions democratic.

CATEGORIES OF SOCIETY

The terms 'society' and 'public' are too amorphous and impenetrable to be meaningful without qualification. For biotechnological, as for any, innovations, some people are more responsive, responsible and/or influential than others. It is thus useful to define some relevant societal categories. Adapting the scheme of Miller (see Ref. 10), we can envisage four groups whose opinions impinge on innovations in animal production: (i) the scientific (biotechnology) community; (ii) the livestock farming community, who stand to gain or lose financially by the innovations; (iii) the 'biotechnologically-attentive' lay public, consumers with a significant degree of knowledge and anticipatory concern; (iv) the 'biotechnologically-inattentive' lay public, who are unlikely to register concern unless or until their own interests are appreciably affected. Group size increases from (i) to (iv), but in global terms there is a fifth, and largest group, (mostly living in less-developed countries) whose poor educational opportunities result in their being totally unaware of biotechnology and thus disenfranchised. The leverage individuals in these groups exert in effecting biotechnological change would seem to be inversely proportional to group size, but as in the mechanical 'law of moments' the relatively large size of a group may compensate for low leverage.

Proponents of biotechnology tend to devote most effort to convincing the most influential individuals (viz, in group (i)) of the value of innovations, but this often leads to resentment among other groups (ii and iii) who consider that their own interests are given insufficient attention. The interests of groups (iv) and (v) are largely unarticulated and thus jeopardized. Prudence suggests that increased attention should be paid to promotion of the value of biotechnology to groups (ii) and (iii): equity requires that they also act as custodians of the interests of others (groups (iv and v)). Ultimately, political decisions have to be taken in accordance with overall public opinion, as demonstrated by the European Commission's ban on the use of steroids in animal production, even before publication of the Lamming Report (21), clearing them of adverse consequences when appropriately used.

These societal groups differ in their competence to assess evidence relating to biotechnology, eg scientists are more aware of certain purely theoretical matters; farmers of the practicalities of recommended procedures; consumers of the factors which will affect their future patterns of food consumption. Broadly speaking, the criteria of acceptability may be considered under two headings (though there is much overlap), viz. physiological, concerning the extent to which the proposed new technology is explicable in terms of current biological theory; and procedural, relating to the practical consequences of implementing the technology on a large scale. Criteria for these two levels of acceptability will be considered in turn.

The range of techniques subsumed by the term 'biotechnology' is very wide and space permits only a partial analysis. Consequently, it is proposed to concentrate on aspects of recombinant-DNA technology, and, consistent with the theme of the conference, to focus on two aspects involving growth hormone (GH) viz. (a) administration of exogenous rGH to lactating (eg Ref. 3) or growing (eg Ref. 11) animals; (b) production of transgenic, or otherwise genetically-modified, farm animals carrying increased gene copy numbers for GH (eg Ref. 39).

PHYSIOLOGICAL CRITERIA

The aim of both exogenous GH treatment and production of transgenic animals with increased copies per cell of the GH gene is to augment supply of GH to target tissues and thereby promote milk secretion or growth of lean body mass. But it is inconceivable that these are the only physiological processes affected. It is thus important to establish as soundly as possible the totality of the effects induced, to define safety limits and ensure predictability and control.

W.B. Cannon in the 1930s first coined the term 'homeostasis' to describe the "coordinated physiological processes which maintain most of the steady-states in the organism" (9), though he owed much to C. Bernard's concept of the 'milieu interieur'. Subsequently, homeostasis was interpreted in terms of general systems theory, and formulated mathematically in accordance with cybernetic principles (eg Ref. 40). Two important characteristics of homeostasis are its inherent complexity, involving the synergistic interaction of numerous components, and its essentially oscillatory nature (40).

Waddington in 1957 introduced a related concept, 'homeorhesis', to describe "an equilibrium (which) is not centred on a static state but rather on a direction or pathway of (developmental) change"; and subsequently the concept received much discussion (40). When, in 1980, Bauman and Currie (4) used the term in connection with the hormonal control of galactopoiesis, they were apparently unaware of its pedigere. They redefined it as the "orchestrated changes for the priorities of a physiological state" and, essentially, equated it with nutrient partitioning. Homeostasis and homeorhesis are alike in that both posit the importance of holistic, goal-seeking processes to maintain or direct metabolism.

Latterly, however, there has been a tendency to describe homeorhesis in reductionist terms, eg "It is interesting that a single endocrine treatment (somatotrophin) can so exquisitely coordinate changes in such an array of tissues and physiological processes .." (29), with reference to the differential effects of GH on different tissues. But since such effects are as much determined by the target tissues as by GH this begs the question of how differential tissue responsiveness is coordinated. As homeostasis and homeorhesis are clearly complex systems, it seems preferable to consider individual hormones as 'agents of coordination' rather than coordinators per se.

An important question concerns the extent to which these agents of coordination may be varied to modulate animal productivity without unduly upsetting the fine-tuning postulated by the homeostatic concept. When does physiology become pharmacology and terminate in pathology? It is clear that answers to such questions can only be obtained from large animal physiological experimentation (eg to determine cardiovascular, renal and metabolic responses) and cannot be accurately assessed simply from productivity responses. Transgenesis raises cognitive problems over and above those encountered in exogenous treatments, eg there is often a need for limitation of gene expression to specific developmental periods (41). According to Wagner, "a deeper understanding of genetic regulation is essential to the successful development of transgenic animals" (32).

There have been some attempts to arrive at an understanding of the galactopoietic action of exogenous GH by a process of induction ie comparing the effects produced by treatment with characteristics of genetically-superior animals (29). Undoubtedly, there are some parallels, but there are also differences. For example, the high milk yields of genetically superior cows are not always associated with significantly higher blood concentrations of GH (15) and it has been stressed that reliable estimates of genetic merit for milk production cannot be derived from measurements of a single hormone (36). Moreover, the normal inverse relationship between GH and insulin concentrations in blood does not obtain in animals treated with exogenous GH (13).

The ultimate limits of plasticity of the homeostatic process are clearly defined by the occurrence of disease states. Reports indicate how narrow the acceptable range might be in the case of pigs treated with pGH. Machlin (24) reported that injections of 0.22 mg/kg/day resulted in "some mortality .. liver and kidney degeneration, haemorrhage of the stomach, edema and arthritis". In a later study pigs receiving a tenth of this dose showed no overt signs of ill health (11).

In summary, physiological explanation forms an important basis for the acceptability of biotechnological innovations. Statements like "We lack knowledge and understanding of the action of somatotropin on the mammary gland" (29), made by acknowledged experts, highlight the necessity to remedy our ignorance.

PROCEDURAL CRITERIA

Popper's prescription for the scientific method, viz. "bold conjectures and ingenious and severe attempts to refute them" (30) provides an invaluable yardstick for technology assessment. In the present context a portmanteau Popperian 'bold conjecture' might be constructed along the lines "This biotechnological innovation will yield significant benefits which will outweigh any significant adverse consequences". The obligation on the proponents of the innovation in question is clear: to subject the assertion to 'ingenious and severe' tests. If the claim patently resists attempted refutation, there can be no rational grounds for prohibiting implementation of the technology.

Acceptability of biotechnology at the procedural level would seem to be dependent on perceptions of risk-benefit analyses (37), where the criteria are defined by anticipated longer-term social, environmental and safety effects as well as economic advantage.

Benefits claimed for rGH technology are: (i) increased profits for pharmaceutical companies; (ii) increased profits for livestock farmers; (iii) cheaper food products for consumers. The validity of claims (ii) and (iii) has been questioned (see below). Prospective benefits need, however, to be assessed in a temporal context, ie for individual farmers relative benefits might accrue from use of new technology by comparison with non-use, if adoption becomes widespread.

A long list of potential risks has been alleged to weigh against these prospective benefits. Some claims are doubtless irrational, whereas others have a more substantial basis: all need to be addressed. Table 1 summarises the alleged biological risks for exogenous rGH treatments under the headings of those affecting (A) the operator; (B) the treated animal; (C) the consumer: it could be used in the manner of a check-list. In addition, various potential risks have been identified pertaining to (D) the agricultural industry; (E) the rural environment; (F) society at large. References cited in Table 1 are biased to the proposed use of rbGH (bovine somatotrophin, bST) in dairying simply because, as the trailblazer for rDNA technology, it has received disproportionate attention; but the approach employed would also seem to be applicable to exogenous growth adjuvants and, more generally, to transgenic animals.

Two dimensions of risk may be identified (18): 'probability of occurrence' and 'consequences of occurrence'. If both are very low the 'risk' can be ignored, but if either is significant its potential impact needs careful evaluation. It is suggested that in evaluating potential risks a four-point scale is used to cover the two dimensions, viz. N, insignificant; *, significant but readily avoidable; **, significant and not readily avoidable but acceptable; X, significant and not readily avoidable and unacceptable. The appearance of an 'X' in risk assessment would seem to be prohibitive, but other combinations might be acceptable. The notion of 'acceptable' is, of course, problematical - but the scheme might well assist rational discussion. The term 'significant' is also contentious. Used by scientists it usually implies 'statistical significance', whereas for most non-scientists it denotes 'importance'. The two interpretations are essentially incommensurable. Statistically-significant effects may well present no important risk, but statistical-insignificance, due eg to innate variability, inadequate experimental method or even inappropriate statistical technique, might obscure important effects. Similarly, the phrase "there is no evidence for (some deleterious effect)" is only reassuring if the effect has been specifically sought. With these general considerations in mind, brief references will be made to each of the alleged risk categories.

A Operator-associated (applicable to person administering exogenous treatments). No claims of serious risk appear to have been made, but effects of accidental or malicious use on humans need to be defined.

B Animal Welfare. The most prominent informed critics are veterinary scientists, Kronfeld (20) and Webster (43). Both identify metabolic disease as the greatest risk in animals already operating at high metabolic rates, ie high yielding dairy cows. Statistically significant increases in the incidence of mastitis (23) and of infertility (44) have been reported in some studies of bST-treated cows, but not in many others (eg Ref. 3). Bauman has asserted that, whatever the prognostications, metabolic diseases do not occur with bST treatment (eg Ref. 3), although admitting that "Examination of subtle health effects will require large numbers of cows over a range of environmental and management conditions" (29). For observers of this debate there is doubtless a lingering suspicion that results achieved with expertly-managed institute and university farm herds may not be transferable to commercial practice as a whole, and Kronfeld (20) claims that even in published results there is a wide variation in response. However, neither Kronfeld nor Webster rules out the use of bST in declining lactation, on purely animal welfare grounds. Similarly, Webster considers that hormonal induction of growth does not encounter risks of metabolic stress (43).

Overt disease is a quantifiable, but in some cases extreme index of reduced animal welfare. Other indices, such as heart-rate, have been recommended (5), and it is surprising that such a simple procedure has not been regularly employed. None of the 51 abstracts of papers on bST presented to the ADSA conference in 1988 (1) mentions heart-rate measurement.

In addition to the common rGH, methionyl somatotrophin, preparation of a Delta-9 analogue, truncated at the N-terminus by 8 residues, has been described (45). Since these are xenobiotics, possible immune responses in treated animals, need

Table 1 Summary (and proposed check-list) of biological risks alleged to be associated with use of exogenous recombinant growth hormone in animal production

All categories below to be assessed in relation to (i) recommended treatment and (ii) graded degrees of misuse

	Categories	Criteria (eg)	Refs (eg)	Risk
A	Operator-associated Accidental/malicious administration	immune reactions	12	
B	Treated animal			
(a)	injection site	inflammation	44	
(b)	physiological responses	stress indices eg heart rate reduced fertility	 5 44	
(c)	behaviour	abnormal	5	
(d)	disease susceptibility	ketosis mastitis liver damage	20 23 24	
+All above categories under unusual circumstances eg heat stress, disease				
C	Consumer			
	(a) normal and (b) abnormal consumption by (c) healthy and (d) diseased (e) adults, (f) children, (g) infants and (h) non-human consumers eg pets of (j) pasteurized/cooked and (k) unpasteurized/uncooked food; ie 32 combinations	Major changes in concentrations of biologically-active food constituents	23 16 17	
+All above categories under specific physiological or environmental conditions eg pregnancy (cf. fetus), lactation, heat/cold stress				

Risk Assessment:

N, insignificant;
*, significant but readily avoidable;
**, significant and not readily avoidable but acceptable;
X, significant and not readily avoidable and unaccepatble.

investigation: 30% of patients receiving met-hGH reportedly experienced antibody responses, though no adverse side effects were apparent (12). Furthermore, GH is highly heterogeneous in vivo (22), so that the implications of augmenting supply of only a single molecular species also demand attention.

Potential welfare problems in transgenic animals (eg Ref. 32) share certain features with exogenously-treated animals (eg relating to raised blood concentrations) and others with conventional breeding practices (where the transgene is incorporated into the germ-line). It seems important in terms of public acceptability to draw a clear distinction between experimental procedures and what would be acceptable for practical application.

The Farm Animal Welfare Council in its recent report (14) assigned a high priority to investigation of potential welfare problems associated with new biotechnology.

C The Consumer. It is likely that perceived risks associated with consumption of food from treated animals constitute the single most important constraint on public acceptability. It would seem to follow that this is where proponents of biotechnology need to demonstrate the most scrupulous attention to detail to avoid the charge of obfuscation. If this was the aim, they cannot be said to have totally succeeded. Teske, of the United States Food and Drug Administration, reporting the basis for the authorization of the sale of milk from bST-treated cows (38), listed four "facts" about bST: (i) it is digested and thus inactivated in the gut; (ii) it is species-specific and inactive in humans even if injected; (iii) it does not cause compositional changes in milk; (iv) consumers have always ingested small amounts of natural bST. However, evidence in the literature calls into question the scientific accuracy, with respect to all potential consumers, of the first three points. Thus, (i) in newborn babies and some adults proteins may be absorbed into the blood stream (17); (ii) chymotrypsin digests of natural bST have been reported to be biologically active in humans (16); and (iii) significant compositional changes in milk have been reported from several laboratories (eg Ref. 34). Quite recently it has been shown (31) that bST increases milk concentrations of insulin-like growth factor-I (IGF-I), a peptide which is biologically active in humans. In theory, IGF-I might affect gut mucosal cells even if not absorbed into blood. McBride et al. (23) claim that "some concern arises as to the possibility of abnormal levels bIGF-I in the milk of rbGH-treated cows, and with it, consumer health".

The listing of these statistically significant effects does not, as discussed above, imply that they are necessarily important risks: concentrations may well be below any conceivable level of danger. But public acceptability is unlikely to be promoted by anything less than explicit discussion of the issues.

Space permits only summary treatment of alleged socioeconomic risks (D-F).

D Agricultural Industry. Claims have been made for likely adverse consequences in relation to (i) income and employment in the livestock industry (28); (ii) ease of livestock management (20); (iii) genetic evaluation of livestock (8).

E Rural Environment. Reduction of cow numbers and dependence on concentrate feeds following widespread use of bST has been predicted to lead to loss of public amenity as grazing pastures are put to alternative uses (43).

F Social Consequences. Apart from employment (D), there are at least two major dimensions: economic and cultural. Economic effects are related to possible surplus production, reliance on imported feeds, and the effects of both of these on less-developed countries (7). Cultural constraints are undoubtedly important, pertaining as they often do to the perceived hubris of scientists, arrogating to themselves the role of 'creator' (19). Limits need to be defined to exorcise unjustified fears (35).

RECOMMENDATIONS

If the benefits of biotechnology are to be realised, the growing numbers of the scientifically-literate lay public will need to be convinced of its net value. As recognized by the Royal Society, public understanding of science is crucial to further social development (6). Aware of this, the Monsanto Company recently held a series of public meetings to foster "informed debate" on the proposed use of bST in dairy farming (2). They faced a formidable problem: credibility is compromised when advocates of any proposal are seen to have a vested interest in its adoption. Thus, where technological innovation is largely dependent on commercial initiatives public acceptability is likely to be constrained not only by innate conservatism but also by a measure of cynicism. This emphasises the role of independent (ie government-financed) scientists (26): and they are needed, at least in the near-market phase, not as 'collaborators' with industry (in which their neutrality is peceived to be compromised) but as 'honest brokers'. Regrettably, reduced government funding jeopardizes the effectiveness of this role (27). Such considerations lead to the following recommendations.

1 An agreed set of criteria, along the lines discussed (A-F), should be publicized and form the basis of decisions on proposed biotechnological innovations.

2 Evaluation (physiological, economic, etc) should be performed, where possible as 'blind trials', by independent research establishments, data being collated by an expert committee (akin to the Veterinary Products Committee, but with a wider remit, eg to include economic assessment) who would publish the full results together with their recommendations (expert report).

3 The licencing authority (eg in the U.K., Minister of Agriculture) should invite comments on the expert report from other interested groups, eg farmers' and consumers' organizations.

4 The House of Lords Select Committee currently considering the future of agricultural R&D should specifically examine the role of universities and government research institutes in evaluating biotechnological innovations.

5 A Committee should be set up, with a status equivalent to that of the Warnock Committee (42), to consider the social and cultural implications of transgenesis in animals.

CODA

In 1963, 'Objections to Christian Belief', a book written by four Cambridge theologians, was published (25). Their aim was not, of course, to recant their religious convictions but to face, with intellectual honesty, the counter-claims of atheism and agnosticism. (A companion volume 'Objections to Humanism', played the same role for four agnostics). In the same way, this paper has sought to identify the objections to biotechnology in animal production, that the strength of the arguments in its favour might be better articulated.

> "Another error is an impatience of doubting and a blind hurry of asserting without a mature suspension of judgement ... if we begin with certainties, we shall end in doubts; but if we begin with doubts and are patient in them, we shall end in certainties"
>
> Francis Bacon (1605)

REFERENCES

1 ADSA (1988). American Dairy Science Association Annual Conference. Abstracts of presented papers. Journal of Dairy Science **71** (Suppl. 1).

2 Anon (1988). Monsanto Goes to the Country. Animal Health Matters (National Office of Animal Health) **8,** 2.

3 Bauman, D.E. (1987). Bovine Somatotropin: The Cornell Experience. In Proceedings of the National Invitational Workshop on Bovine Somatotropin, pp 46-56. St. Louis, Missouri, USDA Extension Service.

4 Bauman, D.E. and Currie, W.B. (1980). Partitioning of nutrients during pregnancy and lactation: a review of mechanisms involving homeostasis and homeorhesis. Journal of Dairy Science **63,** 1514-1529.

5 Broom, D.M. (1988). 'The scientific study of farm animal welfare'. Robinson Memorial Lecture, delivered at the Faculty of Agricultural Science, Nottingham University, February 17th 1988.

6 Bodmer, W. (1987). The public understanding of science. Science and Public Affairs **2,** 69-89.

7 Bunders J. (1988). Appropriate biotechnology for sustainable agriculture in developing countries. TIBTECH **6,** 173-180.

8 Burnside, E.B. (1987). Impact of somatotropin and other biochemical products on sire summaries and cow indices. Journal of Dairy Science **70,** 2444-2449.

9 Cannon, W.B. (1939). 'The Wisdom of the Body' revised, enlarged edition. Published in Norton Library (1963). W.W. Norton, New York.

10 Cantley, M.F. (1987). Democracy and Biotechnology. Swiss Biotechnology **5,** 5-15.

11 Chung, C.S. Etherton, T.D. and Wiggins, J.P. (1985). Stimulation of swine growth by porcine growth hormone. Journal of Animal Science **60,** 118-130.

12 Crawford, M. (1987). Genentech sues FDA on Growth Hormone. Science **235,** 1454-1455.

13 Davis, S.R., Gluckman, P.D., Hart, I.C. and Henderson, H.V. (1987). Effects of injecting growth hormone or thyroxine on milk production and blood plasma concentrations of insulin-like growth factors I and II in dairy cows. Journal of Endocrinology **114,** 17-24.

14 F.A.W.C. (1988). Report on priorities in animal welfare research and development. Farm Animal Welfare Council, Tolworth, Surbiton, Surrey.

15 Flux, D.S., Mackenzie, D.D.S and Wilson, G.F. (1984). Plasma metabolite and hormone concentrations in Friesian cows of different genetic merit measured on two feeding levels. Animal Production **38,** 377-384.

16 Forsham, P.H., Li, C.H., Diraimondo, V.C., Kolb, F.O., Mitchell, D. and Newman, S. (1958). Nitrogen retention in man produced by chymotrypsin digests of bovine somatotropin. Metabolism **7,** 762-764.

17 Hemmings, W.A. (ed.) (1978). Antigen absorption by the gut. MTP Press Ltd., Lancaster, England.

18 Irwin, A. (1985). Risk and the control of technology. Manchester University Press, Manchester.

19 Jackson, D.A. and Stich, S.P. (eds.) (1979). The Recombinant DNA Debate. Prentice-Hall, New Jersey.

20 Kronfeld, D.S. (1988). Biologic and economic risks associated with the use of bovine somatotropins. Journal of the American Veterinary Medical Association June 15, 1701-1704.

21 Lamming, G.E., Ballanini, G., Baulieu, E.E., Brookes, P., Elias, P.S., Ferranda, R., Galli, C.L., Heitzman, R.J., Hoffman, B., Karg, H., Meyer, H.H.D., Michel, G., Poulsen, E., Rico, A., van Leeuwen, F.X.R. and White, D.S. (1987). Scientific report on anabolic agents in animal production. Veterinary Record October 24, 389-392.

22 Lewis, U.J. (1984). Variants of growth hormone and prolactin and their posttranslational modifications. Annual Review of Physiology **46,** 33-42.

23 McBride, B.W., Burton, J.L. and Burton, J.H. (1988). The influence of bovine growth hormone (somatotropin) on animals and their products. Research and Development in Agriculture **5,** 1-21.

24 Machlin, L.J. (1972). Effect of porcine growth hormone on growth and carcass composition of the pig. Journal of Animal Science **35,** 794-800.

25 MacKinnon, D.M., Vidler, A.R., Williams, H.A. and Bezzant, J.S. (1963). Objections to Christian Belief. Constable, London.

26 Mepham, T.B. (1987). Changing Prospects and Perspectives in Dairy Research. Outlook on Agriculture **16,** 182-188.

27 Mepham, T.B. (1988). Questions of validity in mammary physiology: methodology and ethics. Proceedings of the Nutrition Society (in press).

28 Mix, L.S. (1987). Potential impact of the growth hormone and other technology on the United States dairy industry by the year 2000. Journal of Dairy Science **70,** 487-497.

29 Peel, C.J. and Bauman, D.E. (1987). Somatotropin and lactation. Journal of Dairy Science **70,** 474-486.

30 Popper, K.R. (1979). Objective Knowledge (revised edition) Oxford University Press, Oxford.

31 Prosser, C.G., Fleet, I.R., Hart, I.C. and Heap, R.B. (1987). Changes in concentration of insulin-like growth factor I (IGF-I) in milk during bovine growth hormone treatment in the goat. Journal of Endocrinology **112,** 65.

32 Radke, K. and Lagarias, D. (1986). Altering the genetic makeup of animals. Biotechnology **4,** 14.

33 Ravetz, J.R. (1986). Usable knowledge, usable ignorance: incomplete science with policy implications. In Sustainable development of the biosphere (Clark, W.C. and Munn, R.E., eds.). Cambridge University Press, London.

34 Richard, A.L., McCutcheon, S.N. and Bauman, D.E. (1985). Responses of dairy cows to exogenous bovine growth hormone administered during early lactation. Journal of Dairy Science **68,** 2385-2389.

35 Rollin, B.E. (1986). The 'Frankenstein Thing': The moral impact of genetic engineering of agricultural animals on society and future science. In Genetic Engineering of Animals (Evans, J.W. and Hollaender, A., eds.) pp 285-297. Plenum, New York.

36 Sejrsen, K. and Løvendahl, P. (1986). Criteria identifying genetic merit for milk production. In Exploiting new technologies in animal breeding (Smith, C., Boking, J.W. and McKay, J.C., eds.) pp 142-152. Oxford University Press, Oxford.

37 Starr, C., Rudman, R. and Whipple, C. (1976). Philosophical basis for risk analysis. Annual Review of Energy **1,** 629-662.

38 Teske, R.H. (1987). Recombinant DNA: a regulatory perspective. In Proceedings of the National Invitational Workshop on Bovine Somatotropin, pp 24-27. St. Louis, Missouri, USDA Extension Service.

39 Vize, P.D., Michalska, A.E., Ashman, R., Lloyd, B., Stone, B.A., Quinn , P., Wells, J.R.E. and Seamark, R.F. (1988). Introduction of a porcine growth hormone fusion gene into transgenic pigs promotes growth. Journal of Cell Science **90,** 295-300.

40 Waddington, C.H. (ed.) (1967). Towards a theoretical biology. I. Prolegomena. Edinburgh University Press, Edinburgh.

41 Wagner, T.E. (1986). Introduction and regulation of cloned genes for agricultural livestock improvement. In Genetic engineering of animals (Evans, J.W. and Hollaender, A., eds.), pp 151-161. Plenum, New York.

42 Warnock, M., et al. (1984). Report of the Committee of Enquiry into Human Fertilization and Embryology. HMSO Cmnd. 9314, London.

43 Webster, A.J.F. (1988). Farm Animal Protection - Somatotropin. In Farm animal protection - the practical way forward (Carter, V. and Carter, H., eds.), pp 80-88. European Conference Group on the Protection of Farm Animals, Horsham, U.K.

44 Whitaker, D.A., Smith, E.J., Kelly, J.M. and Hodgson-Jones, L.S. (1988). Health, welfare and fertility implications of the use of bovine somatotrophin in dairy cattle. Veterinary Record May 21, 503-505.

45 Wingfield, P.T., Graber, P., Buell, G., Rose, K., Simona, H.G. and Burleigh, B.D. (1987). Preparation and characterization of bovine growth hormones produced in recombinant Escherichia Coli. Biochemical Journal **243,** 829-839.

POSTERS

INDEX OF POSTERS

STRUCTURE AND REGULATION OF THE RAT GROWTH HORMONE RECEPTOR

W.R. Baumbach, D.L. Horner and J.S. Logan

Molecular and Cellular Biology Group, Agricultural Research Division, American Cyanamid Company, Princeton, New Jersey 08540, U.S.A.

Growth hormone (somatotropin) is a 191 amino acid serum polypeptide which profoundly affects postnatal vertebrate growth and metabolism. The molecular mechanisms and specific target tissue involved in growth hormone action are not well understood, although it is hypothesized that growth hormone induced IGF-I (somatomedin C) synthesis in the liver mediates some though not all of the effects of growth hormone. Studies of the proteins to which growth hormone binds may provide insight into its mechanisms and sites of action. Using probes derived from the published nucleotide sequence of the rabbit growth hormone receptor (1), we have isolated and sequenced cDNA clones encoding the full length transmembrane receptor in rat, which is a system well suited to physiological studies of growth hormone action. Preliminary Northern analysis has shown that as expected, liver is the most abundant source of the 4.75 kb mRNA encoding the full length growth hormone receptor. Surprisingly, hypophysectomized animals (lacking endogenous growth hormone) demonstrate high levels of the growth hormone receptor mRNA in liver, and these levels do not appreciably change in response to exogenous GH, suggesting that low observed levels of binding in these animals may reflect regulation at a post-transcriptional level. During normal rat development, levels of receptor mRNA in liver undergo a steady increase from barely detectable at birth to near adult levels at four weeks of age. A tissue distribution analysis of GH receptor resulted in the following observations: both skeletal muscle and subcutaneous adipose tissue contain relatively high levels of GH receptor mRNA (3-4 fold less than liver); heart exhibits somewhat lower levels of receptor mRNA; and low but detectable levels of GH receptor message are found in other tissues such as thymus, brain, testes and lactating mammary gland. Future studies in transgenic mice may lead to insights regarding the direct action of growth hormone upon peripheral tissues.

REFERENCE

1 Leung, D.W., Spencer, S.A., Cachianes, G., Hammonds, R.G., Collins, C., Genzel, W.J., Barnard, R., Waters, M.J. and Wood, W.I. (1987). Nature **330,** 537.

CLONING AND ANALYSIS OF EXPRESSION OF RAT GROWTH HORMONE RECEPTOR

L.S. Mathews, B. Enberg and G. Norstedt

Center for Biotechnology, Karolinska Institute, Huddinge University Hospital F 82, 141 86 Huddinge, Sweden

The growth hormone receptor (GHR) was cloned from a rat liver cDNA library using a pair of oligonucleotides complementary to the human sequence as a probe. Seven overlapping clones were obtained spanning a total of 3.2 kb. These contain the entire coding sequence, as well as 400 bp of 5′ and 850 bp of 3′ untranslated sequences. Nucleotide sequence identity to the rabbit and human GHRs is 76% and 72%, respectively; identity of the translated amino acid sequences is 75% and 70%.

A 560 bp restriction fragment was cloned into an RNA expression vector and used to generate a uniformly labeled RNA probe. This was used in a solution hybridization assay to quantitate GHR mRNA expression. Expression was detected in most tissues, with the highest levels being observed in the liver. In liver, kidney, heart and brain expression was found to be lower in newborn as compared to adult rats, rising to adult levels through neonatal development.

CONFORMATIONAL ANALYSIS OF BOVINE GROWTH HORMONE FRAGMENTS WHICH CORRESPOND TO HELICAL REGIONS OF THE INTACT PROTEIN

S.R. Lehrman, J.L. Tuls and M. Lund

The Upjohn Company, Kalamazoo, MI 49001, U.S.A.

Porcine growth hormone has been shown to be an antiparallel four helix bundle protein (1). The strong homology between porcine and bovine growth hormones suggests that the two proteins have similar tertiary structure. We are interested in determining the secondary structure of bovine growth hormone (bGH) fragments which form α-helices within the intact protein. Previously, we demonstrated that synthetic and proteolytic fragments of bGH, which include residues 109 to 133, form α-helices in aqueous solution (2). We now report the synthesis and conformational analysis of peptide fragments which correspond to helices 2 and 4 of the putative four helix bundle, and fragments which correspond to non-helical regions of bGH. We find that [164-Ser-153-180]-bGH (derived from helix 4) has a greater propensity for α-helix formation in aqueous alcohol than [78-97]-bGH (derived from helix 2), or peptides which are derived from non-helical regions of the protein. The conformational properties of these bGH fragments have been studied as a function of temperature, and trifluoroethanol, peptide and denaturant concentration.

REFERENCES

1 Adbel-Meguid, S.S., Shieh, H.-S., Smith, W.W., Dayringer, H.E., Violand, B.N. and Bentle, L.A. (1987). Proceedings of the National Academy of Sciences, U.S.A. **84,** 6434.

2 Brems, D.N., Plaisted, S.M., Kauffman, E.W., Lund, M. and Lehrman, S.R. (1987). Biochemistry **26,** 7774.

BIOLOGICAL ACTIVITY OF AMINO-TERMINAL AMINO ACID VARIANTS OF BOVINE SOMATOTROPIN

G.G. Krivi, G.M. Lanza, W.J. Salsgiver, N.R. Staten, S.D. Hauser, E. Rowold, T.R. Kasser, T.C. White, P.J. Eppard, L. Kung, R.L. Hintz, K.C. Gleen and D.C. Wood

Biological Sciences and Animal Sciences Division, Monsanto Co., St. Louis, Missouri 63198, U.S.A.

Pituitary bovine somatotropin (bST) is comprised of a mixture of protein species including two 191 amino acid derivatives (N-terminal alanine with leucine or valine at position 127), two 190 amino acid derivatives (N-terminal phenylalanine with leucine or valine at position 126), and several deletion derivatives missing the amino terminal 3 or 4 amino acids. In this study, recombinant DNA techniques were used to produce individual analogues of the bST species found in pituitary preparations. The purified analogues were tested to determine the biological activity of the individual components of pituitary bST preparations. Oligonucleotide-directed, site-specific mutagenesis of the bST structural gene described in Seeberg et. al. (2) was used to prepare genes coding for the following 6 analogues of pituitary bST species: ala(1) leu(127) bST; ala(1) val(127) bST; met(1) leu(127) bST; met(1) val(127) bST; deletion(aa 1-4) leu(127) bST; and deletion(aa 1-4) val(127) bST. The purified 191 amino acid proteins (ala(1) leu(127) bST, ala(1) val(127) bST, met(1) leu(127) bST, and met(1) val(127) bST, showed equivalent activity when tested in a bovine liver radioreceptor assay, in an anti-insulin assay utilizing 3T3L1 adipocytes, and in rat weight gain assays. When administered to dairy cows as a 25 mg daily intramuscular injection, all 4 somatotropins increased 3.5% fat-corrected milk (FCM) production above control ($P < 0.05$). N-terminal methionine and alanine bST's increased FCM to the same extent (met, 7.7 kg/day versus ala, 7.2 kg/day). Somatotropin variants with valine at position 127 had greater FCM response ($P < 0.05$) than the leucine counterparts (val, 8.5 kg/day versus leu, 6.5 kg/day); but this difference occurred mainly with the N-alanine derivatives. In contrast to the results seen with the different 191 amino acid somatotropins, the N-terminal deletion variants showed reduced potency in bovine liver radioreceptor assays and in intact female rat weight gain assays. When tested in dairy cows, the N-terminal deletion variants increased milk production significantly ($P < 0.05$) above controls (3.3 kg/d) but their galactopoietic potency was less ($P < 0.05$) than the activity of 191 amino acid analogues. These data suggest that the reduced potency of pituitary bST when tested in cows for galactopoietic activity (1) is due in major part to the presence of the N-terminal deletion species in the pituitary bST preparations.

REFERENCES

1 Bauman, D.E., Eppard, P.J., DeGeeter, M.J. and Lanza, G.M. (1985). Journal of Dairy Science **68**, 1352.

2 Seeburg, P.H., Sias, S., Adelman, J., DeBoer, H.A., Hagflick I., Jhurani, P., Goeddel, D.V. and Heyneker, H.L. (1983). DNA **2**, 37.

PURIFICATION AND PROPERTIES OF TWO RECOMBINANT DNA-DERIVED OVINE GROWTH HORMONE VARIANTS EXPRESSED IN ESCHERICHIA COLI

O.C. Wallis and M. Wallis

Biochemistry Laboratory, School of Biological Sciences, University of Sussex, Brighton BN1 9QG, U.K.

We have previously prepared a series of E.coli clones containing pUC8-based plasmids in which the coding region of the cDNA for ovine growth hormone (oGH) is expressed under the control of the lac promoter (4). These clones vary in the sequence of the DNA coding for the N-terminus of the expressed protein and in the level of expression.

Two clones, which have now been studied in detail, synthesise GH-related proteins at very high levels (up to 25% of the total cell protein). The recombinant plasmids, pOGHe101 and pOGHe102, encode the proteins oGH1 and oGH2 respectively. In each case the proteins are found as insoluble inclusion granules in induced bacteria. Purification of oGH1 from these granules has been previously reported (3) by a method involving solution of the granules in 6 M guanidium chloride followed by careful renaturation and ion exchange and gel filtration chromatography. oGH2 was found to have a higher pI than oGH1 and ion exchange chromatography was carried out at pH 9.0 instead of pH 8.0 when this protein was purified. Final yields of the proteins were about 100 mg per litre of bacterial culture. The purified proteins ran as single bands on SDS-polyacrylamide gel electrophoresis with apparent M_r of approximately 22000 for oGH1 and 21500 for oGH2. N-terminal sequence analysis demonstrated that oGH1 was identical to pituitary-derived oGH except that the N-terminal alanine was replaced by the sequence: Thr Met Ile Thr Asn Ser Gly Asp. In oGH2 the last two amino acids of this sequence and the first two of the oGH sequence (Phe Pro) were missing. The amino acid sequence of oGH1 was confirmed by determining the nucleotide sequence in plasmid pOGHe101. Restriction fragments were subcloned in plasmid pJBS633 (1) and their sequences were determined by the chain-termination procedure of Sanger (2). Comparison of the properties of the recombinant hormones with pituitary-derived bovine GH indicated that oGH1 showed the same potency as bovine GH in a radioimmunoassay and in a weight gain assay in hypophysectomized rats but a slightly higher potency in a radioreceptor assay. The potency of oGH2 in the radioreceptor assay was close to that of bovine GH. Our results indicate that high yields of oGH variants can be obtained by our procedure and that the presence of an N-terminal extension had little effect on the properties of the hormone in 3 different assay systems.

REFERENCES

1 Broome-Smith, J.K. and Spratt, B.G. (1986). Gene **49,** 341.

2 Sanger, F., Nicklen, S. and Coulson, A.R. (1977). Proceedings of the National Academy of Sciences, U.S.A. **74,** 5463.

3 Wallis, O.C. and Wallis, M. (1987). Journal of Endocrinological Investigation **10** (Suppl. 4), Abstract 12.

4 Wallis, O.C., Warwick, J.M. and Wallis, M. (1986). Journal of Endocrinology **108** (Suppl.), Abstract 223.

GH TREATMENT OF HYPOPHYSECTOMIZED RATS: HORMONE AND NUTRIENT INTERACTIONS ON TISSUE METABOLISM

P.C. Bates[1], P.A. Donachie[1], D. Schulster[2] and D.J. Millward[1]

[1] Nutrition Research Unit, London School of Hygiene and Tropical Medicine, 4 St. Pancreas Way, London NW1 2PE, U.K.
[2] National Institute of Biological Standards, South Mimms, Herts EN6 3QG, U.K.

Recombinant-derived human growth hormone (hGH) has been shown to induce positive nitrogen balance even on low calorie intakes. Although this was assumed to reflect maintenance of muscle protein mass the mode of action and the effects of altered food intake have not been examined fully. We report here a study of the interaction of nutrition and GH on tissue growth in a GH-sensitive animal model: hypophysectomized (H_x) rats treated with hGH.

Nineteen male rats, weighing 190 g, were hypophysectomized and then weighed daily to ensure completeness of H_x from cessation of body weight growth. Fourteen days later 4 rats were killed (H_xO) and the rest were divided into three groups which were treated for 7 days with saline (H_x7), hGH (60 mU/rat/day; GH) and fed ad libitum, or hGH but pair-fed to the intake of the saline group (GH-PF). At death gastrocnemius muscles, livers and hearts were dissected and weighed from all animals and tissue RNA concentrations were measured.

The GH group increased their food intake such that it was 34% greater ($P < 0.01$) than that for the H_x7 and the GH-PF groups. Body weight increased by 12% in the GH group ($P < 0.01$) but no increase occurred in the GH-PF group (H_xO, 178 ± 3; H_x7, 174 ± 5; GH, 195 ± 8; GH-PF, 176 ± 6 g; mean ± SEM). In contrast, muscle growth did occur in both GH-treated groups (gastrocnemius weight: H_xO, 870 ± 11; H_x7, 835 ± 29; GH, 935 ± 24; GH-PF, 947 ± 35 mg; $P < 0.02$). However, in the pair-fed group, heart growth was restricted (heart weight: H_xO, 488 ± 11; H_x7, 439 ± 16; GH, 511 ± 12; GH-PF, 434 ± 17mg) and liver weight actually decreased (liver weight: H_xO, 6952 ± 292; H_x7, 7134 ± 212; GH, 8466 ± 216; GH-PF, 5435 ± 176mg; $P < 0.001$). The muscle growth occurred in the GH-PF group despite very much reduced concentrations of plasma insulin (H_xO, 727 ± 103; H_x7, 716 ± 54; GH, 1096 ± 165; GH-PF, 300 ± 22ng/ml) and of IGF-I (H_xO, 21 ± 3; H_x7, 22 ± 2; GH, 88 ± 14; GH-PF, 23 ± 2 ng/ml). There was an increase in the muscle RNA concentration in both GH-treated groups (H_x7, 1017 ± 9; GH, 1262 ± 48; GH-PF, 1165 ± 31 ug/g; $P < 0.01$) indicating an increased capacity for protein synthesis.

Thus restriction of the increase in food intake, which occurs with GH-treatment in hypophysectomized rats, significantly modulates the effects on protein metabolism; skeletal muscle growth is maintained and occurs at the expense of other tissues. It is noteworthy that pair-feeding abolished the GH-induced increase in plasma IGF-I concentration which is usually associated with GH treatment during hypopituitary states. Therefore the significance of total plasma IGF-I concentrations as an indicator of the anabolic response to GH is questionable.

HUMAN GROWTH HORMONE TREATMENT ENHANCES DEPOSITION OF COLLAGEN IN RAT SKELETAL MUSCLES

B.H. Moreland[1], C. Ayling[2] and D. Schulster[2]

[1]Division of Biochemistry, UMDS (Guy's Hospital Medical School), London, U.K.
[2]Department of Endocrinology, National Institute of Biological Standards and Control, Potters Bar, Hertfordshire, U.K.

Hind limb muscles - gastrocnemius, soleus and extensor digitorum longus (EDL) were studied for the effects of human growth hormone (hGH) injected daily (x 7) to hypophysectomized (hx) rats. In response to hGH, the weights of these muscles increased (10-16%) and morphological/biochemical changes (2) and fibre type changes (1) were observed. We now report observations on the changing pattern of collagen deposition.

Hx rats were maintained for 14 days before subcutaneous injection with pituitary hGH, recombinant DNA hGH (rDNA hGH) or diluent for 7 days; they were then killed and hind limb muscles were excised. Samples of each muscle were fixed in glutaraldehyde, embedded in Araldite epoxy resin and 1 μm transverse sections cut for staining with toluidine blue. These sections from both diluent-treated hx and normal rats (viewed under the light microscope, x 500) were found to be closely packed with normal, polygonal shaped fibres whereas those from hGH-treated hx rats showed marked increases in both the connective tissue material in the endomysial regions and in the numbers of fibroblasts in discrete, localised areas in the sections. When ultra-thin muscle sections (100 nm thick) corresponding to these regions were viewed by transmission electron microscopy (x 9000), the connective tissue between the fibres showed the characteristic banded appearance of collagen. Muscle samples, frozen in liquid nitrogen and sectioned transversely, were histochemically stained for collagen by the Gomori method. Sections from both diluent-treated hx and normal rats showed the presence of collagen in discrete areas confined to the perimyseum regions, whereas the sections from hGH-treated rats showed larger localized patches of collagen extending into areas between the individual muscle fibres. After SDS polyacrylamide gel electrophoresis of muscle extracts, an additional protein band (Coomassie blue stained) of high molecular weight was consisently evident in samples obtained from pituitary hGH-treated or rDNA hGH-treated hx rats. This band corresponded to a protein band present in a standard collagen marker (MW 100-150 kDa) and was not evident in samples obtained from diluent-treated hx rats.

REFERENCES

1 Ayling, C.M., Schulster, D. and Moreland, B.H. (1987). Journal of Endocrinology **115,** Abstract 89.

2 Schulster, D., Wood, R.D., Ayling, C.M. and Moreland, B.H. (1987). International Symposium on Growth Hormone; Serono Symposium, June 1987, Tampa, Florida, U.S.A. Abstract 53.

METABOLIC AND ENDOCRINE CHALLENGE OF SOMATOTROPIN TREATED PIGS

K.-V. Brenner, J. Novakofski, P.J. Bechtel and R.A. Easter

University of Illinois, Urbana 61801, U.S.A.

Pigs treated with somatotropin (pST) are more efficient, have reduced carcass fat and increased muscle. Besides paracrine and endocrine regulation of IGF, pST may have several effects. These include direct effects on adipose tissue and muscle, somatic effects that indirectly alter tissue accretion and/or modulation of other endocrine axes regulating metabolism. The objective of our experiments was to examine indirect effects of pST by a series of substrate and endocrine challenges.

Crossbred pigs (64 kg) were given either pST (3 mg subcutaneous daily) or excipient. Husbandry and performance characteristics of similarly treated pigs have been described (3). During the 5th week of pST treatment, pigs were fitted with jugular cannulas. Pigs were allowed to recover for 4 days and then challenge tests were begun. Feed intake was determined daily and body weight measured every 3 to 4 days to assess health and performance.

Pigs were challenged approximately 2 h after pST treatment with either glucose (0.5 g/kg), insulin (75 mU/kg), epinephrine (10 μg/kg) or TRH (300 mg/pig). Pigs were allowed to recover for 3 days between tests. Blood was collected at -30, -15, 0, 15, 30, 45, 60, 120, 180 and 240 min. pST treated pigs had higher initial insulin level (14 vs 7 μU/ml), a higher initial response to glucose challenge (181 vs 76 μU/ml after 2 min) and a larger maximum insulin response (277 vs 78 μU/ml after 7 min) compared to control pigs. Following epinephrine challenge, changes in free fatty acids (FFA) were larger in pST treated pigs than in controls (1213 vs 227 μM after 10 min). Insulin levels in pST treated pigs changed approximately twice as much as in controls after epinephrine while the FFA increase 30-120 min after insulin was greater in pST treated pigs. The pituitary-thyroid axis was not markedly affected by pST although endogenous somatotropin release after insulin was blunted.

In summary, responses of free fatty acid or responses to epinephrine tended to be accentuated in pST treated pigs. These results suggest that the decreased fatness of treated pigs is at least partially the result of indirect pST modulation of lipolysis. Some results from these experiments have been previously reported as abstracts (1, 2).

REFERENCES

1 Brenner, K.-V., Wagner, W.C., Novakofski, J. and Bechtel, P.J. (1988). Journal of Animal Science **66**, 291.

2 Novakofski, J., Brenner, K., Easter, R., McLaren, D., Jones, R., Ingle, D. and Bechtel, P. (1988). FASEB Journal **2**, A848.

3 Novakofski, J., McKeith, F.K., Grebner, G.L., McLaren, D.G., Brenner, K., Easter, R.A. and Bechtel, P.J. (1988). British Journal of Animal Production **46**, 487.

EFFECT OF SOMATOTROPIN ON NITROGEN AND ENERGY METABOLISM IN GROWING SWINE

D. Wray-Cahen, R.D. Boyd, D.A. Ross and D.E. Bauman

Cornell University, Department of Animal Science, Ithaca, NY, U.S.A.

Dramatic effects on nutrient partitioning have been observed in swine treated with porcine somatotropin (pST). Although these effects are believed to be primarily associated with the post-absorptive use of nutrients, this has not been confirmed in swine. Such knowledge is important in the investigation of mechanisms and for estimating pST's effects on nutrient requirements. Therefore, we examined the effects of pST on nitrogen (N) and energy (E) absorption and retention in growing pigs. Twelve crossbred barrows (65 kg) received daily injections (subcutaneous) of either pituitary pST (120 ug/kg body weight) or an equivalent volume of excipient for 20 day. This level of pST optimised protein accretion and feed efficiency in a previous study (2). Pigs were housed individually in metabolism crates and fed a 19.6% crude protein corn-soyabean diet. Nutrient content of the diet was ≥ 140% of requirements (4) for all nutrients except E. Pigs were fed ad libitum prior to the balance period and received approximately 90% of this intake during the balance period (day 14 to day 19). Urine, faeces, feed, and orts were collected twice daily and analysed for N and E content. Pigs treated with pST retained more N than their controls (38.4 vs 23.0 g/day, $P < 0.01$). They also had a slightly higher apparent digestibility for the N and E consumed (88% vs 84% and 88% vs 85%, respectively, $P < 0.05$). However, this improvement in apparent digestibility was probably a function of the lower feed intake (23% reduction), rather than a direct effect of pST. pST had no affect on the efficiency with which digested E was metabolised (96%). Although pST-treated pigs used digestible N more efficiently than their controls (which received AA in great excess of their requirement), their efficiency of digestible N utilisation (64%) agrees closely with values for pigs fed diets adequate in AA (1). A higher dietary content of AA is required to compensate for both the increase in protein accretion rate and the decrease in feed intake in pST-treated pigs. Overall, pST treatment resulted in greater N retention, without altering the efficiency of E retention. These changes are consistent with the increased carcass N accretion and decreased lipid accretion rates previously reported (2, 3) and support the hypothesis that the effects of pST are due to post-absorptive mechanisms. Changes in AA content and protein:calorie ratio of the diet are required to achieve the maximum performance responses with pST treatment of growing pigs.

REFERENCES

1 Agricultural Research Council (1981). The nutrient requirements of pigs. Commonwealth Agricultural Bureaux, Slough, England.

2 Boyd, R.D., Bauman, D.E., Beermann, D.H., DeNeergard, A.F., Souza, L. and Butler, W.R. (1988). Journal of Animal Science **63** (Suppl. 1), 218.

3 Campbell, R.G., Steele, N.C., Caperna, T.J., McMurtry, J.P., Solomon, M.B. and Mitchell, A.D. (1988). Journal of Animal Science **66**, 1643.

4 National Research Council (19791). Nutrient requirements of swine. National Academic Press, Washington DC, U.S.A.

ANTI-LIPOGENIC EFFECTS OF SOMATOTROPIN ON OVINE ADIPOSE TISSUE

P.A. Sinnett-Smith and J.A. Woolliams

AFRC Institute of Animal Physiology and Genetics Research, Edinburgh Research Station, Roslin, Midlothian EH25 9PS, U.K.

Somatotropin treatment reduces fatness in animals (2). Both lipolytic and anti-lipogenic actions have been attributed to somatotropin (1, 5), however its physiological role in controlling lipolysis has been questioned (1). In this study the physiological role of somatotropin in the control of lipid metabolism in ovine adipose tissue during growth has been investigated in vivo.

Texel sheep were treated with 0.05, 0.10 or 0.20 mg/kg/day recombinant bovine somatotropin (re.BS) or a saline placebo. After approximately 37 days of treatment samples of subcutaneous adipose tissue were obtained by biopsy and the rates of fatty acid synthesis, fatty acid esterification, lipolysis (4) and blood metabolite concentrations (using commercially available test kits) determined.

Treatment with re.BS reduced adipocyte size ($P < 0.10$) consistent with reduced overall fatness ($P < 0.001$). re.BS treatment significantly reduced the rates of fatty acid synthesis ($P < 0.01$) and esterification ($P < 0.001$). No effect on adipose tissue lipolysis was detectable. This was reflected in unchanged blood free fatty acid concentrations. Blood urea concentrations were significantly reduced by re.BS treatment ($P < 0.01$) consistent with increased protein deposition as previously reported (3).

		re.BS Treatment (mg/kg/day)			
	Saline	0.05	0.10	0.20	C.V.
Number of animals	4	4	4	4	
Fatty acid synthesis*	15.65	9.12	4.61	3.01	0.69
Fatty acid esterification*	0.721	0.446	0.335	0.347	0.40
Urea concentration (mmol/l)	9.18	6.36	5.91	5.84	0.25

*μmol/h/10^7 cells

The study has confirmed the ability of re.BS to reduce fat deposition in sheep but casts further doubt on its lipolytic action. The potential role of re.BS as a repartitioning agent, decreasing fat deposition (by decreasing synthesis) and increasing protein deposition has been emphasised.

REFERENCES

1 Goodman, H.M. and Grichtig, G. (1983). Endocrinology **113,** 1697.

2 Johnsson, I.D., Hathorn, D.J., Wilde, R.M., Treacher, T.T. and Butler-Hogg, B.W. (1987). Animal Production **44,** 405.

3 Pell, J.M. and Bates, P.C. (1987). Journal of Endocrinology **115,** R1.

4 Sinnett-Smith, P.A. and Woolliams, J.A. (1988). Animal Production **47** (in press).

5 Vernon, R.G. (1982). International Journal of Biochemistry **14,** 255.

INCREASES IN SHEEP ADIPOCYTE β-RECEPTOR NUMBER ON EXPOSURE TO GROWTH HORMONE IN VITRO

P.W. Watt, R.A. Clegg, D.J. Flint and R.G. Vernon

Hannah Research Institute, Ayr, Scotland

Mobilization of lipids during lactation has been observed in a variety of species, however the mechanisms and individual factors responsible have not been elucidated completely. One factor associated with an increased rate of catecholamine-stimulated lipolysis in adipose tissue is an increase in the number of β-adrenergic receptors (2). A second factor also observed during lactation is an increase in the level of plasma growth hormone (1). The possible role of growth hormone (GH) in the regulation of the number of β-adrenergic receptors in sheep adipocytes during lactation has thus been investigated.

Adipose tissue explants from non-lactating and lactating sheep were incubated in Medium 199 containing sodium acetate (2 mM) and antibiotics for 0 h (Do) and 48 h ± GH (100 ng/ml) as described previously (4). Adipocytes were prepared from the explants and their plasma membranes isolated essentially as described by Malbon et. al. (3). The number of β-adrenergic receptors was measured using (-)-[^{3}H]dihydroalprenolol as described by Malbon et al. (3).

	β-receptor binding [fmol[^{3}H]-dihydroalprenolol/mg protein]	
	Non-lactating	Lactating
Do	28.0 ± 5.1	50.9 ± 15.6
48 h culture	33.1 ± 7.0	77.4 ± 13.0
48 h culture + GH	51.6 ± 10.2	98.8 ± 19.8

Results are means ± SEM of 6 observations

Analysis of variance of log-transformed data showed that the number of β-receptors in adipose tissue from lactating sheep is significantly greater ($P < 0.001$) than for non-lactating animals regardless of additions or length of incubation. In addition the number of receptors increased ($P < 0.02$) after 48 h of culture in the presence of GH. These observations are consistent with GH being the factor responsible for the increased number of β-receptors of adipocytes during lactation.

REFERENCES

1 Hart, I.C., Bines, J.A., Norant, S.V. and Ridley, J.L. (1978). Journal of Endocrinology **77,** 333.

2 Jaster, E.H. and Wegner, T.N. (1981). Journal of Dairy Science **64,** 1655.

3 Malbon, G.C., Moreno, F.J., Cabelli, R.J. and Fain, J.N. (1978). Journal of Biological Chemistry **253,** 671.

4 Robertson, J.P., Faulkner, A. and Vernon, R.G. (1982). Biochemical Journal **206,** 577.

THE EFFECT OF GROWTH HORMONE ON HIND-LIMB MUSCLE METABOLISM IN GROWING LAMBS

F. Lakehal, L.A. Crompton and M.A. Lomax

Department of Physiology and Biochemistry, University of Reading, Whiteknights, PO Box 228, Reading RG6 2AJ, U.K.

Growth hormone (GH) has been shown to have lipolytic and diabetogenic, as well as growth promoting biological actions. The purpose of this experiment was to establish whether the diabetogenic actions of GH on muscle metabolism were related to a change in the supply of oxidisable substrates as a result of the increase in lipolytic rate.

Hind-limb muscle metabolism was measured using arterio-venous difference and blood flow rate procedures in animals which had catheters inserted into the carotid artery and deep femoral vein. Growing wether lambs were treated with either saline (S) or ovine GH (0.25 mg/kg/day) for 11 days. Muscle metabolism was measured in 3 control and 3 GH treated lambs on day 11 of treatment. Results are shown in the table.

	Arterial Concentration (mM)			Net hind-limb uptake (nmol/min/g)		
	S	GH	SED	S	GH	SED
Glucose	5.07	4.88	0.24	50	20*	9
Lactate	0.98	0.93	0.12	4	-3	60
NEFA	0.15	0.30	0.05	1	3	1
Acetate	1.45	1.63	0.08	80	80	10
3-hydroxybutyrate	0.86	0.81	1.17	20	10	8

Significantly different from saline control: *$P < 0.05$

GH treatment significantly decreased hind-limb muscle glucose uptake without any change in arterial glucose concentration or lactate output. Plasma NEFA concentrations were increased by GH treatment, but this did not lead to a major change in either NEFA or 3-hydroxybutyrate uptake by the hind-limb.

It is concluded that the diabetogenic effect of GH on muscle is not related to changes in arterial glucose concentration or the supply of oxidisable substrates measured. In this experiment GH also stimulated plasma insulin concentrations and caused resistance to the hypoglycaemic effects of insulin injection. It is therefore possible that the diabetogenic action of GH on muscle tissue is due to the direct inhibition of insulin action rather than a change in substrate supply to muscle.

THE EFFECT OF GROWTH HORMONE ON MUSCLE ENZYME ACTIVITY AND LACTATE DEHYDROGENASE ISOENZYME PATTERN

F. Lakehal, M.J. Hannah, L.A. Crompton and M.A. Lomax

Department of Physiology and Biochemistry, University of Reading, Whiteknights, PO Box 228, Reading RG6 2AJ, U.K.

Anabolic agents have been shown to both increase the proportion of red to white type muscle fibres (1) and to promote protein deposition to a greater extent in red type muscle fibres (2). Red type muscle fibres differ from white type by having a higher oxidative capacity, ie higher citric acid cycle enzymes and a lactate dehydrogenase (LDH) isoenzyme pattern similar to heart. This study examines the effect of growth hormone (GH) on muscle LDH isoenzyme pattern and the activities of glycolytic and citric acid cycle enzymes.

Six growing wether lambs were treated with either saline (S) or ovine GH (0.25 mg/kg/day) for 5 days. Muscle biopsy samples were taken from a red type muscle (biceps femoris) on the last day of treatment.

GH treatment caused a 52% increase in the activity of the citric acid cycle enzyme, citrate synthase (GH, 4.44 $\pm$ 0.18; S, 2.92 $\pm$ 0.34 μmol/min/g, $P < 0.05$) but had no effect on the activities of the glycolytic enzymes phosphofructokinase, pyruvate kinase and LDH. There was a small but significant increase in the proportion of LDH 1, the isoenzyme which predominates in heart muscle (GH, 10.6 $\pm$ 1.7; S, 5.3 $\pm$ 0.8%, $P < 0.05$). Plasma lipoproteins may be the source for the increased oxidative metabolism, since muscle lipoprotein lipase activity was significantly stimulated by GH treatment (GH, 26.40 $\pm$ 3.69; S, 17.75 $\pm$ 1.76 μeq/h/g, $P < 0.05$).

It is concluded that a short period of GH administration to growing lambs causes an increase in the oxidative capacity of a red type skeletal muscle and a shift towards the heart muscle enzyme pattern. This may indicate an increase in the proportion of red type muscle fibres.

REFERENCES

1 Beerman, D.H., Fishell, V.K., Hogue, D.E., Ricks, C.A. and Dalrymple, R.H. (1985). Journal of Animal Science **61,** Suppl. 1, 254.

2 Pell, J.M. and Bates, P.C. (1987). Journal of Endocrinology **115,** R1.

THE EFFECTS OF CLOSE-ARTERIAL INFUSION OF CIMATEROL INTO THE HIND-LIMB OF GROWING LAMBS

J. Brown, L.A. Crompton and M.A. Lomax

Department of Physiology and Biochemistry, University of Reading, Whiteknights, PO Box 228, Reading RG6 2AJ, U.K.

The β_2-agonist cimaterol has been shown to increase the amount of lean tissue in the carcass of farm animals (1). Whether this effect is due to direct or indirect mechanisms remains unclear.

The direct effects of cimaterol on hind-limb muscle metabolism were studied in growing wether lambs by close-arterially infusing either saline (S), a low dose of cimaterol (LC) (0.1-0.2 mg/day) or a high dose of cimaterol (HC) (2.5-2.7 mg/day) into one femoral artery for 5 days. The other leg was close-arterially infused with saline (LS or HS) for 5 days as a control.

Hind-limb muscle metabolism was studied in both limbs on day 5 of treatment. Early observations on blood flow, oxygen uptake and carbon dioxide output across the hind-limb muscle are shown in the table below.

Treatment	S	LS	LC	HS	HC	
n	4	6	5	4	6	
	Mean	Mean	Mean	Mean	Mean	SED
Blood flow (ml/min)	231	247	432**	315	373*	31
Oxygen uptake (μmol/min)	420	552	977*	828	972*	143
Carbon dioxide output (μmol/min)	342	414	640	806*	868*	117
Respiratory quotient	1.25	1.43	1.67	1.19	1.32	0.24

Sigificantly different from saline (S): *$P < 0.05$, **$P < 0.01$

Close-arterially infused cimaterol, at a low dose, had a local direct effect on hind-limb muscle, significantly stimulating blood flow and oxygen uptake. The other leg (low dose saline) did not show this effect. In the high dose treatment, recirculation of the agonist appears to have occurred as shown by the increased values in both treated and saline legs compared to saline only infusions. Thus it appears that direct effects of cimaterol on muscle are important in its action.

REFERENCE

1 Beermann, D.H., Hogue, D.E., Fishell, V.K., Dalrymple, R.H. and Ricks, C.A. (1986). Journal of Animal Science **62**, 370.

MANIPULATION OF GROWTH IN ENTIRE MALE SHEEP BY THE BETA-ADRENERGIC AGONIST CIMATEROL AT TWO LEVELS OF DIETARY PROTEIN

H. Galbraith and L.A. Sinclair

Department of Agriculture, University of Aberdeen, 581 King Street, Aberdeen, U.K.

A major objective in animal production remains the regulation of growth to optimise lean tissue deposition in meat-producing animals. Current evidence suggests that these processes may be manipulated in sheep most effectively by the administration of certain beta-adrenergic agonists in the presence of endogenous sex-hormones in entire males. Nutritional status may also determine the response of these animals to manipulators of growth. The effect of differences in dietary protein was examined in the present study.

The response to the beta-agonist cimaterol (Boehringer Ingelheim, Vetmedica GmbH) was studied using 24 entire male Suffolk x Blackface lambs aged approximately 11 wk and weighing 24 kg. They were allocated 42 days before slaughter in a 2 x 2 factorial design to receive to appetite high quality isoenergetic diets containing an estimated 11.9 MJ ME per kg dry matter (DM). Two diets were formulated to contain either 120 [diet L] or 180 [diet H]g crude protein per kg DM. The diets were either unsupplemented (U) or supplemented (C) with cimaterol at a final concentration of 2 ppm.

Mean values [with SEM] for treatment UL, UH, CL and CH respectively and significance (*P < 0.05; **P < 0.01; ***P < 0.001) of main effects (P = protein; Ci = cimaterol; I = interaction) for selected characteristics are as follows: live weight gain, kg, 15.9, 17.0, 17.2, 18.7 (1.69; NS); food conversion ratio, 3.36, 3.23, 3.07, 2.98 (0.23; NS); cold carcass weight, kg, 16.4, 17.6, 17.7, 18.6 (0.64; P*, Ci*); liver (g/kg empty body weight [EBW]), 25.0, 28.5, 22.0, 25.6 (1.75; P*, Ci*); gastro-intestinal tract (g/kg EBW), 103, 106, 93, 98 (8.0; NS); m. longissimus dorsi, at 10th rib, cross sectional area, cm^2, 17.8, 19.6, 21.6, 22.6 (1.57; Ci**), and depth, cm, 3.02, 3.50, 3.52, 3.58 (0.12; P**, Ci**, I*); carcass crude protein (g/kg), 155, 169, 171, 175 (3.4; P**, Ci***); carcass fat (g/kg), 242, 201, 194, 193 (16.1; Ci*); RNA:DNA ratio in m. gluteus homogenate, 1.42, 1.43, 1.41, 1.52 (0.17; NS), blood plasma urea (day 28), μg/l, 411, 685, 157, 670 (61.4; P**, Ci**, I*); blood testosterone (day 28), μg/l, 2.07, 1.41, 1.00, 1.25 (0.43; NS).

The data indicate the effectiveness of cimaterol in increasing the weight and crude protein concentration and decreasing the fat concentration of the carcass - similar effects to those produced by elevated levels of dietary protein. Cimaterol also reduced the weight of the liver and the concentration of urea in blood plasma at day 28 - opposite effects to those produced by the increased level of protein, but did not affect significantly the weight of the digesta-free tract or the ratio of DNA to RNA in homogenate prepared from m. gluteus. The absence of a significant statistical interaction for the effects of both cimaterol and additional dietary protein on carcass weight and the proportion of crude protein suggests that these effects may be additive and independent. However, the significant interaction for the m. longissimus dorsa width measurement and plasma urea concentration indicates that certain responses to cimaterol may also depend on the level of protein in the diet.

GROWTH HORMONE RELEASE IN CALVES SELECTED FOR HIGH AND LOW DAIRY MERIT

P. Løvendahl, J.A. Woolliams, P.A. Sinnett-Smith and K. Angus

AFRC Institute of Animal Physiology and Genetics Research, Edinburgh Research Station, Roslin, Midlothian EH25 9PS, U.K.

The possible association between growth hormone (GH) release and dairy merit (1) was studied using four provocative tests in British Friesian calves (21 males + 28 females) from lines selected for high (n = 23) and low (n = 26) milk yield (line difference approximately 20%). GH release following acute intravenous administration of arginine 0.1 g/kg bw, hpGRF1-29NH_2 0.2 μg/kg bw, TRH 0.2 μg/kg bw and GRF plus TRH 0.2 μg + 0.2 μg/kg bw was determined. Animals were tested in 5 batches with each calf receiving 3 of 4 tests. Blood was sampled via jugular cannulae at -15, -5, 5, 10, 15, 20, 30, 60, 90, 120 min relative to injection. Serum was kept frozen until assayed by RIA for GH. Response was measured as the mean value during 5, 10, 15, 20 min following injection of GRF, TRH and GRF + TRH and 15, 20, 30, 60 min following arginine injection. Data were analysed in logarithmic scale and adjusted for effects of sex, batch and diet.

Preinjection baseline values on the high vs low line were, before arginine: 7.7 ng/ml vs 8.3 ng/ml SED 1.2 ng/ml ($P > 0.1$); GRF 7.2 ng/ml vs 6.0 ng/ml SED 1.1 ng/ml ($P > 0.1$); TRH 10.1 ng/ml vs 9.8 ng/ml SED 1.7 ng/ml ($P > 0.1$), GRF + TRH 9.6 ng/ml vs 6.7 ng/ml SED 1.6 ng/ml. The response in high vs low line was following arginine: 20.3 ng/ml vs 15.5 ng/ml SED 4.2 ng/ml ($P > 0.1$), GRF 38.6 ng/ml vs 24.0 ng/ml SED 6.8 ng/ml ($P < 0.05$), TRH 22.7 ng/ml vs 18.0 ng/ml SED 4.8 ($P > 0.01$), GRF + TRH 119.9 ng/ml vs 120.1 ng/ml SED 33.6 ng/ml ($P > 0.1$).

The difference in GH response was consistently higher in the high animals although the difference was only significant after the GRF injection. High animals also tended to have higher baseline concentrations confirming earlier findings (1). These findings indicate that GH release may be an indicator trait of breeding value for milk yield and thereby be of special value in MOET selection schemes (2).

REFERENCES

1 Barnes, M.A., Kazmer, G.W., Akers, R.M. and Pearson, R.E. (1985). Journal of Animal Science **60,** 271.

2 Woolliams, J.A. and Smith, C. (1988). Animal Production **76,** 333.

THE INFLUENCE OF BOVINE SOMATOTROPIN (BST) ON LEUKOCYTE INSULIN RECEPTORS IN LACTATING COWS AND BEEF STEERS

B.W. McBride[1], R.O. Ball[1], A.D. Kennedy[2] and J.H. Burton[1]

[1] Department of Animal and Poultry Science, University of Guelph, Guelph, Ontario, Canada

[2] Department of Animal Science, University of Manitoba, Winnepeg, Manitoba, Canada

Somatotropin has been used to quantitatively improve milk production in dairy cattle and increase rate of growth and amount of lean tissue in beef cattle perhaps because of its ability to alter the partitioning of nutrients among the various bovine tissues. A possible mechanism for this redistribution of nutrients has been presented in the suggestion that bST may reduce tissue responsiveness to insulin (1). To investigate this possibility blood samples were taken from 24 lactating cows in their second or later lactations and from 18 rapidly growing beef steers. Blood samples were taken at approximately 09.00 h, prior to the cows receiving their daily injections. All cows were between weeks 24 and 28 of lactation when sampled. Steers were sampled immediately prior to slaughter.

Insulin receptor analysis was conducted on mononuclear leukocytes (MNL) isolated according to the procedures outlined by Kennedy et al. (2). This analysis was performed according to published methods (2, 3).

Results of the analyses (Table 1) show no significant treatment effect on receptor concentration or affinity for insulin in the lactating cows. Similar results were found in the steers with the exception that both receptor site concentration and low affinity sites approached significance. Average daily gain for the bST treated steers was increased by 15% over controls.

Table 1 MNL insulin receptor characteristics in control and bST treated dairy and beef cattle

Animal	Treatment	N	Sites/cell (x 10^{-3})	KE (x 10^{8} M^{-1})	KF (x 10^{-6} M^{-1})	ADG (kg/d)
	Control	8	44.7 ± 12.6	0.76 ± 0.28	0.81 ± 0.36	
Dairy	10.3 mg/d	7	29.1 ± 5.0	0.54 ± 0.09	0.53 ± 0.13	
	20.6 mg/d	9	55.8 ± 17.3	0.37 ± 0.11	0.42 ± 0.24	
Level of significance ($P \leq$)			0.40	0.33	0.57	
Beef	Control	9	15.9 ± 2.5	0.71 ± 0.27	0.35 ± 0.12	1.3
	20.6 mg/d	9	45.9 ± 20.2	0.48 ± 0.22	0.13 ± 0.06	1.5
Level of significance ($P \leq$)			0.16	0.51	0.13	0.05

KE - high afinity; KF - low affinity; ADG - average daily gain

REFERENCES

1 Bauman, D.E. and McCutchen, S.N. (1986). Proceedings of the VI International Symposium on Ruminant Physiology pp 436, Prentice-Hall, New Jersey.

2 Kennedy, A.D. et al. (1987). Canadian Journal of Animal Science **67,** 721.

3 Pedersen, O. and Beck-Nielsen, H. (1976). Endocrinologica Scandinavica **83,** 556.

BLOOD CONSTITUENTS AND SUBCUTANEOUS ADIPOSE TISSUE METABOLISM OF DAIRY COWS AFTER ADMINISTRATION OF RECOMBINANT BOVINE SOMATOTROPIN (bST) IN A PROLONGED RELEASE FORMULATION

J. Skarda, P. Krejci, J. Slaba and E. Husakova

Institute of Animal Physiology and Genetics, Czechoslovak Academy of Sciences, Prague 10-Uhrineves, Czechoslovakia

In the first experiment 25 primiparous and multiparous dairy cows (Friesian and crossbred Friesian x Bohemian spotted) were randomly assigned to bST treatments (subcutaneous injections at 28 day intervals). Blood samples were collected at day 1, 8, 14 and 22 post injection and plasma was analyzed for bST concentration (using homologous radioimmunoassay) and for clinical parameters. Blood plasma bST concentration increased in primiparous cows from a basal level of 1.4 ng/ml to 10.7, 14.5, 27.0 ng/ml and in multiparous cows from a basal level of 1.2 ng/ml to 6.4, 11.7, 10.5 ng/ml at 24 h post injection of 320, 640 or 960 mg bST respectively and declined during the second week to control level at day 14 post injection. Plasma glucose, free fatty acids (FFA) and serum Ca, Na and K were not affected by treatment.

In the second experiment the profiles of bST in the plasma from the intensive sampling on five multiparous crossbred (Ayrshire x Bohemian spotted) cows showed that as early as during 1 h after the subcutaneous hormone (640 mg) administration, bST concentration in plasma increased to 22 ng/ml and then varied between 17 to 24 ng/ml during the first 5 days post injection. By day 7 bST decreased to 12 ng/ml and levels on day 14 were not different compared to control.

The incorporation of ^{14}C-acetate into total lipids of subcutaneous adipose tissue explants was significantly (+) lower in cows 12-15 days after third injection of 640 mg of bST. Release of FFA from adipose tissue explants was not significantly affected by treatment.

Substrate	Lipid synthesis nmol/h/g tissue		FFA release nmol/h/g tissue	
	Control	bST	Control	bST
Acetate (A): 4 mmol/l	135	44+	8177	8700
A + norepinephrine (NE): 0.016 mmol/l	125	29+	10199	10052
A + glucose (G): 3.5 mmol/l	1601	676+	8119	8922
A + G + NE	889	371+	9916	8472

The suppression of lipogenesis in adipose tissue of bST treated cows support the concept that bST partitions fatty acids from adipose tissue and circulation for preferential use by mammary tissue.

THE EFFECT OF RECOMBINANT BOVINE SOMATOTROPIN (bST) IN A SUSTAINED RELEASE VEHICLE ON THE PERFORMANCE OF DAIRY COWS

J. Skarda[1], P Krejci[1], J Vesely[2] and K Molnarova[1]

[1] Institute of Animal Physiology and Genetics, Czechoslovak Academy of Sciences, Prague 10-Uhrineves, Czechoslovakia
[2] JZD Agrokombinet Slusovice, Czechoslovakia

Numerous studies have reported the effect of daily injections of bST on dairy cow performance but the effect of injections of a prolonged release formulation of bST has not been published.

In the first experiment 56 Friesian and crossbred (Friesian x Bohemian spotted) primiparous and multiparous cows were assigned to one of four treatments at 78-89 day postpartum after 21 day preliminary period. Cows were injected (subcutaneously) with 0, 320, 640 or 960 mg of recombinant bST (Lilly/Elanco) at 29 day intervals in an 84 day experiment (June-September). A 21 day preliminary period was used to collect covariate data to generate least-square means for treatments.

Daily milk yield of control cows gradually decreased during experimental period. Cows treated with bST showed a cyclic pattern of daily milk yield within each injection cycle. Milk production increased to a maximum at 4 to 8 day post injection and then slowly declined to 28th day. Milk, 4% fat-corrected milk (FCM) and fat yields were significantly ($P < 0.01$) improved by each dose of bST. Percent milk fat and protein were not altered.

	bST (mg/28 days)			
Variable	0	320	640	960
Milk, kg/day	19.58	20.85+	21.54+	21.89+
% increase		6.5	10.0	11.8
FCM, kg/day	18.78	20.06+	20.55+	21.11+
Fat, kg/day	0.73	0.78	0.80	0.82
Fat %	3.75	3.74	3.70	3.75
Protein %	2.98	3.05	2.92	3.02

In the second experiment 40 crossbred (Ayrshire x Bohemian spotted) multiparous cows from two commercial herds were randomly assigned to one of two treatments. Ten cows of each herd received 640 mg of bST at 28 day intervals for 112 day (December-April) and 10 cows were controls. Treated cows produced 11.8 and 15.2% more actual milk than control cows. Pregnancy rates and mastitis incidence were not affected by bST.

In all our commercial herds recombinant bST in sustained release vehicle increased milk yield with no adverse effect upon health and reproduction.

CHANGES IN MAMMARY UPTAKE OF ESSENTIAL AMINO ACIDS IN LACTATING JERSEY COWS IN RESPONSE TO EXOGENOUS BOVINE PITUITARY SOMATOTROPIN

F.M. Fullerton[1], T.B. Mepham[1], I.R. Fleet[2] and R.B. Heap[2]

[1] University of Nottingham, Faculty of Agricultural and Food Sciences, Sutton Bonington, Loughborough, Leicestershire LE12 5RD, U.K.
[2] AFRC Institute of Animal Physiology and Genetics Research, Cambridge Research Station, Babraham, Cambridge CB2 4AT, U.K.

Lactating Jersey cows (31-35 wk) were given daily injections of pituitary bovine somatotropin (bST), 30 mg day^{-1}, subcutaneous, 1.4 units mg^{-1} (kindly donated by Dr I.C. Hart) for 7 days. Each cow had been previously surgically prepared, under halothane-oxygen anaesthesia, for carotid arterial (A) and milk vein (V) sampling and mammary blood flow (MBF) measurement by thermodilution. On days -1 (control day), 1, 7, 10 and 21, MBF was measured and 5 pairs of A and V blood samples were taken for chemical analysis.

On day 7 of bST treatment there were increases in milk yield (149.6 ± 7.2% of control values, n = 6, P < 0.01) and mammary plasma flow (MPF, 147.8 ± 18.3%, n = 6, P < 0.05) which had returned to control values by day 21 as previously reported (1). Mean milk protein concentration decreased by 6.6 ± 0.7%, n = 4 (P < 0.05) during bST treatment. Free plasma A concentrations and A-V differences for essential amino acids showed no significant changes in response to bST treatment except for the plasma A concentration of histidine which was significantly reduced on day 7 (64.3 ± 11.6% of control values, n = 5, P < 0.05). The effect of bST treatment on the uptakes of essential amino acids by the mammary gland was only statistically significant on day 1 (Table 1).

Table 1 Mammary uptakes of essential amino acids in response to bST treatment in 5 Jersey cows (mean ± SEM, half udder)

	Control day $mg\ min^{-1}$	Day 1 % of control		Control day $mg\ min^{-1}$	Day 1 % of control
Met	1.57 ± 0.39	149.7 ± 18.4*	Thr	3.32 ± 0.69	132.5 ± 11.6*
Tyr	3.51 ± 0.73	146.0 ± 18.2*	Ile	4.13 ± 0.97	174.7 ± 37.8
Phe	2.81 ± 0.36	133.0 ± 12.5*	Leu	6.33 ± 1.13	149.5 ± 15.1*
His	1.32 ± 0.25	147.7 ± 16.1*	Lys	6.29 ± 1.11	146.2 ± 15.1*
Val	5.44 ± 1.54	146.7 ± 20.0	Arg	5.02 ± 0.69	162.6 ± 39.0

Asterisks denote significant changes from control (P < 0.05, unpaired Student's t test)

Treatment with bST stimulated milk secretion and MPF. The uptakes of essential amino acids were significantly increased on day 1 of bST treatment, in excess of the increase in MPF (119.2 ± 7.9% NS).

REFERENCE

1 Fullerton, F.M., Fleet, I.R., Heap, R.B., Hart, I.C. and Mepham, T.B. (1989). Journal of Dairy Research (in press).

BIOCHEMICAL RESPONSES TO THE USE OF RECOMBINANT BOVINE SOMATOTROPIN (SOMETRIBOVE) IN DAIRY CATTLE IN RELATION TO PRODUCTION AND WELFARE

D.A. Whitaker, E.J. Smith and J.M. Kelly

University of Edinburgh, Department of Veterinary Clinical Studies

Recombinant bovine somatotropin (sometribove) injected into dairy cows results in increases in milk production. This study assessed some of the biochemical consequences and their welfare implications.

22 cows and 16 first lactation heifers were paired, with one of each pair injected at 80 days (± 7) after calving with sometribove in prolonged release formulation or placebo. Injections, subcutaneously, were repeated at two week intervals during the winter feeding period. Measurements made were daily milk yield, twice weekly bodyweight and condition score, weekly milk composition and blood levels of beta-hydroxybutyrate (BHB), urea-nitrogen (urea N), albumin, globulin, calcium, magnesium, inorganic phosphate and plasma glucose.

Treated cows produced on average 4.5 kg more milk per day than controls and heifers 2.5 kg more. Yield increased for 3-4 days after each injection, plateaued and decreased for 2-3 days before the next injection. There were no significant differences in milk quality. Over 12 weeks, treated cows gained on average 6.8 kg bodyweight and lost 0.1 of a condition score. Controls gained 18.8 kg and 0.3 score. Heifers over eight weeks lost 3.0 kg and 0.1 score, controls gaining 6.4 kg and 0.2 score.

Significantly higher BHB and inorganic phosphate levels in blood and lower urea N were found in treated animals on average. There were no other practical or significant differences in the biochemical measurements except, midway between treatments, urea N dropped in cows and BHB levels rose in heifers.

Production responses were similar to other reported trials (1). Biochemical measurements fell within acceptable ranges suggesting no unusual or unreasonable production stress nor any welfare concern in animals treated at 80 days after calving. Differences in urea N suggest better protein utilisation in treated animals. Higher inorganic phosphate implies some alteration in mineral metabolism.

REFERENCE

1 Peel, C.J. and Bauman, D.E. (1987). Journal of Dairy Science **70**, 474.

ROLE OF FEVER ON THE ENDOTOXIN-INDUCED SOMATOTROPIN RELEASE IN THE LACTATING GOAT

C. Burvenich, G. Vandeputte-Van Messom, J. Fabry and A.M. Massart-Leen

Faculty of Veterinary Medicine, Physiological Department, University of Ghent, Casinoplein 24, B-9000 Ghent, Belgium

Somatotropin (STH) blood concentrations increased during the fever of infection and artificial hyperthermia in man (Bunner et al., 1984). Fever is observed after administration of bacterial endotoxin to humans and experimental animals. In goats febrile reactions to intravenous (i.v.) injection of endotoxins are accompanied by elevations of plasma STH levels (Burvenich et al., 1984). In the present experiments we investigated the role of fever in the endotoxin or lipopolysaccharides (LPS)-induced changes in plasma STH. The effects of i.v. LPS were compared in goats before and after pretreating the animals with the non-steroidal, anti-inflammatory agent flurbiprofen (The Boots Company). Plasma levels of non-esterified fatty acids (NEFA) were also measured.

Fifty six experiments were carried out on 6 goats weighing between 49 and 63 kg. Goats were kept in invididual standings and were well adapted to experimental manipulations. In a first series of control LPS experiments goats were randomly injected (7 experiments) with one of the 4 following doses of Escherichia coli LPS: 0, 25, 100 or 500 ng/kg (Difco Laboratories, Detroit, U.S.A. 026: B6). In a second series of antipyretic experiments LPS was injected as in the first series but goats were i.v. pretreated with 2 mg/kg of flurbiprofen, 30 min before LPS injection. A second dose of the antipyretic agent was given 6 h later. Blood samples were taken hourly via a catheterized external jugular vein until the 10th h as well as at 4.5, 5.5 and 6.5 h after LPS injection and rectal temperature (RT) was monitored. Plasma STH was measured by RIA (Fabry et al., 1982) and NEFA was assayed according to the method of Smith (1975).

I.v. injection of increasing doses of LPS to goats caused fever, characterised by a striking biphasic response after the highest LPS dose, with peaks occurring at 1.5 and 4.5 h after injection. Fever response was dose-dependent, but only to low (40.3°C) and moderate (40.6°C) doses of LPS. High doses of LPS (500 ng/kg) also induced fever, but peak values were below those seen after a dose of 100 ng/kg. Fever was accompanied by clinical symptoms including piloerection, depression, miosis and loss of appetite. Endotoxin also caused a dose-dependent rise in plasma NEFA. Peak values were observed 2 h after injection and occurred before fever peak. Plasma STH levels significantly increased in a dose-dependent way. Peak effects of 68%, 72% and 128% above preinjection level were attained 4 to 5 h after LPS injection and coincided with RT peaks. I.v. injection of NaCl 0.99% (dose of 0 ng/kg LPS) did not induce any significant change in RT, NEFA or STH plasma levels.

I.v. pretreatment with flurbiprofen blocked fever, as well as the increase in plasma NEFA and STH induced by i.v. endotoxin. Fever response and clinical symptoms were completely abolished. RT even slightly decreased below control level. The increase in NEFA was either completely inhibited (100 ng/kg) or partly for 58% (500 ng/kg; $P < 0.05$). STH response to LPS was also significantly depressed for 48% (25 ng/kg), 53% (100 ng/kg) and 66% (500 ng/kg) by flurbiprofen. The onset as well as time of appearance of peak STH values were delayed for at least 1 h in comparison to control experiments.

It is generally accepted that the sympathetico-adrenomedullary system is involved in fever development. Peripheral vasoconstriction and increased metabolic rate would help to bring about the rise in body temperature. In the present experiments the increase in plasma NEFA started immediately after LPS injection. Peak values were observed during the increment phase of fever. During that period an elevation of catecholamines in plasma has been observed (Bernabé et al., 1986). The effect of endotoxin on plasma NEFA probably can be ascribed to sympathetic activation. Inhibition of these effects by flurbiprofen further supports this conclusion. Antipyretics are known to reverse vasoconstriction and to reduce fever. It is believed that the subsequent and sudden drop in the concentration of plasma NEFA caused the rise in plasma STH during LPS fever. The increase in plasma STH was suppressed by flurbiprofen. It thus seems that the changes in plasma during endotoxin fever are indirectly mediated by an early activation of the sympathetic nervous system.

ACKNOWLEDGEMENTS

This study was supported by a grant of the Belgian N.F.W.O.

REFERENCES

1 Bernabé, J., Vandeputte-Van Messom, G. and Burvenich, C. (1986). Archives Internationales de Physiologie et de Biochimie **94**, P67.

2 Bunner, D.L., Morris, E. and Smallridge, R.C. (1984). Metabolism **33**, 337.

3 Burvenich, C., Reynaert, R., Vandeputte-Van Messom, G. and Peeters, G. (1984). Archives Internationales et de Biochimie **93**, P25.

4 Fabry, J., Oger, R. and Reynaert, R. (1982). Bulletin des Recherches Agronomiques de Gembloux **17**, 331.

5 Smith, S.W. (1975). Analytical Biochemistry **67**, 531.

EFFECTS OF RECOMBINANT PORCINE GROWTH HORMONE IN GERMAN PIG BREEDS DURING GROWTH AND LACTATION

F. Ellendorff[1], E. Kallweit[1], E. Hüster[1], D. Ekkel[1], R.E. Ivy[2] and D.J. Meisinge[2]

[1] Institut für Tierzucht und Tierverhalten (FAL), Mariensee, 3057, Neustadt 1, Federal Republic of Germany
[2] Pitman-Moore Inc., Terre Haute, IN 47808, U.S.A.

German breeds of pigs have a long record of selection for performance characteristics. It was thus tempting to test, whether or not the use of recombinant porcine growth hormone (re.pGH) would influence production characteristics, carcass quality and meat quality in a similar fashion as in breeds that have undergone less rigid selection procedures.

German Large White (Deutsches Edelschweing, n = 22) and German Pietrains (n = 22) were included in the study. Starting with 50 kg of body weight (47-52) treated animals (n = 10/breed) received 5 mg re.pGH (Pitman-Moore) daily intramuscularly. Sham animals (n = 5/breed) received the solvent and 6 animals/breed served as controls. Seventy days (approximately 100 kg body weight) after onset of treatment the animals were slaughtered. Mean daily gain (50-100 kg) ranged from 774 to 918 g/day between different groups with no significant differences. Similarly, significant differences for food intake or food conversion were distributed between groups in a pattern that did not allow the interpretation of rpGH having a significant influence.

The results relating to carcass composition were very different. Even in the lean Pietrain breed a further highly significant ($P < 0.001$) reduction in fat measurements was clearly attributable to re.pGH-treatment. Equally, the percent muscular tissue was significantly ($P < 0.05$ to 0.01) increased in Large White and Pietrains; so were a number of specific muscles. Criteria of meat quality such as pH, colour, rigor mortis, conductivity, tissue composition as well as water binding, uptake and capacity were not significantly affected.

It is concluded, that in pigs highly selected for production re.pGH does improve carcass composition (ie reduce fat and increase muscular tissue), but it does not alter performance charcteristics (eg growth, feed conversion) or meat quality.

SEMISYNTHESIS OF INSULINS WITH MODIFIED GROWTH PROMOTING PROPERTIES

A.N. McLeod, J.E. Pitts, A. Auf der Mauer and S.P. Wood

Laboratory of Molecular Biology, Department of Crystallography, Birbeck College, Malet Street, London WC1E 7HX, U.K.

Insulin and IGF-I have a good deal of sequence and probable structure homology (1). They also exhibit some ability to bind each other's receptors. They also exhibit some ability to bind each other's receptors. Since some residues important for insulin activity are known (2) and some insulin sequence variants have been shown to have different mitogenic properties (3), there exist several guidelines on how to manipulate the properties of insulin. Here we report progress in the semisynthesis of insulins modified in the B chain C-terminal region. These compounds should help in understanding the basis of specificity for insulin and IGF receptors.

REFERENCES

1 Blundell, T.L. and Humbel, R.E. (1980). Nature **287,** 781.

2 Blundell, T.L., Pitts, J.E. and Wood, S.P. (1982). Critical Review of Biochemistry **13,** 141.

3 King, G.L. and Kahn, C.R. (1981). Nature **292,** 644.

IDENTIFICATION OF THE PORCINE INSULIN-LIKE GROWTH FACTOR (IGF)I GENE PROMOTER REGION

M.C. Dickson, J.C. Saunders, K. Kwok, J.R. Miller, N.S. Huskisson and R.S. Gilmour

Biochemistry Department, AFRC Institute of Animal Physiology and Genetics Research, Cambridge Research Station, Babraham, Cambridge CB2 4AT, U.K.

It is thought that growth hormone acts at the gene level to increase IGF-I polypeptide synthesis (1) and by analogy with other genes that the 5′ promoter region of the gene is responsible for regulating this response. As a first step towards understanding the mechanisms involved we have cloned the porcine IGF-I gene and isolated a 10 kilo base stretch of DNA which contains the first coding exon and extensive upstream sequence. About 2.5 kilobase of this DNA has been sequenced including exon 1. A comparison with published sequence for the human (2) and rat IGF-I (3) shows a remarkable degree ($>70\%$) of homology not only in the coding region but also in about 1.5 kilobase of untranslated 5′sequence. Thereafter homology breaks down rapidly. The strong interspecies homology in untranslated DNA might suggest either a highly conserved, transcribed leader sequence which precedes the coding region or common control elements in an untranscribed promoter structure. Analysis of the sequence does not reveal the usual motifs associated with polymerase binding and the initiation of transcription. Current work is directed at mapping the extent of transcription from the 5′ sequences by hybridization to IGF-I mRNA.

REFERENCES

1 Isaksson, O.G.P., Jansson, J.O. and Gause, I.A.M. (1982). Science **216,** 1237.

2 Rotwein, P., Pollock, K.M., Didier, D.K. and Krivi, G.G. (1986). Journal of Biological Chemistry **261,** 4828.

3 Shimatsu, A. and Rotwein, P. (1987). Journal of Biological Chemistry **262** 16, 7894.

STUDIES ON THE SECRETION OF HUMAN PROINSULIN AND INSULIN-LIKE GROWTH FACTORS I AND II IN E.COLI

S.E. Lawler and J.E. Pitts

Laboratory of Molecular Biology, Department of Crystallography, Birkbeck College, Malet Street, London WC1E 7HX, U.K.

Our aim is to study structure-function relationships in the hormones of the insulin family by examining the properties of engineered recombinant insulin and IGF's (1). To achieve this the expression vector pIN-III-OmpA$_2$ (2) is being used. This vector has been successfully used previously to direct a number of recombinant proteins into the periplasm of E.coli (3, 4, 5).

Fragments containing the genes for proinsulin and both the IGF's have been subcloned into pIN-III-OmpA$_2$. Site-specific deletion of the non-coding DNA between the signal peptide coding sequence and the inserted gene is currently being carried out using synthetic oligonucleotides.

The secretion of a proinsulin analogue containing 5 additional amino acids at the N-terminus coded for by DNA yet to be deleted is being examined. SDS-PAGE of cell fractions has shown that a protein of the correct molecular weight is present in the periplasm of induced cells, and phase-contrast microscopy has shown that morphological changes also occur. We aim to purify this analogue, determine its activity and carry out N-terminal sequencing to check that the OmpA signal peptide has been correctly cleared.

REFERENCES

1 Blundell, T.L., Pitts, J.E. and Wood, S.P. (1982). Critical Reviews of Biochemistry **13,** 141.

2 Ghrayeb, J., Kimura, H., Takahara, M., Hsiung, H.M., Masui, Y. and Inouye, M. (1984). EMBO Journal **3,** 2437.

3 Hsiung, H.M., Mayne, N.G. and Becker, G.W. (1986). Bio/Technology **4,** 991.

4 Takahara, M., Hilber, D.W., Barr, P.J., Gertt, J.A. and Inouye, M. (1985). Journal of Biological Chemistry **260,** 2670.

5 Takahara, M., Sagai, H., Inouye, S. and Inouye, M. (1988). Bio/Technology **6,** 195.

INSULIN-LIKE GROWTH FACTOR (IGF) I AND II mRNA LEVELS IN THE TISSUES OF THE OVINE FETUS AND NEONATE

J.C. Saunders and R.S. Gilmour

Biochemistry Department. AFRC Institute of Animal Physiology and Genetics Research. Cambridge Research Station. Babraham. Cambridge CB2 4AT. U.K.

Two distinct forms of IGF have been identified in many species (1. 3). IGF-I. which is highly growth hormone dependent and is regulated by nutritional status. and IGF-II which is closely related but less growth hormone dependent. Although these peptides are present in the circulation and may act by endocrine mechanisms (2. 6). it is widely believed they also act by autocrine or paracrine mechanisms. To list this systematically. total RNAs were extracted from ovine fetus and neonate tissues and analysed for IGF-I and II mRNA content by quantitative dot blot hybridisation against cDNA probes specific for each IGF.

For fetal tissues collected between 28-145 days gestation IGF-II mRNA predominates over that of IGF-I. Episodic synthesis of IGF-II mRNA is seen in some tissues (liver. spleen. skeletal muscle) while continuous. elevated levels occur in kidney and lung: the latter shows the highest values for tissues examined. IGF-I mRNA levels are much lower and almost absent in some tissues: however synthesis in fetal liver is constant (5 x less than IGF-II) throughout gestation. This suggests a paracrine rather than an endocrine role for IGF-II in the fetus; however. the possibility exists that for IGF-I the fetus may respond to hepatic synthesis in an endocrine manner.

In the neonate. IGF-I synthesis increases dramatically under the influence of growth hormone (4). IGF-I synthesis in the liver is regarded as the main source of serum IGF-I (5). Nevertheless. it can be seen from mRNA analysis that all tissues examined synthesise IGF-I mRNA to varying degrees and therefore have the potential for IGF synthesis. This supports the notion that in the adult IGF-I may exert both endocrine and paracrine functions and that these may differ in response depending on the site of synthesis.

REFERENCES

1 Daughaday. W.H. (1982). Proceedings of the Society for Experimental Biology and Medicine **170**, 257.

2 D'Ercole. A.J.. Applewhite. G.T. and Underwood. L.E. (1980). Developmental Biology **75**, 315.

3 Froesch. E.R.. Schmid. C.. Schwander. J. and Zapf. J. (1985). Annual Review of Physiology **47**, 443.

4 Isaksson. E.G.P.. Jansson. J.O. and Gause. I.A.M. (1982). Science **216**, 1237.

5 Phillips. L.S. et. al. (1976). Endocrinology **98**, 606.

6 Rinderknecht. E. and Humbel. R.E. (1978). FEBS Letters **89**, 283.

LEVELS OF INSULIN-LIKE GROWTH FACTOR I IN SHEEP TISSUES

K.M. Crimp, M.A. Lomax and D. Savva

Department of Physiology and Biochemistry, University of Reading, Whiteknights, PO Box 228, Reading RG6 2AJ, U.K.

The plasma concentrations of insulin-like growth factor I (IGF-I) are known to respond to growth hormone treatment and nutritional status in sheep and cattle (1, 4). It has been demonstrated that the abundance of IGF-I messenger RNA (mRNA) is regulated by growth hormone in a variety of rat tissues (3).

We are carrying out studies to measure IGF-I mRNA levels in sheep tissues in response to hormonal and nutritional treatments. However, a sheep IGF-I cDNA has not been characterised yet and we have therefore been using a human IGF-I cDNA probe to measure the levels of IGF-I mRNA in sheep tissues.

Total RNA was isolated from a variety of sheep tissues using the one-step guanidine thiocyanate-phenol procedure (2). This procedure yields RNA that is free from any contaminating DNA and is relatively undegraded.

We have employed dot and Northern hybridisation to compare the levels of IGF-I mRNA in tissues taken from sheep treated with growth hormone. Northern hybridisation analysis of total RNA indicates that the human IGF-I cDNA probe hybridises to a mRNA species whose size is approximately 1 Kb.

REFERENCES

1 Breier, B.H., Gluckman, P.D. and Bass, J.J. (1988). Journal of Endocrinology **118,** 243.

2 Chomczynski, P. and Sacchi, N. (1987). Analytical Biochemistry **162,** 156.

3 Hynes, M.A., von Wyk, J.J., Brooks, P.J., D'Ercole, A.J., Jansen, M. and Lund, P.K. (1987). Molecular and Cellular Endocrinology **1,** 233.

4 Pell, J.M., Gill, M., Beever, D.E., Jones, A.R. and Cammell, S.B. Proceedings of the Nutrition Society (in press).

MODULATION OF INSULIN-LIKE GROWTH FACTOR I AND II mRNA DURING LOCALIZED GROWTH OF RAT SKELETAL MUSCLE AND BROWN ADIPOSE TISSUE

D. DeVol[1], J. Novakofski[1], P. Bechtel[1] and P. Rotwein[2]

[1] University of Illinois, Urbana, IL 61801, U.S.A.
[2] Washington University School of Medicine, St. Louis, MO 63110, U.S.A.

We have evaluated the expression of the insulin-like growth factor I and II (IGF-I and IGF-II) genes during localized, tissue specific growth using a specific solution-hybridization nuclease protection assay to detect IGF-I and IGF-II mRNA (3). Several tissue specific growth models have been used including compensatory hypertrophy of rat skeletal muscle, cold induced hyperplasia of rat brown adipose tissue (BAT), and administration of rat growth hormone (GH) to hypophysectomized rats.

Compensatory hypertrophy of the plantaris (P) and soleus (S) muscles of the rat hind limb was induced by unilateral excision of the distal gastrocnemius tendon, with the contralateral leg serving as a control. By 8 days post-surgery, hypertrophy in P and S muscles had increased to 120% of control values. Hypertrophy muscle IGF-I mRNA levels rose to 2-3 fold above control values by 4 days post-surgery in both the P and S ($P < 0.05$). By 4 days post-surgery, hypertrophy muscle IGF-II mRNA levels were 5- ($P < 0.05$) and 7- ($P < 0.025$) fold higher than control values in the P and S muscles, respectively. In addition, systemic administration of GH to hypophysectomized rats increased IGF-I levels 2-3 fold in both the P and S, while IGF-II mRNA levels were increased only in the S muscle (3-fold increase over controls) (1). We are currently evaluating the effects of hypophysectomy on skeletal muscle IGF-I and IGF-II levels during compensatory hypertrophy.

In order to investigate the potential involvement of IGF-I and/or IGF-II in the localized growth of another tissue type, hyperplasia of BAT was induced by placing rats in a cold environment (4°C). After 24 h of cold exposure, [^{3}H]-thymidine incorporation into BAT was increased from 76 to 684 dpm/μg DNA, and by 96 hours of cold exposure, a significant increase in BAT pad weight had occurred (354 to 443 mg; $P < 0.05$). There was no increase in BAT IGF-I mRNA levels as a result of cold exposure; however, IGF-II mRNA levels increased 4-6 fold after 8 and 24 h of cold exposure. IGF-II levels decreased to 2-fold above control values by 96 h of cold exposure. Furthermore, systemic administration of GH to hypophysectomized rats indicated that BAT IGF-I mRNA levels, but not IGF-II mRNA levels, were under control of GH (2).

The above results support the hypothesis that both locally produced and systemic growth factors may regulate tissue specific growth.

REFERENCES

1 DeVol, D., Bechtel, P., Novakofski, J. and Rotwein, P. (1988). Endocrinology **122,** (Abstract Supplement), 74.

2 DeVol, D., Rotwein, P., St. Denis, S., Bechtel, P. and Novakofski, J. (1988). The FASEB Journal **2,** A1416.

3 Shimatsu, A. and Rotwein, P. (1987). Journal of Biological Chemistry **262,** 7894.

IDENTIFICATION OF IGF-I mRNA IN RABBIT MAMMARY GLAND AND EVOLUTION DURING PREGNANCY AND LACTATION

H. Jammes, M. Duclose and J. Djiane

Unite d'Endocrinologie Moleculaire, INRA, 78350 Jouy en Josas, France

We have recently demonstrated the presence of specific receptors of IGF-I and IGF-II in the mammary gland from pregnant rabbit and ewes (1, 2). Moreover, we have performed an autoradiographic identification of IGF-I receptors on frozen mammary gland sections and observed a distribution of IGF-I receptors exclusively in epithelial tissue.

In the present study the possible production of IGF-I by the rabbit mammary gland itself was investigated by analysing the presence of IGF-I mRNAs in the mammary tissue at various stages of pregnancy and lactation. RNAs were extracted from mammary gland using the guanidinium isothiocyanate/lithium chloride method. The fractions were enriched for Poly(A) + RNA by two cycles of oligoT cellulose affinity chromatography. The Poly(A) + RNA was separated by electrophoresis in an agarose (1.4%) formaldehyde gel and transferred to hybond C paper. After prehybridization for 4 h at 42°C, the hybridization was performed using a human ^{32}P-IGF-I cDNA probe in 20% formamide.

The Northern blot analysis of mRNAs from rabbit mammary tissue at 18, 22, 26 and 29 days of pregnancy and at 6, 15 and 26 days of lactation demonstrated a specific hybridization with the IGF-I cDNA probe. In the mammary gland from pregnant rabbit, IGF-I cDNA probe hybridized with 3 groups of transcripts of 7, 1.8 and 0.8-1 kb; the predominant form being the 0.8-1 kb. In mammary gland from lactating rabbit, only the 0.8-1 kb was observed. Differences by initiation, polyadenylation sites or alternative splicing of a common primary transcript could contribute to explain these differences between the IGF-I mRNA patterns in pregnancy and lactation.

The concentrations of IGF-I mRNAs were estimated by dot blot hybridization. An increase of IGF-I mRNA is positively correlated with the state of pregnancy and with the mammogenesis: epithelial cell multiplication and onset of cell differentiaton. However, during lactation at the time of full milk protein synthesis, the relative level of IGF-I mRNAs is decreased.

The presence of IGF-I mRNAs in the mammary tissue provide further support for the production of IGF-I in situ and for a possible autocrine action of this factor on the mammary cell. The hormonal control of IGF-I production by the mammary gland remains to be determined.

REFERENCES

1 Disenhans, C., Belair, L. and Djiane, J. (1988). Reproduction, Nutrition and Development **28**, 241.

2 Duclose, M., Servely, J.L., Houdebine, L.M. and Djiane, J. (1987). Journal of Endocrinology **115** (Suppl.), Abstract 87.

TISSUE-SPECIFIC AND DEVELOPMENTAL EXPRESSION OF INSULIN-LIKE GROWTH FACTOR-I IN THE PREGNANT AND LACTATING PIG

F.A. Simmen[1], R.C.M. Simmen[1], L.R. Letcher[1], A. Tavakkol[1] and F.W. Bazer[2]

[1] The Ohio State University, Department of Animal Science and Laboratories of Molecular and Developmental Biology, Wooster, Ohio, U.S.A.
[2] University of Florida, Department of Animal Science, Gainesville, Florida, U.S.A.

Insulin-like growth factor-I (IGF-I) is a polypeptide growth factor of 70 amino acids believed to be the major in vivo mediator of growth hormone action and also implicated in the oestrogen-promoted growth of reproductive tissues. However, its role in embryonic, fetal and neonatal development is less clear. We have therefore initiated a survey of maternal fluids and maternal, embryonic, and fetal tissues of the pig for the presence of IGF-I peptide and messenger RNAs. Immunoreactive IGF-I was quantified by radioimmunoassay of acid-ethanol extracted biological fluids or acetic-acid extracts of tissues (1). Tissue levels of IGF-I mRNAs were compared after hybridization of pig IGF-I complementary DNAs (cDNAs) (2) to dot-blots or Northern blots containing tissue RNA.

Mammary gland secretions contained significant quantities of IGF-I (colostrum: 50-300 ng/ml; milks corresponding to days 5-14 of lactation: 4-14 ng/ml). Maternal sera obtained at parturition displayed a range of IGF-I content (20-91 ng/ml) which was typically higher than that of newborn pig sera (12-16 ng/ml). Uterine luminal fluids obtained from sows during early pregnancy also contained immunoreactive IGF-I (day 11 of pregnancy: 100-300 ng/uterine horn; day 14 of pregnancy: 5-50 ng/uterine horn). In contrast, allantoic fluids were devoid of detectable IGF-I. Acid extraction of uteri and mammary glands obtained from pregnant sows also yielded immunoreactive IGF-I. At any given stage of pregnancy, uterine IGF-I levels exceeded those of the mammary gland. In addition uterine tissue IGF-I levels were maximal during early pregnancy and declined as pregnancy proceeded.

IGF-I mRNAs were identified in multiple tissues of sows and porcine fetuses. Uteri of early pregnant sows exhibited the highest levels of IGF-I mRNAs among those tissues surveyed to date. In contrast, IGF-I mRNA levels in maternal liver, spleen and mammary gland were 10-50 fold lower. IGF-I mRNAs were also detected in liver, intestine and stomach of late-gestation fetuses. The identification of this growth factor and its mRNA in both reproductive and non-reproductive tissues of maternal and fetal origin and in biological fluids suggests a generalized role for IGF-I in early embryonic, fetal and neonatal development.

REFERENCES

1 Simmen, F.A., Simmen, R.C.M. and Reinhart, G. (1988. Developmental Biology (in press).

2 Tavakkol, A., Simmen, F.A. and Simmen, R.C.M. (1988). Molecular Endocrinology **2**, 674.

IGF-BINDING PROTEINS FROM PIG PRE-IMPLANTATION BLASTOCYSTS

A.N. Corps, C.J. Littlewood and K.D. Brown

Department of Biochemistry, AFRC Institute of Animal Physiology and Genetics Research, Cambridge Research Station, Babraham, Cambridge CB2 4AT, U.K.

Insulin-like growth factor (IGF)-binding proteins, generally inhibiting interactions of IGFs with their cellular receptors, are a major secretory product of endometrium (1), are secreted by mouse embryo fibroblasts in culture (2), and are present in the serum of fetal pigs (3). We are investigating the involvement of growth factors during the late pre-implantation period in the pig, and the aim of the present study is to characterise the IGFs, their receptors and binding proteins at this stage.

Blastocysts were flushed from the uterine horns of gilts between 16-18 days of pregnancy using Dulbecco's Modified Eagle's Medium (DMEM). Cell suspensions were prepared by mechanical disruption of whole blastocysts or of trophoderm after the removal of vascularised embryo-associated material. The cells were washed and incubated in DMEM for 24 h at 37°C; the supernatant medium was collected, centrifuged and stored frozen until required. Specific binding of ^{125}I-IGF-I to the cells was determined as described previously for Swiss 3T3 fibroblasts (2). To test for IGF binding proteins, dilutions of the 24 h supernatant were incubated with ^{125}I-IGF-I for 30 min at 37°C, cooled on ice, and incubated at 4°C with confluent monolayers of Swiss 3T3 cells; specific binding was then determined (2). Additionally, the ^{125}I-IGF-I/supernatant mixtures were incubated with the cross-linking agent disuccinimidyl suberate (DSS), followed by polyacrylamide gel electrophoresis (2).

The results obtained demonstrate that pig blastocyts possess specific binding sites for IGFs, and that supernatants from the cultured pig blastocyts inhibited the binding of ^{125}I-IGF-I to the type 1 IGF receptors on Swiss 3T3 cells. The potency of the inhibition was similar whether or not the cultures were depleted of embryo-associated material. Chemical cross-linking using DSS revealed a ^{125}I-IGF-I labelled complex of about 45000 kDa, the labelling of which could be blocked by IGFs but not insulin.

We conclude that cells of the pre-implantation trophoderm secrete an IGF-binding protein that can modulate IGF binding to cellular receptors, which could therefore affect IGF-stimulated growth and development of the early blastocyst.

REFERENCES

1 Bell, S.C. (1988). Research in Reproduction **20** (2), 3.

2 Corps, A.N. and Brown, K.D. (1988). Biochemical Journal **252,** 119.

3 McCusker, R.H., Campion, D.R. and Clemmons, D.R. (1988). Endocrinology **122,** 2071.

INSULIN-DEPENDENCE OF THE SMALL IGF-BINDING PROTEIN (IGF-SBP)

J.M.P.Holly[1], D.B.Dunger[2], J.A. Edge[2], C.P.Smith[1], S.A.Amiel[1], R.A.Biddlecombe[1], M.O. Savage[1] and J.A.H. Wass[1]

[1] St Bartholomew's Hospital, London, U.K.
[2] John Radcliffe Hospital, Oxford, U.K.

Evidence is accumulating that the IGF-SBP may be an important acute modulator of IGF activity. That the IGF-SBP is more than a mere carrier protein is suggested by reports that its circulating levels undergo a marked circadian variation and are related to those of insulin. We have investigated this relationship with insulin both within and between individuals.

The relationship within individuals was examined in profiles of IGF-SBP and insulin measured over 12 or 24 h periods in 14 normal subjects (10 adolescents and 4 adults) and in 11 adolescents with diabetes mellitus. In 5 of the diabetics overnight profiles were repeated with euglycaemia maintained with a glucose clamp. The relationship between individuals was studied in a cross-sectional study of 116 normal subjects (aged 5 to 48 years) in whom fasting levels of IGF-SBP and insulin were measured. IGF-SBP was measured with a specific RIA.

In normal adults and late pubertal adolescents (Tanner stages 3-5) IGF-SBP levels were very low except for a nocturnal peak of 30-69 ng/ml occurring between 0.300-10.00 h coinciding with the overnight low in insulin levels. However, in 2 prepubertal subjects much higher levels were observed at night (127 and 131 ng/ml). In the prepubertal and pubertal adolescents a strong negative correlation was obtained between IGF-SBP and insulin ($r = -0.69$, $n = 53$, $P < 0.001$). In diabetic adolescents similar circadian rhythms were again found but peak levels appeared to depend on the insulin regime followed, with levels of 410-460 ng/ml being observed in subjects receiving inadequate insulin and who experienced nocturnal hyperglycaemia. Maintenance of euglycaemia overnight failed to suppress the IGF-SBP peak although the shape of the peak was changed reflecting a change in free insulin levels with which a strong negative correlation was again found ($r = -0.67$, $n = 77$, $P < 0.001$).

In the cross-sectional study of IGF-SBP levels were initially high in early childhood and fell throughout puberty as fasting insulin levels rose. These changes could be accounted for almost entirely by pubertal development rather than age. IGF-SBP correlated with fasting insulin in children < 16 years ($r = -0.63$, $n = 60$, $P < 0.001$) but not in the adults. After puberty fasting insulin decreased to prepubertal levels but there was no accompanying rise in IGF-SBP.

We have shown the IGF-SBP to be closely inversely related to insulin in both normals and subjects with diabetes mellitus. In the diabetics IGF-SBP levels were generally high consistent with the insulin deficiency. Very high concentrations were observed in poorly controlled subjects, although a simple relationship with the level of acute metabolic control could not be established. In puberty the rise in IGF-I is accompanied by a rise in insulin and a fall in IGF-SBP. This may represent an integrated mechanism contributing to the pubertal growth spurt. In addition this raises the possibility of an acute modulator of somatogenic activity linked, through insulin, to nutritional intake.

TRANSFER OF PLASMA IGF-I INTO LYMPH

M.D. Baucells, I.R. Fleet and C.G. Prosser

AFRC Institute of Animal Physiology and Genetics Research, Cambridge Research Station, Babraham, Cambridge CB2 4AT, U.K.

In vitro experiments suggest free unbound IGF-I can be transferred intact across the capillary endothelium (1). However, in the intact animal 95% or more of the circulating IGF-I is bound to specific high molecular weight binding proteins which are thought to restrict the passage of IGF-I out of the vascular space. In the present study we have examined the transfer of [^{125}I] labelled IGF-I from blood into lymph of male kid goats as a model for its movement into the extracellular fluid space.

[^{125}I]IGF-I was injected into the jugular vein of 8 anaethesised male Saanen goats (3.5-6 wk old; 7-12.5 kg). Anaesthesia was induced and maintained with Na pentobarbitone. Lymph was collected continuously from the subclavian duct over 10 min intervals for 240 min. Blood was taken every 10 min from a carotid arterial catheter. The amount of lymph collected varied from 60 to 1540 mg/10 min between animals but was constant within animals.

Plasma levels of [^{125}I]IGF-I declined rapidly following its intravenous injection. At 4 min the amount of radioactivity in plasma was 26 $\pm$ 4.3 cpm/mg. By 20 min 70.1 $\pm$ 2.7% (mean $\pm$ SEM) of this amount remained in circulation. Thereafter, levels declined more slowly with 53.1 $\pm$ 3.1% of the initial total radioactivity remaining at 120 min and 41.5 $\pm$ 3.0% at 240 min. The integrity of [^{125}I]labelled material in plasma remained unchanged throughout the experiments. Total radioactivity in lymph reached a peak value of 4.7 $\pm$ 1.3 cpm/mg at 120 min. However, at 20 min only 65.3 $\pm$ 2.2% of the [^{125}I]labelled material in lymph was precipitable by 15% TCA and this fell to 49.9 $\pm$ 2.8% by 120 min. Thus the amount of intact [^{125}I]IGF-I in lymph reached its maximum by 20 min and was maintained at this level until at least 120 min. Thereafter, it declined slowly, reflecting the decline in plasma levels. The ratio of intact [^{125}I]IGF-I in plasma to lymph was approximately 6:1 at 120 min and was not significantly altered for the remainder of the experiment.

Sephacryl S-200 column chromatography of plasma and lymph collected between 50 and 90 min showed that 6.6 $\pm$ 0.8% and 23 $\pm$ 4.1% of the intact [^{125}I]IGF-I remained in the free form. The rest was predominantly associated with the 150 kDa binding protein. The degraded material eluted with or before the salt peak indicated the presence of peptide fragments and free [^{125}I]. The ratio of free [^{125}I]IGF-I in plasma to lymph was 2:1; the ratio for bound [^{125}I]IGF-I in plasma relative to lymph was 8:1.

These experiments provide in vivo evidence that IGF-I is rapidly transferred intact from the vascular space into lymph. As a significant proportion of [^{125}I]IGF-I in lymph is in the free, unbound form it would seem that it is this component that is more readily transferred. The results further suggest that IGF-I enters extracellular fluid and that tissues ae exposed to IGF-I from the circulation.

REFERENCE

1 Bar, R.S., Boes, M. and Yorek, M. (1986). Endocrinology **118**, 1072.

GROWTH HORMONE (bST), INSULIN, INSULIN LIKE GROWTH FACTOR I (IGF-I) AND IGF-I BINDING PROTEINS DURING LACTATION IN CATTLE

D. Schams, U. Winkler, R. Einspanier, M. Theyerl-Abele and B. Graule

Institut für Physiologie, Techn. Universität München, 8050 Freising-Weihenstephan, Federal Republic of Germany

There is clear evidence for the growth hormone family being essential for galactopoiesis in cattle. In high-yielding dairy cows marked endocrine changes occur that are considered responsible for the shift from an anabolic state during late pregnancy towards a primarily catabolic state in early lactation and again to an anabolic state during mid- and late-lactation. The aim of the work was to study the interrelationship between bST and some anabolic hormones during lactation.

Blood was collected from 9 cows (Brown Swiss) by needle puncture from the jugular vein 3 times/wk, beginning during late pregnancy until 44 wk post partum (pp). Additionally frequent samplings (10 h, 30 min interval) by means of a permanent cannula were performed at certain intervals. In further experiments with 5 cows IGF-I and activity of binding proteins after injection of exogenous bST during early lactation (wk 1-8 pp) were examined. bST, IGF-I, insulin were determined by RIA. IGF-I binding proteins were revealed in cow plasma by competitive tracer binding and size exclusion chromatography.

Concentrations of bST increased after parturition for about 8 wk, decreased thereafter and reached a lower plateau 18 wk pp and remained at that level for the rest of the lactation. The increase of bST after parturition was related to an enhanced episodic secretory activity of bST. bST levels were positively correlated to milk yield contrary to concentrations of IGF-I and insulin which were negatively correlated. The IGF-I response to exogenous bST was diminished immediately after parturition and increased continuously thereafter.

The 45 kDa binding protein of IGF-I seems to be constant throughout lactation. The 150 kDa binding protein was lower during the first weeks after parturition and increased parallel with IGF-I thereafter continuously. An increase of binding activity was only measured for the 150 kDa protein after injection of exogenous bST during mid-lactation indicating its growth hormone dependency. It is assumed that low IGF-I and insulin levels during peak lactation may support the lipolytic action of bST, which antagonises the shortage of energy and/or provides precursors for milk synthesis as the main action of bST. Later on during the anabolic phase growth hormone may preferably act indirectly via IGF-I originating from the liver.

THE EFFECT OF NUTRITIONAL STATUS AND GROWTH HORMONE (GH) TREATMENT ON THE IN VITRO MITOGENIC RESPONSES OF MAMMARY GLAND TISSUE IN LAMBS

S.D. Wheatley, J.M. Pell and I.A. Forsyth

AFRC Institute for Grassland and Animal Production, Hurley, Maidenhead, Berkshire SL6 5LR, U.K.

Both restricted rates of liveweight gain (LWG) and exogenous GH administration stimulate mammary gland development in prepubertal female lambs. We have studied the effects of nutritional status and GH treatment in vivo on the subsequent response of mammary gland explants to mitogenic stimuli in vitro. Sixteen lambs were randomly allocated to four dietary treatments: 12 (L) or 20 (H)% crude protein offered at either ad libitum (A) or restricted (R; 3% of liveweight) rates. Within each diet lambs were given daily injections of buffered saline (S) or GH (0.1 mg/kg/d). Treatments were imposed from 9 weeks of age for a period of 10 weeks. The in vitro mitogenic responses were measured by determining the incorporation of [^{3}H]thymidine into the explant DNA over the first 24 h of culture in serum-free medium, supplemented with no hormone (NH), insulin (I) or insulin-like growth factor-I (IGF-I).

Main effects
dpm/μg DNA (n = 8)

Stimulus	R	A	S	GH	L	H	SED
NH	1117	571	891	797	518	1171	314.2
I: 10 ng/ml	1140	690	824	1006	728	1102	414.2
10 μg/ml	1775	583*	1000	1357	845	1513	513.1
IGF-I: 1 ng/ml	1445	492**	866	1071	648	1289	308.6
100 ng/ml	1854	489***	1009	1334	1151	1193	232.9
Live weight gain g/day	166	368***	252	282	265	269	20.7

*P < 0.05, **P < 0.01, ***P < 0.001

In all treatments, apart from ad libitum feeding, there was a dose-dependent response of DNA synthesis to both I and IGF-I. Basal and stimulated response was greatest in mammary tissue from restricted lambs. Growth hormone treatment and high dietary protein were also associated with enhanced DNA synthesis. The results suggest that observed nutritional and hormonal effects on prepubertal mammogenesis may be mediated by alterations in tissue sensitivity or proliferative capacity in response to mitogens.

COMPARATIVE ASPECTS OF SELECTED HORMONES AND GROWTH FACTORS ON PROTEIN METABOLISM IN ADULT AND FETAL OVINE PRIMARY MUSCLE CULTURES

J.A. Roe, C.M. Heywood, J.M.M. Harper and P.J. Buttery

Department of Applied Biochemistry and Food Science, University of Nottingham School of Agriculture, Sutton Bonington, Loughborough, Leicestershire LE12 5HD, U.K.

In order to study the control of lean deposition in ruminants the effects of hormones and growth factors on protein synthesis and degradation in muscle cell cultures derived from sheep have been investigated. Cells from fetal tissue are easier to prepare than those from the adult. We have compared the responses of the two systems to selected hormones and growth factors.

No significant effects of physiological concentrations (10^{-10} M) of insulin have been seen in either the adult or fetal systems. However, supra-physiological concentrations (10^{-6} M) of insulin have been shown to stimulate ($P < 0.001$) protein synthesis by 36% $\pm$ 2.2 (n = 4) (mean $\pm$ SEM (no. of observations)) and inhibit ($P < 0.001$) protein degradation by 17% $\pm$ 1.0 (n = 4) compared to control values in the adult system. This compares with the 6% increase in protein synthesis and the 4% inhibition of protein degradation seen in fetal cells (1).

Both cell types were more sensitive to insulin-like growth factor I (IGF-I) although the adult cells were again more responsive than the fetal cells; 10^{-8} M IGF-I increased ($P < 0.001$) protein synthesis by 36% $\pm$ 3.3 (6) and inhibited ($P < 0.01$) protein degradation by 14% in the adult system whereas 1.3 x 10^{-8} M IGF-I in the fetal system increased protein synthesis by only 16% and inhibited protein degradation by 7% (1).

In contrast adult cells were slightly less sensitive to epidermal growth factor (EGF) than were fetal cells; EGF (10^{-9} M) increased ($P < 0.001$) protein synthesis by 15% $\pm$ 6.1 (3) and inhibited ($P < 0.05$) protein degradation by 7% $\pm$ 2.1 (3). In comparison Harper et al. (1987) reported a 24% stimulation of protein synthesis and an 8% inhibition of protein degradation in fetal cells treated with 1.7 x 10^{-9} M EGF.

Bovine growth hormone had no direct effects on either adult or fetal muscle cells in culture.

We conclude that there are quantitative differences between the two cell types.

REFERENCE

1 Harper, J.M.M., Soar, J.B. and Buttery, P.J. (1987). Journal of Endocrinology **112**, 87.

INDUCTION OF LACTOGENIC RECEPTORS IN HYPOPHYSECTOMIZED RATS TREATED WITH BOVINE GROWTH HORMONE-MONOCLONAL ANTIBODY COMPLEXES

R. Thomas[1], I.C. Green[1], M. Wallis[1] and R. Aston[2]

[1] Biochemistry Laboratory, School of Biological Sciences, University of Sussex, Brighton, Sussex BN1 9QG, U.K.
[2] Coopers Animal Health Ltd., Berkhampsted, Hertfordshire HP4 2QE, U.K.

In vivo administration of growth hormone (GH) complexed with monoclonal antibodies (MAbs) enhances the somatogenic and lactogenic properties of the hormones (2). The mechanisms behind this enhancement are unclear. Recently, it has been shown that hypophysectomized (hypox) rats treated with bovine (b)GH-MAb complexes exhibit significantly raised somatomedin C levels as well as body weight, compared to animals treated with hormone alone (3). Here we report changes in the binding sites for human (h)GH in liver microsomal preparations from hypox rats treated with bGH-MAb complexes.

Male hypox rats were injected daily for up to 12 days with normal mouse IgG (control) or with bGH (50 μg) alone or in complex with MAbs OA13, OA14 or OA17 (500 μg) (1). Every second day individual microsomal preparations were made from livers of 2-3 rats per treatment group. Binding studies using [^{125}I]-hGH were performed on these microsomal preparations.

Treatment of hypox rats with bGH stimulated body weight gain over the 12 day treatment period but specific binding of [^{125}I]-hGH to the microsomes of these animals was not significantly increased. OA13 and OA14 complexed to bGH significantly enhanced its effects on weight gain of hypox rats but OA17-bGH showed no such enhancement effect. Animals treated with bGH-MAb complexes showed significantly raised hepatic binding of [^{125}I]-hGH after 2 days treatment irrespective of the MAb used. Binding increased to a maximum by day 8 and began to fall by day 12. The binding sites were lactogenic in nature, the bound [^{125}I]-hGH being displaced by ovine prolactin but not by bovine or porcine GH. The specificity of the binding sites was similar to that in microsomal preparations from pregnant rats.

The results confirm that some MAbs can potentiate the somatogenic action of bGH in male hypox rats and that both growth-enhancing and non-growth-enhancing bGH-MAb complexes can enhance the concentration of lactogenic receptors in the livers of these animals.

REFERENCES

1 Aston, R., Holder, A.T., Ivanyi, J. and Bomford, R. (1987). Molecular Immunology **24,** 143.

2 Holder, A.T., Aston, R., Preece, M.A. and Ivanyi, J. (1985). Journal of Endocrinology **107,** R9.

3 Wallis, M., Daniels, M., Ray, K.P., Cottingham, J.D. and Aston, R. (1987). Biochemical Biophysical Research Communications **149,** 187.

POTENTIATION OF GROWTH HORMONE ACTIVITY USING A POLYCLONAL ANTIBODY OF RESTRICTED SPECIFICITY

J.M. Pell[1], C. Elcock[1], A. Walsh[1], T. Trigg[2] and R. Aston[2]

[1]AFRC Institute for Grassland and Animal Production, Hurley, Berkshire, U.K.
[2]Peptide Technology Ltd., 4-10 Inman Road, Deewhy, NSW 2099, Australia

A polyclonal antibody was raised against a specific region of bovine growth hormone (bGH) in sheep. When complexed to bGH, this antibody was found to enhance [^{35}S]sulphate uptake into costal cartilage of Snell dwarf mice by approximately two-fold, when compared to stimulation by bGH alone (50 μg + control Ig, 2135 ± 197; 50 μg bGH + antipeptide Ig, 3900 ± 209 dpm per mg; mean ± SEM, n = 6 per group, $P < 0.001$). The effects of the antipeptide were investigated in growing lambs. Ten week old lambs were randomly allocated to one of four groups (n = 7 per group) and were offered a concentrate diet (16% crude protein) ad libitum. They were subjected to a 21 day control period and were then treated with either: control Ig, 10 mg bGH, 10 mg antipeptide or 10 mg bGH pre-complexed to 10 mg antipeptide Ig. Treatments were administered as a subcutaneous injection every other day for 16 days. On days -6 and +8 of treatment, 18 hourly jugular blood samples were removed for the determination of plasma hormone and metabolite concentrations. On days -4 and +10 of treatment, an insulin tolerance test (ITT) was performed. At slaughter, samples of subcutaneous fat were removed for the determination of rates of lipogenesis, oxidation and lipolysis in vitro.

	Control Ig	Antipeptide Ig	bGH	Antipeptide Ig + bGH	SE	Significance
Liveweight gain (g/day)	235	300	367	341	35	NS
Liver weight (g/kg liveweight)	23.5	25.0	25.9	25.2	0.7	NS
IGF-I conc. (ng/ml)	273[a]	243[a]	398[b]	371[b]	27	0.001
ITT*	-2.70[a]	-1.22[a]	4.88[b]	6.65[b]	1.23	0.001
Fat (μgmol/2 h/g):						
lipogenesis	45.2[a]	32.3[a,b]	18.3[c]	19.7[b,c]	4.7	0.01
oxidation	2.69[a]	1.81[b]	1.11[b,c]	1.01[c]	0.25	0.001
lipolysis	5.79	6.03	6.93	7.36	0.94	NS

*change in glucose concentration (mmol/l) one hour after insulin administration on day -4 minus that for day +10 of treatment. Values with same superscript not significantly different.

These preliminary data demonstrate that the antipeptide Ig may enhance the activity of endogenous GH in growing lambs. This observation may provide the basis for further work on growth enhancing peptide vaccines in production animals. It is likely that additional enhancement of exogenous bGH activity by antipeptide Ig was not observed because, in this particular experiment, the exogenous bGH was a potent stimulator of most of the physiological processes which were investigated.

PROTEINS COVALENTLY LINKED TO HUMAN GROWTH HORMONE WILL ENHANCE ITS ACTIVITY IN VIVO

D.J. Morrell[1], R. Aston[2] and A.T. Holder[3]

[1] Institute of Child Health, Department of Growth and Development, London, U.K.
[2] Peptide Technology, Sydney, New South Wales, Australia
[3] Department of Endocrinology and Animal Physiology, AFRC Institute of Grassland and Animal Production, Hurley, Berkshire, U.K.

Certain monoclonal antibodies (MAbs) are known to enhance the biological activity of human or bovine growth hormone (GH) in vivo (1, 2). It is not clear whether the phenomenon of enhancement requires a specific interaction between GH and a MAb or whether an association between GH and other proteins may achieve the same effect. Therefore, we have investigated this possibility by examining the in vivo biological activity of human growth hormone (hGH) when covalently linked to ovalbumin or bovine serum albumin (BSA).

Growth promoting activity was assessed by measuring the incorporation of [^{35}S]-sulphate into costal cartilage of hypopituitary Snell dwarf mice (n = 6 per treatment group) after two daily injections (0.1 ml; subcutaneous) of the test substance (3). Human GH was covalently linked to BSA or ovalbumin using glutaraldehyde (final concentration 0.08% v/v; 24 h incubation, r.t.) or dimethylsuberimidate (DMS; final concentration 10 mM; 30 min incubation, r.t.). Covalent linking was carried out in solutions which contained equimolar amounts of hGH and BSA or ovalbumin, the concentration of hGH being 100 mU/0.1 ml (injection volume). Additional treatment groups were hGH alone, hGH alone cross-linked with glutaradehyde or DMS and mixtures of hGH and BSA ovalbumin taken through a mock cross-linking procedure.

There was a significantly greater incorporation of [^{35}S]-sulphate in mice treated with hGH cross-linked to either BSA or ovalbumin than that observed in control animals treated with hGH alone or with hGH mixed with BSA or ovalbumin. The degree of enhancement did not appear to favour any combination of protein and cross-linking agent. Treatment of mice with hGH which had been exposed alone to glutaraldehyde or DMS provoked a significantly greater biological response than treatment with untreated hGH. It is likely that inter-molecular cross-linking of hGH was responsible for this effect.

These results demonstrate that enhancement of GH activity is not a property peculiar to antibodies and that, at least in this case, enhancement of GH action is not dependent upon dissociation of the complex.

REFERENCES

1 Aston, R., Holder, A.T., Ivanyi, J. and Bomford, R. (1987). Molecular Immunology **24,** 143.

2 Holder, A.T., Aston, R., Preece, M.A. and Ivanyi, J. (1985). Journal of Endocrinology **107,** 9.

3 Holder, A.T., Wallis, M., Biggs, P. and Preece, M.A. (1980). Journal of Endocrinology **85,** 35.

THE EFFECT OF IMMUNONEUTRALIZATION OF CRF ON THE GROWTH OF LAMBS

C.M.M. Reynolds[1], P.J. Buttery[1], N.B. Haynes[1] and N.S. Huskisson[2]

[1] Department of Applied Biochemistry and Food Science, University of Nottingham, School of Agriculture, Sutton Bonington, Loughborough, Leicestershire LE12 5RD, U.K.
[2] Microchemical Facility, AFRC Institute of Animal Physiology and Genetics Research, Cambridge Research Station, Babraham, Cambridge CB2 4AT, U.K.

Glucocorticoids are generally thought to be growth inhibiting steroids; thus, manipulation of the glucocorticoid status in growing lambs could be a potential means of improving growth rate. A relationship between glucocorticoids at physiological concentrations and growth rate was demonstrated by Sillence and Rodway (1987). Plasma corticosterone levels were reduced in female rats by the adrenal inhibitor trilostane, with an improvement in growth rate by up to 30%. Purchas (1973) also found a significant negative correlation between plasma corticosterone concentration and the average daily liveweight gain in heifers. Corticotrophin Releasing Factor (CRF) is a 41 amino acid hypothalamic peptide which stimulates the secretion of adrenocorticotrophic hormone (ACTH) from the anterior pituitary gland which in turn stimulates corticosteroid production in the adrenal glands. Antibodies to CRF were raised in ewe lambs by active immunization against a short sequence of ovine CRF (amino acids 28-41) conjugated to a carrier protein, bovine serum albumin. During a 12 wk growth trial a total of 5 immunizations were administered with control animals receiving immunizations against bovine serum albumin only.

The total and free plasma glucocorticoid concentrations were significantly lower in the CRF-immunized animals than in the controls ($P < 0.1$, $P < 0.05$). However the growth rate of the CRF immunized animals was significantly lower ($P < 0.01$) than that of the control animals. The feed intake was also significantly lower than that of the controls ($P < 0.001$). At slaughter there were no significant differences in the carcass composition of the two groups, although significant adrenal atrophy was observed in the CRF-immunized animals.

Thus the immunoneutralization of endogenous CRF successfully reduced the circulating glucocorticoid concentration; however, the effect on growth rate may implicate the involvement of CRF in the secretion of other hormones, especially growth hormone. This is currently being investigated.

REFERENCES

1 Purchas, R.W. (1973). Australian Journal of Agricultural Research **24**, 927.

2 Sillence, M.N. and Rodway, R.G. (1987). Journal of Endocrinology **113**, 479.

EFFECTS OF PASSIVE IMMUNIZATION WITH AN INSULIN-LIKE GROWTH FACTOR-I (IGF-I) MONOCLONAL ANTIBODY (MAB) ON GROWTH AND PITUITARY GROWTH HORMONE (GH) CONTENT IN THE GUINEA PIG

D.E. Kerr[1], B. Laarveld[2] and J.G. Manns[1]

[1] Department of Veterinary Physiological Sciences, University of Saskatchewan, Saskatoon, Canada
[2] Department of Animal and Poultry Science, University of Saskatchewan, Saskatoon, Canada

IGF-I is an important mediator of GH stimulated growth and it may have a feedback role on pituitary (PIT) GH synthesis and secretion. However, the importance of circulating, opposed to locally produced IGF-I, is in question. By using a passive immunoneutralization technique, our objectives were to evaluate the roles of circulating IGF-I in growth and in the regulation of PIT GH content.

A high affinity (K_a = 1.1 x 10^{11} l/mole) IGF-I MAb (DK82-9A), generated in our laboratory, was purified from ascites fluid by caprylic acid-ammonium sulfate precipitation. It has 14% cross-reactivity with hIGF-II and does not cross-react with MSA, insulin or GH. In an RIA, dilutions of guinea pig serum show parallel displacement to recombinant hIGF-I. The MAb also inhibits the binding of hIGF-I to its receptor (human placenta). Twenty-four, 3 wk old, male guinea pigs were divided into 3 groups and weighed and injected (INJ; intraperitoneal) every 3 days for 24 days. Treatments were IGF-I MAb or control MAb (MAb to bovine herpes virus; BHV) at 20 μg/g body weight or vehicle (PBS) alone. For the IGF-I MAb group this dose was 0.45 μl/g body weight of a stock solution that bound 50% of [^{125}I]-IGF-I (~150 pg/400 μl) at a final dilution of 1:4 x 10^7. Blood was obtained on day 23 (48 h post-INJ) by cardiac puncture and on day 25 (24 h post- INJ) by decapitation.

Serum from the IGF-I-MAb treated group bound 38 ± 8 (day 23) and 56 ± 7% (day 25) of tracer at a final dilution of 1:10000. From an IGF-I-MAb standard curve, estimated serum IGF-I-MAb concentrations in this group were 10 ± 3 and 19 ± 4 μg/ml. Of the total IGF-I in 1:100 diluted serum 50 ± 5 and 561 ± 4% could be immunoprecipitated with an excess of rabbit anti-mouse IgG in day 23 and day 25 serum, respectively. These are minimal estimates of the amount of MAb-bound IGF-I because factors such as % bindability of the MAb in serum, dissociation of the MAb-IGF-I complex due to dilution and storage, and interference by guinea pig anti-mouse antibodies were not determined. Both MAb treated groups developed similar anti-mouse IgG titres (ELISA ED_{50} ~ 1:1000). Treatment effects are tabulated (mean ± SE; $^{a b}$ means in a column differ, $P < 0.05$).

	Growth Rate (g/d)	Tibial Width (mm)	Pit wt (mg)	Pit GH (U/mg)	Serum IGF-I (μg/ml)* day 23	day 25
IGF-I-MAb	6.3 ± 0.3^{a}	0.28 ± 0.1	7 ± 1	19 ± 1	0.86 ± 0.13^{a}	1.17 ± 0.16^{a}
BHV-MAb	6.8 ± 0.5^{ab}	0.28 ± 0.1	7 ± 1	21 ± 2	0.51 ± 0.03^{b}	0.59 ± 0.03^{b}
PBS	8.4 ± 0.6^{b}	0.29 ± 0.1	8 ± 1	20 ± 3	0.63 ± 0.07^{ab}	0.71 ± 0.05^{b}

*Total serum IGF-I determined after acid-gel chromatography. Pit, pituitary.

Passive immunization with doses of a high affinity IGF-I-MAb sufficient to bind all circulating IGF-I had little effect on growth rate, tibial epiphyseal width, PIT weight, or PIT GH content. However, this treatment resulted in a large increase in total IGF-I in serum. Whether this reflects a compensatory mechanism, or an extension of the half-life of serum IGF-I was not determined in this study.

CONTROL OF GROWTH HORMONE SECRETION DURING GESTATION AND LACTATION IN GILTS ACTIVELY IMMUNIZED AGAINST GROWTH HORMONE-RELEASING FACTOR

J.D.Armstrong[1],K.L.Esbenshade[1],M.T.Coffey[1],E.Heimer[2],R.Campbell[2],T.Mowles[2] and A. Felix[2]

[1]Department of Animal Science, North Carolina State University, Raleigh, U.S.A.
[2]Hoffmann La Roche, Nutley, NJ, U.S.A.

Changes in pattern of GH secretion during gestation and lactation, as well as possible opioid involvement in the control of GH during lactation, are not well understood in the pig. Immunization against growth hormone-releasing factor (GRF) in the gilt suppresses episodic release of GH and inhibits release of GH in response to stimulation with an opioid agonist (1). The objective of this study was to determine the effects of immuno-neutralization of GRF on changes in GH secretion during gestation and lactation and on responsiveness of GH to an opioid antagonist in pigs.

Gilts that had been immunized for approximately 15 wk (GRF conjugated to HSA, n = 5; HSA, n = 4 (1)) were exposed to mature males and bred. A booster of 1 mg antigen (GRF-HSA or HSA) was administered on d 97 of gestation. Patterns of GH were determined by obtaining blood samples at 15 min intervals for 5 and 10 h on d 110 of gestation and d 21 of lactation, respectively. On d 21 of lactation, sows were administered 0.25 mg/kg naloxone (opioid antagonist) at 15 min intervals from 45 to 210 min. Each sow received naloxone and vehicle (saline) in a switchback design across treatment with 4 sows receiving naloxone and 5 sows receiving saline during the first 5 h.

Antibody titres (expressed as dilution at 50% binding) were 1:12000 $\pm$ 2832 and 1:9100 $\pm$ 1629 on d 110 of gestation and d 21 of lactation, respectively. Frequency of GH release (peaks/5 h) was no different in GRF (0) and HSA (0.3 $\pm$ 0.3) gilts during gestation. Mean GH was greater during lactation when compared to gestation in all gilts. Frequency of release (peaks/5 h) and concentration (ng/ml), respectively, of GH on d 21 of lactation were greater ($P < 0.05$) in HSA (2.0 $\pm$ 0.4 and 4.7 $\pm$ 0.2) than in GRF (0.5 $\pm$ 0.3 and 2.6 $\pm$ 0.1) pigs.

Administration of naloxone during lactation suppressed GH in both GRF and HSA gilts. In HSA gilts, frequency of release (peaks/4 h) and mean GH (ng/ml), respectively, were lower ($P < 0.05$) following naloxone (0 and 3.2 $\pm$ 0.1) than following saline (1.5 $\pm$ 0.3 and 4.8 $\pm$ 0.2). Naloxone lowered average GH in GRF gilts (2.0 $\pm$ 0.1 vs 2.6 $\pm$ 0.1); however, frequency of GH release was not affected by naloxone (0 vs 0.4 $\pm$ 0.2).

In summary, immuno-neutralization of GRF suppressed GH release during lactation, but failed to significantly alter GH during gestation. Administration of naloxone during lactation decreased GH in all gilts. The effect of naloxone on concentrations of GH in GRF-immunized gilts was independent of changes in frequency of release of GH.

REFERENCE

1 Esbenshade, K.L., Armstrong, J.D., Heimer, E., Campbell, R., Mowles, T. and Felix, A. (1988). Journal of Animal Science **66** (Suppl. 1), 408 (Abstr.).

IMMUNIZATION AGAINST GROWTH HORMONE-RELEASING FACTOR SUPPRESSES IGF-I AND ABOLISHES OPIOID AGONIST INDUCED RELEASE OF GROWTH HORMONE IN LACTATING CATTLE

J.D.Armstrong[1],K.E.Lloyd[1],K.L.Esbenshade[1],E.Heimer[2],R.Campbell[2],T.Mowles[2] and A. Felix[2]

[1]Department of Animal Science, North Carolina State University, Raleigh, U.S.A.
[2]Hoffmann La Roche, Nutley, NJ, U.S.A.

We previously demonstrated that lactating beef cows could be actively immunized against growth hormone releasing factor (GRF, 1-29; 1), as evidenced by antibody titres against GRF (percentage binding at 1:100 dilution) of 30.2 ± 2.2%. Immunization against GRF abolished episodic release of GH and suppressed the release of GH following a GRF agonist or arginine (1). The objective of this study was to determine the effects of active immunization against GRF in lactating beef cows on concentrations of IGF-I, milk production, reproductive performance and changes in GH following an opioid agonist and antagonist.

Multiparous, Hereford cows (534 ± 23 kg) were immunized against GRF (1-29) conjugated to human serum albumin (GRF-HSA; n = 3) or HSA (n = 3) at 52 d prior to calving as previously reported (1). Blood samples were collected at 15 min intervals for 9 h on day 19, 47 and 121 post partum (pp). On each day, cows were given an injection (i.v.) of an opioid agonist (FK33-824, 10 μg/kg) at 5 h and an opioid antagonist (naloxone, 0.75 mg/kg) at 7 h. Onset of ovarian actvity was estimated by weekly progesterone concentrations. Milk production was estimated using the weigh-suckle-weigh method on day 80, 86, 96 and 115 pp.

Mean GH and IGF-I (ng/ml) was consistently lower ($P < 0.05$) in GRF than in HSA cows. Concentrations of IGF-I in GRF and HSA cows, respectively, on day 19, 47 and 119 were 13 ± 3 and 49 ± 6, 9 ± 1 and 60 ± 4, and 9 ± 0.3 and 28 ± 4.

Immuno-neutralization of GRF resulted in lower ($P < 0.05$) milk yield in GRF (3.2 ± 0.5) than in HSA (6.5 ± 0.7) cows on day 86 and 96 pp. Interval from calving to elevated progesterone averaged 53 ± 0 and 72 ± 9 day in GRF and HSA cows ($P < 0.12$), respectively.

Immunization against GRF abolished the opioid agonist stimulated release of GH. Administration of FK33-824 elevated GH ($P < 0.05$) in HSA, but not in GRF cows; FK33-824 was effective in elevating GH only on day 47 and 119 pp in HSA cows. A single injection of naloxone failed to alter concentrations of GH in either GRF or HSA cows. In HSA cows, concentration, but not frequency, of GH release was greater on days 47 and 119 than on day 19.

In conclusion, abolishment of GH pulses by active immunization against GRF suppressed IGF-I concentrations, lowered milk yield and reduced the interval from calving to oestrus. These data also demonstrate that opioid peptides stimulate GH in the cow through a mechanism involving GRF.

REFERENCE

1 Lloyd, K.E., Armstrong, J.D., Esbenshade, K.L., Heimer, E., Campbell, R., Mowles, T. and Felix, A. (1988). Journal of Animal Science **66** (Suppl. 1), 406.

GROWTH HORMONE EXPRESSION AND EFFECTS IN TRANSGENIC SHEEP

C.D. Nancarrow[1], C.M. Shanahan[1], R.J. Dixon[2], B. Farquharson[2], K.A. Ward[1], J.D. Murray[1] and J.T.A. Marshall[1]

[1]CSIRO, Division of Animal Production, Blacktown 2148, Australia
[2]University of Sydney, Department of Veterinary Clinical Studies, Camden 2570, Australia

Transgenic sheep were produced to investigate the possibility that high growth hormone (GH) secretion would enhance growth characteristics and aid meat production (3). We report here some physiological parameters and the results of examinations of tissues from these sheep.

Pronuclei of Merino sheep embryos were injected with the fusion gene construct ovine metallothionein Ia-ovine GH (GH9). Energy and nitrogen utilization (1), muscle protein synthesis (4) and plasma GH concentrations (2) were estimated on 2 transgenic ewes. The ewes died at about 10 months of age, weighing 28 and 37 kg. Post-mortem examinations were carried out and tissues were tested for evidence of GH9 expression using Northern blot techniques.

Ovine GH concentrations in plasma ranged from 0.9 to 30 μg/ml throughout the life-span of 2 ewes (2). Growth rates were not increased although the ewes were composed of only 5-7% total body fat compared with 26-33% in normal ewes; methane production and nitrogen retention were normal but heat production was increased by 20 to 50%, resulting in a 30% increase in basal metabolic rate (1). Both oxygen uptake (μmol/kg/min) and protein degradation (μmol Phe/kg/min) were significantly higher in transgenic ewes ($P < 0.05$) while protein synthesis was not significantly greater (4). Tissues in which mRNA was detected were in descending order of significance: kidney, brain, liver (one only), lung, intestine, spleen and uterus. Gross abnormalities of the fore- and hind-limbs could be seen and these were due to abnormal enlargement of joint capsules, eroded articular surfaces and a 45° rotation of the radius and ulna which was manifest in malposition of the carpus and distal extremities of each fore-limb. Although there was some thickening of the endocardium due to fibre hypertrophy and valvular endocarditis, it was considered that each ewe died from degenerative effects in the kidney and liver. Areas of necrosis and haemorrhage occurred in both tissues.

While chronically high rates of secretion of GH result in a positive effect on body composition by lowering the fat content, other degenerative effects on the major organs and limbs suggest that greater control needs to be exerted over the expression of the GH9 fusion genes.

REFERENCES

1 Graham, N.McC. and Margan, D.E. (1988). Nutrition Society of Australia **13** (in press).

2 Nancarrow, C.D. et. al. (1988). Eleventh International Congress of Animal Reproduction and Artificial Insemination **4**, 478.

3 Nancarrow, C.D., Ward, K.A. and Murray, J.D. (1988). Australian Journal of Biotechnology **2**, 39.

4 Oddy, V.H., Warren, H.M. and Ewoldt, C.L. (1988). Nutrition Society of Australia **13**, (in press).

PARTICIPANTS

Adriaens, F. Monsanto plc., Chineham Court, Chineham, Basingstoke, Hampshire RG2 40UL, U.K.

Agargård, N. National Institute of Animal Science, Research Center Foulum, DK 8833 Ørum Sdrl., Denmark

Allen, W.R. T.B.A., Equine Fertility Unit, Animal Research Station, 307 Huntingdon Road, Cambridge CB3 OJQ, U.K.

Armstrong, S.S. Nutrition Division, Rowett Research Institute, Greenburn Road, Bucksburn, Aberdeen AB2 9SB, U.K.

Armstrong, J. Department of Animal Science, North Carolina State University, Box 7621, Raleigh, NC 27695-7621, U.S.A.

Aston, R. Peptide Technology Ltd., 4 - 10 Inman Road, P.O.B. 444, Dee Why, N.S.W., Australia 2099

Aumaitre, L.A. INRA, Saint-Gilles, 35590 L'Hermitage, France

Axelsson, K. KabiVitrum Peptide Hormones, R and D Department, S - 112 87 Stockholm, Sweden

Baile, C.A. Monsanto Company - BB3F, 700 Chesterfield Village Parkway, St. Louis, MO 63198, U.S.A.

Baldwin, D.R. Cyanamid, Rue du Bosquet 15, B1348 Mont-St-Guibert, Belgium

Bates, P.C. Nutrition Research Unit, London School of Hygiene and Tropical Medicine, 4 St. Pancras Way, London NW1 2PE, U.K.

Baucells, M.D. AFRC Institute of Animal Physiology and Genetics Research, Cambridge Research Station, Babraham, Cambridge CB2 4AT, U.K.

Bauman, D.E. 262 Morrison, Cornell University, Ithaca, NY 14850, U.S.A.

Baumbach, W.R. American Cyanamid, P.O. Box 400, Princeton, NJ 08540, U.S.A.

Beardow, A.W. Small Animal Hospital, University of Liverpool, Crown Street, Liverpool L7 7EX, U.K.

Blum, J. Department Nutritional Pathology, Institute Animal Breeding, University of Bremgartenstr. 109a, 3012 Berne, Switzerland

Bonneau, M. INRA, Station de Recherches Porcines, Saint Gilles, 35590 L'Hermitage, France

Bootland, L.H. Department of Genetics, University of Edinburgh, Kings Buildings, West Mains Road, Edinburgh EH9 1NH, U.K.

Boyd, R.D. Department of Animal Science, Morrison Hall, Room 252, Cornell University, Ithaca, NY 14853-4801, U.S.A.

Bouchet, M. Roussel-UCLAF, Service Sante Animale, 102 route de Noisy, 93230 Ronainville, France

Brinklow, B.R. Institute of Zoology, Regent's Park, London NW1 4RY, U.K.

Brown, A.C.G. Lilly Research Company Ltd., Erlwood Manor, Windlesham, Surrey, U.K.

Brown, J. Department of Physiology and Biochemistry, University of Reading, Whiteknights, Reading, Berkshire RG2 2AJ, U.K.

Bruce, C.I. Animal Health Biology, Pfizer Central Research, Ramsgate Road, Sandwich, Kent CT13 9NJ, U.K.

Bruneau, P. Monsanto, Division Sciences Animales, 52 Rue Marcel Dassault, 92514 Boulogne Billancourt Cedex, France

Brunner, E. London Food Commission Ltd., 88 Old Street, London EC4 9AR, U.K.

Buchanan, G. Elanco Products Ltd., Dextra Court, Chapel Hill, Basingstoke, Hampshire RG21 2SY, U.K.

Burton, J.H. Department of Animal and Poultry Science, University of Guelph, Guelph, Ontario, NIG 2WI Canada

Burvenich, C. Rijksuniversiteit Gent, Facultert van De Diergeneeskunde, Laboratorium Voor Fysiologie Van De Huisdieren, Casinoplein 24, 9000 Gent/Belgium

Buttery, P.J. Department of Applied Biochemistry and Food Science, University of Nottingham, School of Agriculture, Sutton Bonington, Nr Loughborough, Leicestershire LE12 5RD, U.K.

Buttle, H. AFRC Institute for Grassland and Animal Production, Hurley, Maidenhead, Berkshire SL6 5LR, U.K.

Carter, A.P.D. Hoechst (UK) Ltd., Animal Health Division, Walton Manor, Walton, Milton Keynes MK47 7AJ, U.K.

Caygill, J. Ministry of Agriculture, Food and Fisheries, Chief Scientist's Group, Room 119, Great Westminster House, Horseferry Road, London SW1P 2AE, U.K.

Clayton, J. Serono Laboratories (U.K.) Ltd., 2 Tewin Court, Welwyn Garden City, Hertfordshire AL7 1AU, U.K.

Coert, A. Intervet International, PO Box 31, 5830 AA Boxmeer, Netherlands

Collier, R.J. Monsanto Company, Mail Zone - BB3F, 700 Chesterfield Village Parkway, St. Louis, MO 63198, U.S.A.

Corps, A.N. AFRC Institute of Animal Physiology and Genetics Research, Cambridge Research Station, Babraham, Cambridge CB2 4AT, U.K.

Cottingham, J.D. Animal Productivity Group, Coopers Animal Health Ltd., Berkhamsted Hill, Berkhamsted, Hertfordshire HP4 2QE, U.K.

Cracknell, B.C. Beecham Pharmaceutical Research Divison, Walton Oaks, Dorking Road, Tadworth, Surrey KT20 7NT, U.K.

Craven, N. Monsanto Company plc., Chineham Court, Chineham, Basingstoke, Hampshire RG2 4OUL, U.K.

Crilly, P.J. BST Department, Hannah Research Institute, Ayr KA6 5HL, U.K.

Crompton, L.A. Department of Physiology and Biochemistry, University of Reading Whiteknights, PO Box 228, Reading RG6 2AJ, U.K.

Cross, B.A. AFRC Institute of Animal Physiology and Genetics Research, Cambridge Research Station, Babraham, Cambridge CB2 4AT, U.K.

Curtis, R.J. Camco, Chesterford Park, Saffron Walden, Essex CB10 1XL, U.K.

David, P. Pig Improvement Company, Fyfield Wick, Abingdon, Oxford OX13 5NA, U.K.

Davis, C. Elanco Products Ltd., Dextra Court, Chapel Hill, Basingstoke, Hampshire RG21 2SY, U.K.

Davis, S.L. Department of Animal Science, Oregon State University, Corvallis, Oregon, U.S.A. 97331-6702

Dawson, J.M. Department of Applied Biochemistry and Food Science, University of Nottingham, School of Agriculture, Sutton Bonington, Nr Loughborough, Leicestershire LE12 5RD, U.K.

De Boer, S. State Institute for Quality Control of Agricultural Products, Department of Toxicology, Postbox 230, 6700 AE Wageningen, Netherlands

De-Netto, L.A. 33 Western Road, Lewes, East Sussex BN7 1RL, U.K.

Deletang, Sanofi Richerche, 195 Route D, Espangene, Paris, France

Demade, I. SmithKline AHP, 287 Avenue Louise, Bte 13 1050 Brussels, Belgium

Disenhaus, C. Institut National Agronomique, Paris-Grigon, Departement aes Sciences Animales, Chaire de Zootechnie, (Centre de Paris) 16 Rue Claude-Bernard, 75231 Paris, France

Duclos, M.J. AFRC Institute of Animal Physiology and Genetics Research, Edinburgh Research Station, Roslin, Midlothian EH25 9PS, U.K.

Dufour, R. Rhone Merieux, 254 rue M. Merieux, 69342 Lyon Cedex 07, France

Dyer, R.G. AFRC Institute of Animal Physiology and Genetics Research, Cambridge Research Station, Babraham, Cambridge CB2 4AT, U.K.

Ealey, P.A. Department of Chemical Pathology, University College Hospital, Gower Street, London WC1E 6AU, U.K.

Einspanier, R. Institute for Physiology, Technical University Munich, D-8050 Freising-Weihenstephan, West Germany

Ekberg, S. Department of Physiology, University of Göteborg, P.O. Box 330 31, 400 33 Goteborg, Sweden

Elcock, C. AFRC Institute of Grassland and Animal Produciton, Hurley, Maidenhead, Berkshire SL6 5LR, U.K.

Ellendorff, F. Institut fur Tierzucht und Tiervenhalten FAL, Mariensee, 3057 Neustadt 1, West Germany

Essex, C.P. Department of Applied Biochemistry and Food Science, University of Nottingham, School of Agriculture, Sutton Bonington, Nr Loughborough, Leicestershire LE12 5RD, U.K.

Etherton, T.D. 301 W.L. Henning Blg., Pennsylvania State University, University Park, PA 16802, U.S.A.

Fagin, K.D. Amgen, 1400 Oak Terrace Lane, Thousand Oaks, CA 91320, U.S.A.

Finley, E. Hannah Research Institute, BMB Department, Ayr KA6 5HL, U.K.

Fleet, I.R. AFRC Institute of Animal Physiology and Genetics Research, Cambridge Research Station, Babraham, Cambridge CB2 4AT, U.K.

Fletcher, J. Unilever Research, Colworth House, Sharnbrook, Bedford MK44 1LR, U.K.

Forsyth, I.A. AFRC Institute of Grassland and Animal Production, Hurley, Maidenhead SL6 5LR, U.K.

Fritz, I.B. AFRC Institute of Animal Physiology and Genetics Research, Cambridge Research Station, Babraham, Cambridge CB2 4AT, U.K.

Galbraith, H. University of Aberdeen, Department of Agriculture, 581 King Street, Aberdeen AB9 1UD, U.K.

Garssen, G.J. Research Institute for Animal Husbandry, "Schoonoord", P.O.B. 501, 3700 AM, Zeist, Netherlands

Gilmour, R.S. AFRC Institute of Animal Physiology and Genetics Research, Cambridge Research Station, Babraham, Cambridge CB2 4AT, U.K.

Glade, M. Department of Pharmacology, Northwestern University, 303 E. Chicago Avenue, Chicago, IL 60611, U.S.A.

Gluckman, P.D. Department of Paediatrics, University of Auckland, New Zealand

Goddard, C. AFRC Institute of Animal Physiology and Genetics Research, Edinburgh Research Station, Roslin, Midlothian EH25 9PS, U.K.

Goldspink, G. Molecular and Cellular Biology, Veterinary Basic Science, Royal Veterinary College, London NW1 OTU, U.K

Gough A. Hannah Research Institute, Ayr KA6 5HL, U.K.

Gray, A.K. Medicines Unit, Central Veterinary Laboratory, New Haw, Weybridge, Surrey KT15 3NB, U.K.

Griffin, H.D. AFRC Institute of Grassland and Animal Production, Poultry Department, Roslin, Midlothian EH25 9PS, U.K.

Guler, H. Metabolic Unit, Department of Medicine, University Hospital, Zurich, Switzerland CH 8091

Haji Baba, A. Department of Applied Biochemistry and Food Science, University of Nottingham, School of Agriculture, Sutton Bonington, Nr Loughborough, Leicestershire LE12 5RD, U.K.

Hannah, M.J. AFRC Institute of Animal Physiology and Genetics Research, Cambridge Research Station, Babraham, Cambridge CB2 4AT, U.K.

Hard, D.L. Monsanto Europe S.A./N.V., Avenue de Tervuren 270-272, Tervurenlaan 270-272, Bte/Bus 1, B-1150 Brussels, Belgium

Harrington, G. Meat and Livestock Commission, PO Box 44, Queensway House, Bletchley, Milton Keynes MK2 2EF, U.K.

Hart, I.C. American Cyanamid Company, Agricultural Research Division, P.O. Box 400, Princeton, NJ 08540, U.S.A.

Heap, R.B. AFRC Institute of Animal Physiology and Genetics Research, Cambridge Research Station, Babraham, Cambridge CB2 4AT, U.K.

Heiman, M.L. Lilly Research Laboratories, Greenfield Laboratories, P.O. Box 798, Greenfield, Indiana 46140, U.S.A.

Heywood, C.M. Department of Applied Biochemistry and Food Science, University of Nottingham, School of Agriculture, Sutton Bonington, Nr Loughborough, Leicestershire LE12 5RD, U.K.

Hildick Smith, G. Johnson and Johnson, One Johnson and Johnson Plaza, New Brunswick, NJ 08933, U.S.A.

Hochberg, Z. Rappaport Family Institute for Medical Research, Department of Pharmacology, P.O.B. 9697, Haifa 31096, Israel

Holder, A.T. AFRC Institute of Grassland and Animal Production, Hurley Research Station, Hurley, Maidenhead, Berkshire SL6 5LR, U.K.

Holly, J. St Bartholomew's Hospital, West Smithfield, London EC1A 7BE, U.K.

James, S. Animal Productivity Group, Coopers Animal Health Ltd., Berkhamsted Hill, Berkhamsted, Hertfordshire HP4 2QE, U.K.

Jammes, H. Unite' d'Endocrinologie, Cellulaire et Moleculaire, INRA, 78350 Jouy-en-Josas, France

Jensen, S. Forum House, 41-52 Brighton Road, Redhill, Surrey, U.K.

Jensen, M.V. National Institute of Animal Science Research in Cattle and Sheep, Forsogsanlaeg Foulum, DK-8833 Orum Sonderlying, Denmark

Jewell, D.E. 1401 S. Henley, St. Louis, MO 63166, U.S.A.

Jones, R. American Cyanamid Company, Agricultural Research Center, P.O.B 400, Princeton, NJ 08540, U.S.A.

Kerr, D. Department of Animal Science, University of Alberta, Edmonton, Alberta T6G 2P5, Canada

Klotz, G. Bayer AG, Institut fur Tierernahrung, Geb. 6700, d-5090 Leverkusen 1, West Germany

Knight, C. Hannah Research Institute, Ayr KA6 5HL, U.K.

Kriel, V. University of Reading, Whiteknights, PO BOX 228, Reading RG6 2AJ, U.K.

Krivi, G.G. Monsanto Company (Mail Code BB3D), 700 Chesterfield Village Parkway, St. Louis, Missouri 63198, U.S.A.

Laarveld, B. Animal and Poultry Science Department, University of Saskatchewan, Saskatoon S7N OWO, Canada

Lammiman, M.J. Department of Applied Biochemistry and Food Science, University of Nottingham, School of Agriculture, Sutton Bonington, Nr Loughborough, Leicestershire LE12 5RD, U.K.

Lamming, G. E. University of Nottingham, Department of Physiology and Environmental Science, School of Agriculture, Nr Loughborough, Leicestershire LE12 5RD, U.K.

Lawler, S.E. Laboratory of Molecular Biology, Department of Crystallography, London WC1E 7HX, U.K.

Lehrman, S.R. The Upjohn Company, Unit 4861-259-12, 7000 Portage Road, Kalamazoo, MI 49001, U.S.A.

Lindsey, T.O. SmithKline Beckman Animal Health, 1600 Paoli Pike, West Chester, Pennsylvania 19380, U.S.A.

Logan, J. American Cyanamid Company, Agricultural Research Division, P.O. Box 400, Princeton, NJ 08540, U.S.A.

Lomax, M.A. Department of Physiology and Biochemistry, University of Reading, Whiteknights, PO Box 228, Reading RG6 2AJ, U.K.

Lorens, J.B. Marine Genetics, Laboratory of Biotechnology, P.O. Box 3152 Aarstad, N-5001 Bergen, Norway

Loudon, A. Institute of Zoology, Regent's Park, London NW1 4RY, U.K.

Løvendahl, P. Danish National Institute of Animal Science, Department Sheep and Cattle, DK8833 Ørum Sønderlyng, Denmark

MacKenzie, N. Coopers Animal Health Ltd., Berkhamstesd Hill, Berkhamstead, Hertfordshire HP4 2QE, U.K.

MacRae, J.C. Rowett Research Institute, Greenburn Road, Bucksburn, Aberdeen AB2 9SB, U.K.

Male, R. Marine Genetics, Laboratory of Biotechnology, P.O. Box 3152 Aarstad, N-5001 Bergen, Norway

Marshall, N.J. Department of Chemical Pathology, University College Hospital, Gower Street, London WC1 6AU, U.K.

Mason, W.T. AFRC Institute of Animal Physiology and Genetics Research, Cambridge Research Station, Babraham, Cambridge CB2 4AT, U.K.

Mathews, L.S. Center for Biotechnology, Huddinge University Hospital F82, 141 86 Huddinge, Sweden

McCarthy, M.I. Medical Unit, London Hospital, Whitechapel, London E1 1BB, U.K.

McKay, J.C. Ross Breeders Ltd., Newbridge, Midlothian EH28 8S2, U.K.

Mepham, T.B. University of Nottingham, Department of Physiology and Environmental Science, School of Agriculture, Sutton Bonington, Nr Loughborough, Leicestershire LE12 5RD, U.K.

Milner, C.K. Beecham Animal Health Research Centre, Walton Oak, Dorking Road, Tadworth, Surrey KT20 7NT, U.K.

Mitchell, E.J. NRC, Biotechnology Research Institute, 6100 Royalmount Avenue, Montreal, Quebec H4P 2R2, Canada

Moreland, B. Division of Biochemistry, UMDS, Guy's Hospital, London Bridge, London SE1 9RT, U.K.

Morrison, C.A. Ciba-Geigy SA, Suisse, Centre de Recherches Agrocoles, CH1566 Saint Aubin, France

Mueller, M. Institut fur Biochemie, Karls Strasse 23, 8000 Munchen 2, West Germany

Nancarrow, C.D. AFRC Institute of Animal Physiology and Genetics Research, Cambridge Research Station, Babraham, Cambridge CB2 4AT, U.K.

Odonkor, P.O. Department of Physiology, University of Ghana Medical School, P.O. Box 4236, Accra, Ghana

Peel, C.J. Monsanto Europe S.A./N.V., Avenue de Tervuren 270-272, Tervurenlaan 270-272, Bte/Bus 1, B-1150 Brussels, Belgium

Pell, J.M. EAP Department, AFRC Institute of Grassland and Animal Production, Hurley Research Station, Hurley, Maidenhead, Berkshire SL6 5LR, U.K.

Pendleton, J.W. AFRC Institute of Animal Physiology and Genetics Research, Babraham, Cambridge CB2 4AT, U.K.

Peters, A.R. British Technology Group, 101 Newington Causeway, London SE1 6BU, U.K.

Pinkert, C. Embryogen Inc., One President Street, Athens, OH 45701, U.S.A.

Pitts, J.E. Laboratory of Molecular Biology, Department of Crystallography, London WC1E 7HX, U.K.

Plastow, G. Dalgety PLC, Group Research Laboratory, Station Road, Cambridge CB1 2JN, U.K.

Plaut, K. 262 Morrison Hall, Cornell University, Ithaca, NY 14853, U.S.A.

Polge, E.J.C. Animal Biotechnology Cambridge Ltd., Animal Research Station, University of Cambridge, 307 Huntingdon Road, Cambridge CB3 OJQ, U.K.

Prosser, C.G. AFRC Institute of Animal Physiology and Genetics Research, Cambridge Research Station, Babraham, Cambridge CB2 4AT, U.K.

Puri, N. Department of Veterinary and Preclinical Sciences, University of Melbourne, Parkville, Victoria 3052, Australia

Pursel, V.G. U.S.D.A. Reproduction Laboratory, Bldg. 200 Room 2, BARC-East, Beltsville, MD 20705, U.S.A.

Purup, S. National Institute of Animal Science, Research Centre Foulum, DK 8833 Orum Sdrl, Denmark

Reynolds, C.M.M. Department of Applied Biochemistry and Food Science, University of Nottingham, School of Agriculture, Sutton Bonington, Nr Loughborough, Leicestershire LE12 5RD, U.K.

Roberts, C.J. Home Office, Mansion House, Whitehall, Monkwood Road, Shrewsbury SY2 5AN, U.K.

Robins, S.P. Rowett Research Institute, Bucksburn, Aberdeen AB2 9SB, U.K.

Robinson, I.C.A.F. N.I.M.R. The Ridgeway, Mill Hill, London NW7 1AA, U.K.

Rodway, R. Department of Animal Physiology and Nutrition, University of Leeds, Leeds LS2 9JT U.K.

Roe, J.A. Department of Applied Biochemistry and Food Science, University of Nottingham, School of Agriculture, Sutton Bonington, Nr Loughborough, Leicestershire LE12 5RD, U.K.

Rolph, T.P. Pitman-Moore Ltd, Breakspear Road South, Harefield, Middlesex UB9 6LS, U.K.

Ronnholm, H. Building 54, KabiVitrum, 11287 Stockholm, Sweden

Scanlon, M.F. Neuroendocrine Unit, Department of Medicine, University of Wales, College of Medicine, Heath Park, Cardiff, Wales, U.K.

Schams, D. Institute for Physiology, Technical University Munich, D-8050 Freising-Weihenstephan, West Germany

Schmidely, Ph. Institut National Agronomique de Paris-Grignon, 78850 Thiverval-Grignon, France

Schulster, D. National Institute Biological Standards, Blanche Lane, South Mimms, Potters Bar, Hertfordshire EN6 3Q6, U.K.

Sejrsen, K. National Institute of Animal Science, Research Center Foulum, DK 8833 Orum Sdrl, Denmark

Sharp, P.J. AFRC Institute of Animal Physiology and Genetics Research, Edinburgh Research Station, Roslin, Midlothian EH25 9PS U.K.

Simmen, F.A. Department of Animal Science, Ohio Agricultural Research and Development Center, Wooster, OH 44691, U.S.A.

Simmen, R.C.M. Department of Animal Science, Ohio Agricultural Research and Development Center, Wooster, OH 44691, U.S.A.

Sinnett-Smith, P.A. AFRC Institute of Animal Physiology and Genetics Research, Edinburgh Research Station, Roslin, Midlothian EH25 9DS, U.K.

Skarda, J. Institute of Animal Physiology and Genetics, Academy of Sciences, 104 00 Prague 10, Uhrineves, Czechoslovakia

Smiley, M. Institute de Selection Animale, Mauguerand, 22800 Quintin, France

Smith, H. Cyanamid of Great Britian, 154 Fareham Road, Gosport, Hanmpshire PO13 OAS, U.K.

Soar, J.B. Department of Applied Biochemistry and Food Science, University of Nottingham, School of Agriculture, Sutton Bonington, Nr Loughborough, Leicestershire LE12 5RD, U.K.

Stansfield, D. Ciba-Geigy Agrochemicals, Hill Farm, Whittlesford, Cambridge CB2 4QT, U.K.

Stevenson, L.Q. BST Department, Hannah Research Institute, Ayr KA6 5HL, U.K.

Stewart, F. AFRC Institute of Animal Physiology and Genetics Research, Cambridge Research Station, Babraham, Cambridge CB2 4AT, U.K.

Sweet, A.T. Department of Applied Biochemistry and Food Science, University of Nottingham, School of Agriculture, Sutton Bonington, Nr Loughborough, Leicestershire LE12 5RD, U.K.

Swift, P.J. Ciba-Geigy SA, CRA, CH-1566 St Aubin (FR), Switzerland

Thomas, H.M. Biochemistry Department, School of Biological Sciences, University of Sussex, Falmer, East Sussex, U.K.

Totland, G. Marine Genetics, Laboratory of Biotechnology, P.O. Box 3152 Aarstad, N-5001 Bergen, Norway

Trigg, T.E. Project Director, Peptide Technology Ltd., P.O. Box 444 Dee Why, N.S.W. 2099, Australia

Turay, L. Molecular and Cellular Biology, Veterinary Basic Science, Royal Veterinary College, London NW1 OTU, U.K.

van Miert, A. Department of Veterinary Pharmacology, State University of Utrecht, Yalelaan 6, PO Box 80.155, 3f508 Utrecht, Netherlands

Vernon, R. Hannah Research Institute, Ayr KA6 5HL, U.K.

Vigneron, P. INRA, Montpellier Centre, Differenciation Cellulaire et croissance, Place Viala, 34060 Montpellier Ledex, France

Wallis, M. The University of Sussex, Biology Building, School of Biological Sciences, Falmer, Brighton, Sussex BN1 0QG, U.K.

Wallis, O.C. School of Biological Sciences, University of Sussex, Falmer, Brighton, Sussex BN1 9QG, U.K.

Wang, American Cyanamid Co., P.O. Box 400, Princeton, NJ 08540, U.S.A.

Wass, J. St Bartholomew's Hospital, West Smithfield, London EC1A 7BE, U.K.

Waters, M.J. Department of Physiology and Pharmacology, University of Queensland, St. Lucia, Queensland 4067, Australia

Watt, P.W. 216 High Street, Ayr, U.K.

Weaver, J.U. Medical Unit, Floor Alex Wing, The London Hospital, Whitechapel Road, London E1, U.K.

Wheatley, S.D. AFRC Institue of Grassland and Animal Production, Hurley, Maidenhead SL6 5LR, U.K.

Wilde, C. Hannah Research Institute, Ayr KA6 5HL, U.K.

Wilkinson, J.I.D. Lilly Research Centre Ltd., Erl Wood Manor, Windlesham, Surrey, U.K.

Williams, J.P.G. Department Biological Sciences, City of London Polytechnic, Old Castle Street, London E1 7NT, U.K.

Wishart, D.F. Beecham Pharmaceuticals Research Division, Walton Oaks, Dorking Road, Tadworth, Surrey KT20 7NT, U.K

Withers, R.M. Department Applied Biochemistry and Food Science, University of Nottingham, School of Agriculture, Sutton Bonington, Nr Loughborough, Leicestershire LE12 5RD, U.K.

Witkamp, R.F. Department of Veterinary Pharmacology, State University of Utrecht, Yalelaan 6, PO Box 80.155, 3f508 Utrecht, Netherlands

Witty, M.J. Animal Health Discovery Department, Pfizer Central Research, Sandwich, Kent CT13 9NJ, U.K.

Wolfrom, G. Pitman-Moore Inc., P.O. Box 207, Terre Haute, IN 47808, U.S.A.

Wollny, C. Monsanto (Deutschland) GmbH, Immermann Str 3, 4000 Dusseldorf, West Germany

Wray-Cahen, D. 248 Morrison Hall, Cornell University, Ithaca, NY 14853, U.S.A.

Wylie, A.R.G. Agricultural Chemistry Division, Department of Agriculture, Newforge Lane, Belfast BT9 5PX, U.K.

Young, I.M. 29 Greetwell Lane, Nettleham LN2 2PN, U.K.

INDEX